S0-CBV-394

Bernard P. Boudreau

Diagenetic Models and Their Implementation

Modelling Transport and Reactions in Aquatic Sediments

Springer
Berlin
Heidelberg
New York
Barcelona
Budapest
Hong Kong
London
Milan
Paris
Santa Clara
Singapore
Tokyo

Bernard P. Boudreau

Diagenetic Models and Their Implementation

Modelling Transport and Reactions in Aquatic Sediments

With 75 Figures and 19 Tables

Springer

Bernard P. Boudreau
Associate Professor
Department of Oceanography
Dalhousie University
Halifax, NS B3H 4J1
Canada

ISBN 3-540-61125-8 Springer-Verlag Berlin Heidelberg New York

Library of Congress Cataloging-in-Publication Data applied for

Die Deutsche Bibliothek – CIP-Einheitsaufnahme
Boudreau, Bernard P.: Diagenetic models and their implementation: modelling transport and reactions in aquatic sediments / Bernard P. Boudreau. - Berlin; Heidelberg; New York; Barcelona; Budapest; Hong Kong; London; Milan; Paris; Santa Clara; Singapore; Tokyo: Springer 1996
ISBN 3-540-61125-8 (Berlin ...)
ISBN 0-387-61125-8 (New York ...)

Printed in Germany

Production: PRO EDIT GmbH, D-69126 Heidelberg
Cover Design: design & production GmbH, D-69121 Heidelberg

SPIN: 10521317 32/3136-5 4 3 2 1 0 - Printed on acid-free paper

To:

Robert A. Berner

advisor and friend, who started it all.

PREFACE

Over the past fifteen years a limited number of monographs concerned with the theory of transport and reaction of substances in sediments (Berner, 1980; Lerman, 1979; and in part, Thibodeaux, 1979) have appeared. If there is one common thread that runs through these texts, it is that they are primarily concerned with explaining the application of this theory to geochemical and environmental problems. There is less stress on the development of the models themselves and still less on methods for their solution (however, see Albarède, 1995). This emphasis on applications leaves a weak link in the chain of reasoning needed to implement a model. Practitioners of diagenetic modelling (as this science has become known) must then resort to books and papers from the fields of applied mathematics and engineering for this missing information. While such sources are irreplaceable, their language and examples are usually unfamiliar and removed from the experience of most students and researchers of geochemistry and environmental science.

This book attempts to correct this imbalance by presenting a detailed account of model formulation and by explaining some useful solution techniques, all within the context of sediment geochemistry. Consequently, the present text constitutes a complement to the application oriented textbooks cited above. At the same time, I do not claim that the choice of material herein is exhaustive with respect to either modelling theory or methods of solution; instead, I have chosen to illustrate methods that are simple to explain and implement, yet powerful enough to attack even the most complicated diagenetic problems, some of which are very difficult indeed. In particular, the material on solving models is presented solely from the point of view of a *user* who is not a mathematician, but needs some down-to-earth advice on solving equations. Consequently, this book is also an adjunct to mathematical texts on differential equations and numerical methods, and is not a replacement.

This book can hardly make up for a deficient foundation in basic mathematics. I must assume the reader has had some mathematical training, which includes linear algebra and college level calculus to the level of ordinary differential equations. However, to refresh the reader, some review material is presented in Chapter 2. Chapter 3 details the construction of transport-reaction models, while Chapter 4 explains how the parameters in these models may be calculated. The transport-reaction models are usually in the form of differential equations, and these cannot be solved without added boundary conditions. Consequently, the theory of boundary conditions relevant to diagenesis is covered in Chapter 5. Once formulated, the models must be solved. Analytical (exact) methods for transport-reaction models as ordinary and partial differential equations are examined and explained in Chapters 6 and 7, respectively. Such models quite often do not admit to analytical solutions, and numerical methods must be employed. This is the topic of Chapter 8.

Part of the struggle facing both neophyte and veteran modellers is the formulation of working computer "code". It is one thing to simply describe a potential solution method and quite another to implement the same on a computer. With this in mind,

I have created a number of example codes in FORTRAN that illustrate the numerical methods. While these codes hardly represent pinnacles of efficient programming, I do believe that they all work and are relatively easy to understand by someone who has attended at least a one-term course in FORTRAN. They can be obtained by the method explained in the Appendix.

Finally, I note that many important contributions to diagenetic modelling predate this monograph. Much inspiration and knowledge are to be gleaned from these sources, and both hardened professionals and beginners are well advised to be familiar with them. If, however, the text appears to be lacking in references in some sections or omits to mention a particularly relevant contribution, I apologize in advance for such oversights of acknowledgment.

This monograph has as its origins my notes prepared for an invited set of lectures given for the Graduierten Kolleg of the University of Bremen, Germany, and the Max-Planck-Institut für marine Mikrobiologie (MPI-MM), also in Bremen. My hosts were extraordinarily generous with their support, and an instructor could not have asked for better students. The work to convert my notes into this monograph was supported by funding from the Natural Sciences and Engineering Research Council of Canada, by a research fellowship provided by the Alexander von Humboldt-Stiftung, and by a sabbatical leave granted by Dalhousie University. Various individuals aided materially in the development of this book: Cindy Lee (SUNY) read and corrected my original lecture notes, David Archer (U. Chicago) kindly supplied his tortuosity data, Bo Barker Jørgensen (director of the MPI-MM) offered the hospitality and support of his institute, and Warwick Kimmins (Dean of Science, Dalhousie U.) greatly facilitated my sabbatical leave. I am equally thankful for the outstanding reviews provided by my friends and colleagues: Rob Raiswell (U. Leeds), Peter Herman, Jack Middelburg and Karline Soetaert (Netherlands Inst. Ecology, Yerseke), and Mike Velbel (MSU). The text also profited from proof-reading by Susan Clarke. The text was prepared in camera-ready form by the author; if there remain problems with this text, they are solely of my own doing.

Last but not least, I must thank my ever-patient wife, Elizabeth, who not only tolerates, but supports the sometimes unconventional behavior of her scientist-husband. The sacrifice of her time made this book possible.

Bernard P. Boudreau
Halifax, Nova Scotia
May 1, 1996

DISCLAIMER OF WARRANTY AND LIMITS OF LIABILITY

The author of this book has used his best efforts in preparing this material. These efforts include the development, research, and testing of the theories, examples, and programs to determine their effectiveness. Neither the author nor the publisher makes warranties of any kind, expressed or implied, with regard to both the programs and the documentation contained in this book, which includes, without limitation, warranties of merchantability or fitness for a particular purpose. No liability is accepted in any event for any damages, including incidental or consequential damages, lost profits, costs or lost data or program material, or otherwise in connection with or arising out of the furnishing, performance, or use of the programs referred to in this book or available through the specified Internet sites.

TABLE OF CONTENTS

1 THE PHILOSOPHY OF DIAGENETIC MODELLING

"In Science - in fact, in most things - it is usually best *to begin at the beginning*. ..."

"No, no! said the Professor. "We haven't got to the *Experiments* yet. And so", returning to his note-book, "I'll give you the Axioms of Science. After that I shall exhibit some Specimens. Then I shall explain a Process or two. And I shall conclude with a few Experiments. ..."

(Lewis Carroll, *The Professor's Lecture in Sylvie and Bruno*, 1889)

1.1 Introduction

The study of sedimentary chemistry and its associated processes is becoming far more mathematical. This new emphasis is being driven by pressures originating from both basic research and field applications. In particular, pressing environmental problems, such as predicting the fate and transfer rates of contaminants in aquatic sediments, now demand numerical answers. Furthermore, there is a growing desire to gain a quantitative understanding of the reasons for the natural chemical changes observed in sediments as they are buried.

Such predictions and explanations are possible through the application of so-called diagenetic models. *Diagenesis* refers to "the sum total of processes that bring about changes in a sediment or sedimentary rock, subsequent to deposition in water. The processes may be physical, chemical, and/or biological in nature ..." (Berner, 1980, p. 3). As discussed later, diagenetic processes include various types of transport and reaction phenomena, both biological and physical in nature. A diagenetic model is an idealized mathematical representation of diagenesis. Textbooks by Berner (1971, 1980), Lerman (1979), Thibodeaux (1979), devoted in part or in whole to diagenetic models, major reviews of such models by Berner (1974), Domenico (1977), Lerman (1977, 1978), Aller (1982, 1988), Duursma and Smies (1982), Reible et al. (1991), and van der Weijden (1992), and literally scores of original scientific papers published over the past 30 years bear witness to the success of this approach.

There remains, however, the question as to why we should resort to a mathematical description of diagenesis. Couldn't scientists and engineers simply go out and measure what they need to know? Arguably the answer is yes, but this purely observational approach is constrained by at least three factors. First, measurements generally do not tell us how the varied processes operating in sediments combine to give observed distributions; they tell us only the net results. One can build physical laboratory models (microcosms, etc.) to obtain some insights into this interplay, but this experimental approach is not always practical or successful. Therefore, a mathematical model presents a very attractive alternative to understanding the interaction between sedimentary and biogeochemical processes.

Secondly, Nature seems to continuously conspire against our search for understanding. In particular, the time scale of many diagenetic processes is so long that an adequate observational/experimental program would last longer than the lifetime of any investigator. The presence of a deep water column can constitute a significant barrier to some direct observations. Furthermore, our sampling techniques often disturb the system we are trying to study, while our methods of analysis may be so time consuming that resolution of transient processes may not be possible.

Finally, prediction of the future behavior of an evolving sedimentary system is a troublesome task from an experimental point of view. On the other hand, modelling allows an investigator to study possible evolutionary scenarios without disturbing the system itself, altering its behavior, or requiring that it be reproduced accurately in the laboratory.

Further, modelling is often a necessary step in data verification. Once some quantity (e.g. a flux) or parameter (e.g. a rate constant) has been measured, it is necessary to determine if its value reflects that in a natural setting or if the methodology has altered the situation to the extent that the measurement no longer represents a truly unperturbed value. This question of validation can be approached with relative ease through diagenetic models; e.g. one can verify a laboratory-measured rate constant by reproducing an observed concentration profile.

Diagenetic modelling, or any type of modelling for that matter, is not simply the manipulation of mathematical symbols, whether on paper, in one's mind or via a computer. Successful modelling is a science in itself. To this aim, investigators from various fields of applied mathematics have elucidated the necessary steps involved in sound modelling (Lin and Segel, 1974; Noble, 1982). These apply equally well to diagenetic modelling, and they consist of:

1) *Identification* of the problem to be addressed; i.e. what is the precise question you wish to answer?

2) *Formulation* of the appropriate model; i.e. what are the correct equations for the problem?

3) *Solution* of the model; i.e. by what means do you obtain an accurate answer to the question?

4) *Interpretation* of the results in the context of diagenesis; i.e. does the model describe the phenomena under consideration?

These steps do not normally constitute a unidirectional recipe, and there is usually a feedback loop between steps 4 and 2, in which a model must be refined or modified before it is truly useful.

Contrary to the impression some novices may have about modelling, each of the above steps is equally important, and the solution step is not necessarily the hardest. A good modeller places as much weight on valid model construction as on a verifiable solution. However, this book deals solely with points 1 through 3, as interpretation is case specific. Furthermore, the books by Berner (1980), Lerman (1979), Thibodeaux (1979), and Phillips (1991) adequately address this aspect. As such, it is worthwhile at this point to briefly discuss the first three steps.

1.2 Problem Identification

Proper identification of the problem is essential. Diagenetic modelling is a *tool,* and not an end unto itself. There must be a reason for using the modelling tool, and the chosen instrument must be suited to the task. For example, one would not use a hammer to slice a tomato. This example may seem trite, but it "cuts" to the very heart of the problem. To a properly trained modeller, misuse of a model can appear as outrageous, and it will certainly lead to no better results (if somewhat less mess). Conversely, it is equally important to realize that modelling is not an activity reserved for the mathematical *illuminati.* Any competent scientist or engineer should be capable of creating and solving basic diagenetic models, once the mysticism has been dispelled.

The number of questions that modelling can address is infinite, but some for which it is particularly well-suited include:

- What processes can account for an observed distribution, and which of those processes dominates the phenomenon?

- Does an independently determined parameter (e.g. a rate constant as measured in the laboratory or derived from an independent fundamental theory) explain an observed profile?

- How do different data sets compare based on a common fitting parameter in a diagenetic model, e.g. rate constants, irrigation constants, etc.

- Can fluxes from sediments be accurately calculated from concentration profiles? Do these match independently measured values?

- How simple need a model be in order to explain an observed (complex) phenomenon, i.e. the diagenetic version of Occam's Razor?

- If the observed sediment system is perturbed in a specific way, how might it respond and on what time scale?

At the same time, there are many reasons that are not good justifications for modelling. Currently fashionable pedagogy would recommend that specific examples should not be cited, but the abuses are sufficiently common that they need to be identified explicitly so that the student can recognize them for what they are: a) I need to make my paper look more profound; b) my advisor/supervisor/director/reviewers/peers told me to add a model (why?); c) modelling is "neat" (or any other trendy qualifier); and d) someone else is doing it. The bottom line is that there is only one legitimate rationale for modelling:

You must have a well-defined problem/question, and one for which modelling is useful, if not necessary, in providing an answer.

1.3 Model Formulation

Formulation of the appropriate model is sometimes a daunting task for beginners. In large part, schools and colleges don't seem to train people adequately in conceptualizing real problems in mathematical terms. Yet, once you have an appropriate question, you need to obtain a suitable model. This can be done by building the model yourself, which is preferable, or locating an appropriate model constructed by someone else for a sufficiently similar purpose. Either way, you still need to understand how the model was constructed, i.e.

- What processes should be or are included or excluded?
- How are these represented mathematically?
- What approximations and what assumption can be or were adopted?

Without the answers to these questions, you will never know if the adopted model is truly appropriate to your task and if its results can be trusted. For good reason, this book devotes a significant portion of its length to the process of model formulation (see Chapters 3, 4 and 5).

1.4 Solution

> "If a meaningful physical problem is correctly formulated mathematically as a differential equation, then the mathematical problem should have a solution. In this sense an engineer or scientist has some check on the validity of his mathematical formulation."
>
> (Boyce and DiPrima, 1977, 3rd ed., p. 4)

This is, in my opinion, a remarkable statement about models of physical phenomena. One certainly needs to be liberal in one's interpretation of the meaning of the word "solution"; yet, in spite of this caveat, it indicates that proper modelling (i.e. applied mathematics) has a logical consistency that makes it a tool of extraordinary power. As Richard Feynman wrote,

> "Mathematics is language plus reasoning; it is like a language plus logic. Mathematics is a tool for reasoning."
>
> (*The Character of Physical Law*, 1965, p. 40.)

And applied mathematics is a tool for reasoning about the real world, in this case, sediments. This is the inspiration of this book and the reason why it's worth going through the intellectual labor (and possible discomfort) of learning the mathematical techniques needed to solve diagenetic models.

The variety of methods available to solve diagenetic models is literally staggering. Applied mathematics has been a "growth industry" since the 18th century, and mathematicians have been clever at devising new ways to solve problems, often many different ways for the same problem. The applied mathematics literature is consequently immense, and the successful modeller eventually needs to become familiar with the most relevant parts. Additionally, finding a solution often requires some discretionary decisions about the technique to be utilized, e.g. numeric versus analytical, the type of numerical approximation, or the "flavor" of the analytical treatment. There are some guides to help the modeller towards an optimal choice. For example, some problems simply cannot be solved analytically (e.g. most so-called nonlinear partial differential equations), and they must be dealt with numerically. But once the fundamental restrictions have been addressed, there is often more than one method available. The choice becomes one of preferences, and I readily admit that, in this respect, the material in this book is strongly colored by my own prejudices about what is both easy and effective. Eventually, only increased knowledge and experience will teach the student which method is personally simplest to use. The remainder of this book is devoted to helping the reader obtain this knowledge and gain some experience.

2 MATHEMATICAL BACKGROUND

"Can you do Addition?" the White
Queen asked. "What's one and one and
one and one and one and one and one and
one and one and one?"

"I don't know," said Alice. "*I lost count.*"

(Lewis Carroll, *Through the Looking-Glass*, 1871)

2.1 Introduction

Before a diagenetic model can be properly constructed and eventually solved, certain mathematical skills must be mastered. There is no avoiding this necessity. Fortunately, the essential material is largely well-honed 19th century mathematics. I present here the minimum required to properly *formulate* a diagenetic model. The question and techniques of solution are considered in later chapters.

2.2 Variables and Parameters

Any discussion of mathematical models must begin with the knowledge of what you intend to model and, consequently, what the model describes. This means identifying the variables and parameters that will appear in the model. There is much confusion in the geochemical literature with regard to what constitutes a variable versus a parameter. In principle, the difference between these various quantities is well defined. This may appear to be only a matter of semantics, but incorrect usage of terms can make a problem quite incomprehensible to others, especially to neophyte modellers.

Quantities that change continuously in a model are called variables, of which there are two types: *independent variables*, those that dictate the values of all the other variables and that don't depend on each other, i.e. space (x, y, and z) and time (t); and *dependent variables*, the measured or observed quantities that are functions of space and time. The dependent variable in the models considered below is some form of concentration, e.g. concentration of a solute or solid species per unit volume of total sediment. This dependent variable will be designated C, for a generic concentration. The other quantities that appear in diagenetic equations are the *parameters*, which are user-defined constants that do not change during a given model calculation. These include, molecular diffusion coefficients, rate constants, etc.

2.3 Types of Equations

A diagenetic model is not one equation or even one type of equation; instead, it is the sum of whatever type and number of equations are needed to describe a chosen diagenetic situation. Diagenetic models may, therefore, include the following types of equations.

2.3.1 Algebraic Equations

Algebraic equations are mathematical equations that contain only dependent and independent variables, i.e. C, x, t, plus parameters, but no derivatives or integrals (see below). They can originate as primary descriptions of diagenesis (e.g. thermodynamic equations) or as solutions to differential diagenetic models. Two sub-types of algebraic equations are normally distinguished, related to their ease of solution.

i) *Linear* algebraic equations can be written as:

$$C = ax + b \qquad (2.1)$$

This equation is linear because it is a simple polynomial where x and C are raised only to the first power. It contains, in addition, two parameters, a and b.

ii) *Nonlinear* algebraic equations involve functions of the variables; for example,

$$C = ax^2 + bx + c \qquad (2.2)$$

is a nonlinear quadratic in x, while

$$C = A\,exp\left(\frac{vx}{D}\right) + B - \frac{\kappa\,exp(-\alpha x)}{\alpha(D\alpha + v)} \qquad (2.3)$$

contains nonlinear functions of x, i.e. the exponential, *exp*( ). Finally,

$$C + log(C) = ax + b \qquad (2.4)$$

is a nonlinear equation in C because of the *log*( ) function. The various constants in Eqs (2.2) through (2.4), i.e. a, b, c, v, A, B, D, and α, are all parameters (constants).

2.3.2 Differential Equations

Consider a concentration profile that depends only on depth in a sediment, x; that is to say, it does not change with time, t (e.g. Fig. 2.1). The change in concentration over the distance (depth) between two arbitrarily chosen points x_1 and x_2 is:

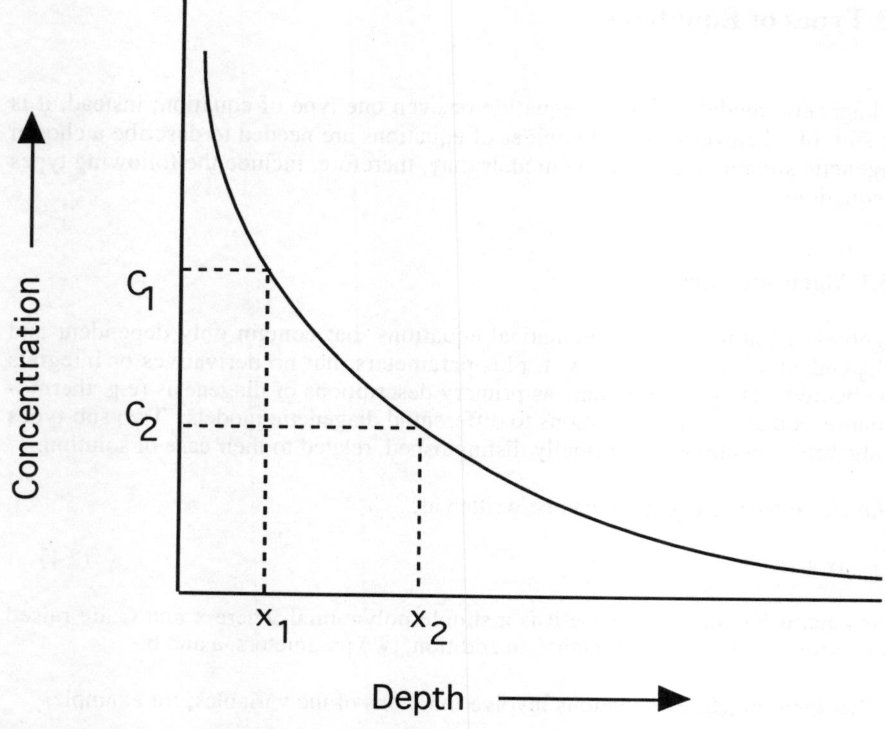

Figure 2.1. A hypothetical concentration profile in a sediment, used in defining the slope of this profile.

$$\frac{\Delta C}{\Delta x} = \frac{C_2 - C_1}{x_2 - x_1} \tag{2.5}$$

The quantity $\Delta C/\Delta x$ is also an approximation of the slope of $C(x)$ at the mid-point, $(x_1+x_2)/2$.

Now, let us make the distance (x_2-x_1) smaller and smaller; the ratio $\Delta C/\Delta x$ does not disappear. This ratio simply becomes a better and better approximation to the change in $C(x)$, i.e. the slope, at the mid-point, $x = (x_1+x_2)/2$; thus, we can define,

$$\underset{\Delta x \to 0}{\text{limit}} \frac{\Delta C}{\Delta x} \equiv \frac{dC}{dx} \tag{2.6}$$

if there is no time dependence. Taking a derivative means taking the difference over an interval and letting that interval approach zero, i.e. $\Delta x \to 0$, but not $\Delta x = 0$. The same can be done with time, t, if C depends only on t (and not on x). We call dC/dx or dC/dt the *ordinary derivative* of C with respect to depth or time, respectively.

Remember that the ordinary derivative occurs when there is only one independent variable.

We can also take the derivative of a derivative:

$$\underset{\Delta x \to 0}{\text{limit}} \frac{\left.\frac{\Delta C}{\Delta x}\right|_{x_2} - \left.\frac{\Delta C}{\Delta x}\right|_{x_1}}{\Delta x} = \underset{\Delta x \to 0}{\text{limit}} \frac{\Delta\left(\frac{\Delta C}{\Delta x}\right)}{\Delta x} \equiv \frac{d}{dx}\left[\frac{dC}{dx}\right] = \frac{d^2C}{dx^2} \tag{2.7}$$

Therefore, the second derivative is a calculation of the slope of dC/dx, which itself is the slope of C(x).

There exist, in addition, *partial derivatives* if there is more than one independent variable, e.g. time and depth. The derivative of a dependent variable with respect to one of the independent variables (or even one dependent variable with respect to another dependent variable) accounts for only part of the change seen in the dependent variable. For example, one can have the first-order partial derivative of concentration with respect to time,

$$\underset{\substack{\Delta t \to 0 \\ x = \text{const.}}}{\text{limit}} \left\{\frac{\Delta C}{\Delta t}\right\} = \frac{\partial C}{\partial t} \tag{2.8}$$

or, the second-order partial derivative with respect to depth,

$$\underset{\substack{\Delta x \to 0 \\ t = \text{const.}}}{\text{limit}} \left\{\frac{1}{\Delta x}\left(\left.\frac{\Delta C}{\Delta x}\right|_{x_2} - \left.\frac{\Delta C}{\Delta x}\right|_{x_1}\right)\right\} = \frac{\partial^2 C}{\partial x^2} \tag{2.9}$$

Given these definitions, an *ordinary differential equation* (ODE) is an equation that involves ordinary derivatives and, perhaps, algebraic terms. An ODE may or may not explicitly contain the independent variables x or t (except as dx or dt).

ODEs are *linear* if they don't contain functions (square, cubes, square root, cube root, sines, cosines, exponential, etc.) of either the dependent variable C or its ordinary derivatives. Functions of x or t don't affect the linearity of the ODE. (Mathematics often uses the same term to indicate different, but related concepts. The term "linear" is an example of this type of usage. Linear in the context of algebraic equations is different from linear in the context of ODEs. Beware of this, for it can be the cause of much confusion.)

ODEs are classified according to their *order* and whether they are linear or not. The order is the highest derivative of concentration in a given ODE. An ODE that contains only first-order derivatives of C is a first-order equation; one that contains one or more second derivatives is second-order; etc. Examples of *first-order* ODEs include:

i) $$\frac{dC}{dt} = 0 \tag{2.10}$$

which is a *linear first-order* ODE that says that the concentration, C, does not change with time;

ii) $$\frac{dC}{dt} = -kC \tag{2.11}$$

which is also linear and says that the concentration changes with time in direct proportion to its own magnitude, with a proportionality constant k, called a first-order decay constant, i.e. this is the famed "decay equation";

iii) $$v(x)\frac{dC}{dx} = -kC \tag{2.12}$$

which is again linear with first-order decay, but now with a proportionality function k/v(x), i.e. the so-called "advection-decay equation";

iv) $$\frac{dC}{dt} = -kC^2 \tag{2.13}$$

This equation is nonlinear, as it says that the concentration changes with time in proportion to the square of its own magnitude, with a proportionality constant k having different units than before. This equation is nonlinear because C is squared. Finally,

v) $$\left[C + \frac{dC}{dx}\right]^2 = \sqrt[3]{\frac{C}{K+C}} \tag{2.14}$$

which is decidedly nonlinear, with concentration changing with depth in a very complex and non-obvious way. In fact, this is an *implicit* equation because one cannot transform it so that only dC/dx appears linearly on the left-hand side. It is also called a *second-degree* ODE because dC/dx is raised to a second power.

Second-order ODEs involve second derivatives of the dependent variable, C, and they may be linear or nonlinear, as in the case of first-order ODEs. A later chapter of this book describes and illustrates how such equations arise from the principles governing diagenesis. Examples of second-order ODEs include:

i) The so-called *linear* diffusion-advection equation (see Chapter 3),

$$D\frac{d^2C}{dx^2} - v\frac{dC}{dx} = 0 \tag{2.15}$$

ii) The so-called linear advection-diffusion-reaction equation (ADR),

$$D\frac{d^2C}{dx^2} - v\frac{dC}{dx} - kC = 0 \tag{2.16}$$

iii) A nonlinear ADR equation with a *power-law* source,

$$D\frac{d^2C}{dx^2} - v\frac{dC}{dx} + k_s\,(C_s - C)^{2.5} = 0 \tag{2.17}$$

Here, the nonlinearity originates from the presence of the 2.5 exponent in the reaction term; and

iv) A linear ADR equation with a spatially distributed source,

$$D\frac{d^2C}{dx^2} - v\frac{dC}{dx} + R_o\,exp(-\alpha x) = 0 \tag{2.18}$$

This last equation is still linear in the sense of differential equations, even though there is an exponential in x, the independent variable. Note that in equations (2.15) through (2.18), the constants D, v, k, C_s, R_o and α are all parameters.

Second-order ODEs are said to be *homogeneous* if the independent variable, say x, does not appear explicitly, e.g.

$$D\frac{d^2C}{dx^2} - v\frac{dC}{dx} - kC = 0 \tag{2.19}$$

Such equations are *inhomogeneous* if x (or t) appears explicitly, e.g.

$$D\frac{d}{dx}\left(exp(-\beta x)\frac{dC}{dx}\right) - v\frac{dC}{dx} + R_o\,exp(-\alpha x) = 0 \tag{2.20}$$

which is inhomogeneous, both because of the exponential in the diffusion term and because of the exponential source term. Whether an ODE is homogeneous or inhomogeneous is quite important if we are looking for an analytical solution (Chapters 6 and 7), but less so for a numerical solution (Chapter 8).

Partial differential equations (PDEs) result when a dependent variable, like concentration, is a function of two or more independent variables. Differential equations involving this dependent variable will then contain partial derivatives. Examples include: i) the time-dependent linear diffusion equation,

$$\frac{\partial C}{\partial t} = D\frac{\partial^2 C}{\partial x^2} \tag{2.21}$$

which is first-order in time and second-order with respect to depth; and ii) the time-dependent nonlinear diffusion-reaction equation,

$$\frac{\partial C}{\partial t} = D\frac{\partial^2 C}{\partial x^2} - kC^2 \tag{2.22}$$

A concept that is related to differential equations and is essential to the concept of diagenesis is that of a *total* or *material derivative.* Mathematically speaking, the total change in the concentration of a species in an observed small mass of either porewater or solid sediment, $(\Delta C)_{total}$, is related to changes in all the independent variables. In the case of a concentration that depends on both depth, x, and time, t,

$$(\Delta C)_{total} = \left(\frac{\Delta C}{\Delta t}\right)_{x\text{-constant}} \Delta t + \left(\frac{\Delta C}{\Delta x}\right)_{t\text{-constant}} \Delta x \tag{2.23}$$

where the bracketed quantities $(\Delta C/\Delta x)$ and $(\Delta C/\Delta t)$ are the changes of C per unit change in x and t, respectively. Equation (2.23) assumes that there is no intervening or intermediate variable that links C to x and t. Now, divide this last equation by Δt and take the limit $\Delta t \rightarrow 0$ and $\Delta x \rightarrow 0$,

$$\frac{DC}{Dt} = \frac{\partial C}{\partial t} + \frac{\partial C}{\partial x}\frac{dx}{dt} \tag{2.24}$$

where,

$$\lim_{\substack{\Delta x \rightarrow 0 \\ \Delta t \rightarrow 0}} \frac{(\Delta C)_{total}}{\Delta t} \equiv \frac{DC}{Dt} \tag{2.25}$$

Equation (2.24) expresses the total or material derivative of C. But, what is this dx/dt term in equation (2.24)? The x is the position of the small mass being observed. Let the observer sit on this position. If he/she does not move with time in the chosen coordinate system, then the position label x of the observer does not change with time, i.e.

$$\frac{dx}{dt} = 0 \tag{2.26}$$

However, if the mass and observer move with time with reference to the coordinate system, then the position label x of the mass and observer would appear to change with time, i.e.

$$\frac{dx}{dt} \neq 0 \tag{2.27}$$

Therefore, this derivative is the position change with time or the *velocity*, v, of the mass in the chosen reference frame, i.e.

$$\frac{dx}{dt} = v \tag{2.28}$$

The total derivative is then the total change that occurs in a chosen small mass (unit mass) as it is advected through the sediment medium.

2.3.3 Integrals and Integral Equations

Integrals also make an appearance in the construction of some diagenetic models. In many ways, integration is the opposite of differentiation. Consider, for example, the concentration profile, C(x), as shown in Fig. 2.2.

Let us say we are interested in the total amount of this species present between depths x = a and x = b. This is equivalent to asking, what is the area between the curve C(x) and the x-axis between the points x = a and x = b? That area can be approximated by first dividing the depth interval x = a to x = b into n segments of size Δx. Next, draw rectangles of width Δx and height $C(a+(i-1)\Delta x) = C_i$, where i is the i-th interval, as shown in Fig. 2.2. Now the area, A, between the curve and the x-axis is roughly the sum of the areas of these rectangles:

$$A \approx C_1 \, \Delta x + C_2 \, \Delta x + C_3 \, \Delta x + ... + C_n \, \Delta x \tag{2.29}$$

or

$$A \approx \sum_{i=1}^{n} C_i \Delta x \tag{2.30}$$

If we let the number of divisions between x=a and x=b approach infinity, $n \to \infty$, then we obtain the exact area, and the summation is transformed into an integral:

$$\underset{n \to \infty}{\text{limit}} \left(\sum_{i=1}^{n} C_i \Delta x \right) \equiv \int_a^b C(x) dx \tag{2.31}$$

Equation (2.31) shows up in the derivation of some diagenetic models and in some numerical methods. Concentration is not the only function we can integrate. Virtually any diagenetic quantity can be integrated. For example, a depth integration of the reaction rate per unit volume produces the total amount of reaction per unit area, which is a quantity of great significance in diagenesis.

An *integral equation* defines a broad class of equations in which a function or variable is defined by an expression involving an integral. For example, at steady state in a sediment, the flux of a radioactive tracer to the sediment-water interface, F_o, must be equal to the total amount of decay within the sediment:

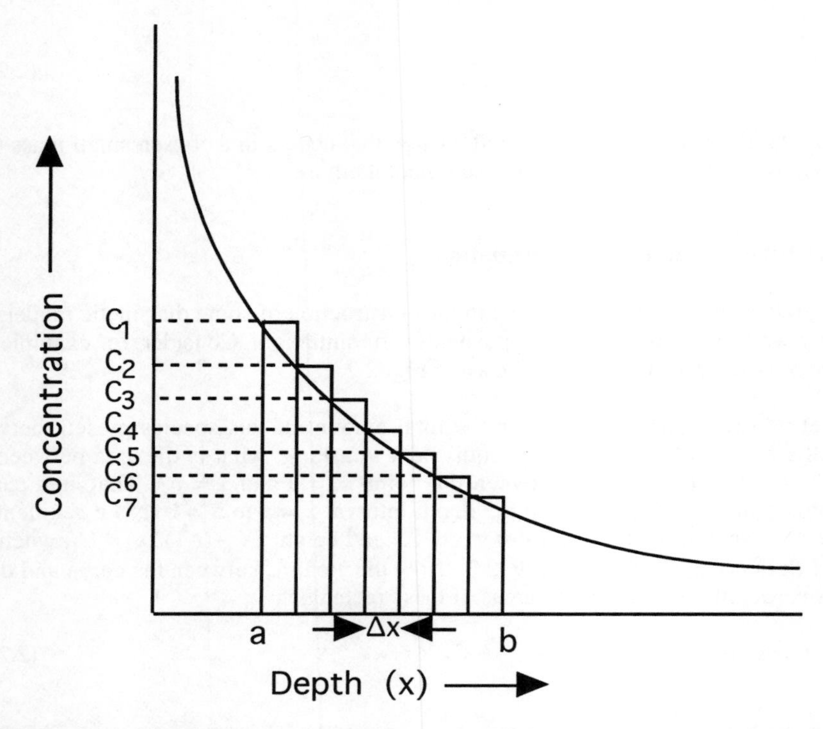

Figure 2.2. Hypothetical concentration profile used to illustrate the concept of an integral as the area between the curve and the x-axis from x=a to x=b, i.e. Eq. (2.31).

$$F_o = \lambda \int_0^\infty C(x)dx \tag{2.32}$$

where λ is the decay constant. The limits of integration need not be constants, e.g.

$$F_x = \lambda \int_x^\infty C(x)dx = \lambda \int_0^\infty C(x)dx - \lambda \int_0^x C(x)dx \tag{2.33}$$

defines the net steady state flux of a tracer across any plane at depth x in the sediment. Integral equations are not common players in diagenetic models, but they do sometimes appear in their solutions.

2.3.4 Integro-Differential Equations

There exist composite mathematical creatures that contain both differentials and integrals. For example (Boudreau and Imboden, 1987),

$$\frac{\partial C}{\partial t} = -v\frac{\partial C}{\partial x} - (k + K_e)C(x) + \frac{K_e}{L}\int_0^L C(x)dx \tag{2.34}$$

where v is an advective velocity, k is a decay constant, K_e is an exchange constant, and L is a mixing depth. This is called an *integro-differential equation*; this particular equation relates the change in the concentration with time to the difference between the change in the advective flux, disappearance due to decay and removal, and the appearance due to transport from other points. The derivation and solution of such equations are considered in Chapters 3 and 6, and need not worry us now.

2.4 Series Approximation and Expansion

Series approximations and expansions are vital tools in the derivation and solution of diagenetic models. A series is the sum of an orderly set of known functions. With a series, it is possible to represent many known and unknown functions. Below are some useful series in this regard.

2.4.1 Power Series

A *power series* of concentration as a function of depth is of the form:

$$C(x) = a_o + a_1 x + a_2 x^2 + a_3 x^3 + a_4 x^4 + ... \tag{2.35}$$

where a_o, a_1, a_2, a_3, etc., are appropriately chosen parameters. The evaluation of the values for these parameters represents the crucial step in using this series. In fact, it is most often the different methodologies used to calculate these parameters that discriminate different types of series; however, this point need not be discussed immediately.

2.4.2 Taylor Series

A *Taylor series* relates the value of a function, say the concentration C at a chosen point and/or time to the value at another point and/or time. If C depends only on x, the concentration $C(x_1)$ at can be related to $C(x_o)$ via:

$$C(x_1) = C(x_o) + \left.\frac{dC}{dx}\right|_{x_o} (x_1 - x_o) + \frac{1}{2!}\left.\frac{d^2C}{dx^2}\right|_{x_o} (x_1 - x_o)^2 + ... \tag{2.36}$$

where the derivatives are evaluated at x_o, and the exclamation mark (!) indicates a factorial, i.e. n! = n(n-1)(n-2)...3·2·1. Comparison with Eq. (2.35) shows that Eq. (2.36) is a special power series that tells us explicitly how its coefficients must be calculated. As an example, the exponential function, C = *exp*(-αx), has the Taylor series expansion with $x_o = 0$ and $x_1 = x$,

$$\left.\frac{dC}{dx}\right|_{x_0} = -\alpha \tag{2.37}$$

$$\left.\frac{d^2C}{dx^2}\right|_{x_0} = \alpha^2 \tag{2.38}$$

etc., so that

$$exp(-\alpha x) = 1 - \alpha x + \frac{\alpha^2 x^2}{2} - \frac{\alpha^3 x^3}{6} + ... \tag{2.39}$$

The special case of a Taylor series where $x_o = 0$ is often called a *Maclaurin series.* If C is a function of both x and t, and we wish to capture this behavior, then C can be expanded as a double Taylor series:

$$C(x_1,t_1) = C(x_o,t_o) + \left.\frac{\partial C}{\partial x}\right|_{x_o}(x_1 - x_o) + \left.\frac{\partial C}{\partial t}\right|_{t_o}(t_1 - t_o) + \frac{1}{2!}\left.\frac{\partial^2 C}{\partial x^2}\right|_{x_o}(x_1 - x_o)^2$$

$$+\frac{1}{2!}\left.\frac{\partial C}{\partial x}\right|_{x_o}\left.\frac{\partial C}{\partial t}\right|_{t_o}(x_1 - x_o)(t_1 - t_o) + \frac{1}{2!}\left.\frac{\partial^2 C}{\partial t^2}\right|_{t_o}(t_1 - t_o)^2 + ... \tag{2.40}$$

2.4.3 Binomial Expansion

Another series can result from application of the *Binomial Expansion* this particular expansion can be stated as (Spanier and Oldham, 1987),

$$(a \pm x)^\nu = a^\nu\left[1 \pm \binom{\nu}{1}\frac{x}{a} + \binom{\nu}{2}\frac{x^2}{a^2} \pm \binom{\nu}{3}\frac{x^3}{a^3} + ...\right] \tag{2.41}$$

for |x| < |a| (i.e. bars indicate absolute values), and where

$$\binom{\nu}{m} = \left(\frac{\nu - m + 1}{1}\right)\left(\frac{\nu - m + 2}{2}\right)\left(\frac{\nu - m + 3}{3}\right)...\left(\frac{\nu}{m}\right) \tag{2.42}$$

and the arbitrary constant ν need not be an integer. The series given by Eq, (2.41) contains a finite number of terms if ν is a positive integer and an infinite number otherwise. In more explicit form, the series can be written as:

$$(a \pm x)^{\nu} = a^{\nu} \pm \nu a^{\nu-1}x + \frac{\nu(\nu-1)}{2!}a^{\nu-2}x^2 + \frac{\nu(\nu-1)(\nu-2)}{3!}a^{\nu-3}x^3 + \dots \tag{2.43}$$

for $x^2<a^2$. From this formula we can generate the following example cases:

$$(1 \pm x)^4 = 1 \pm 4x + 6x^2 \pm 4x^3 + x^4 \tag{2.44}$$

$$(1 \pm x)^{3/2} = 1 \pm \frac{3}{2}x + \frac{3}{8}x^2 \pm \frac{1}{16}x^3 + \frac{3}{128}x^4 \pm \frac{3}{256}x^5 + \dots. \tag{2.45}$$

$$(a \pm x)^{-1} = \frac{1}{a \pm x} = \frac{1}{a}\left(1 \mp \frac{x}{a} + \frac{x^2}{a^2} \mp \frac{x^3}{a^3} + \dots\right) \tag{2.46}$$

$$(1 \pm x)^{-1/2} = 1 \mp \frac{1}{2}x + \frac{3}{8}x^2 \mp \frac{5}{16}x^3 + \frac{35}{128}x^4 \mp \frac{63}{256}x^5 + \dots \tag{2.47}$$

$$(1 \pm x)^{-3/2} = 1 \mp \frac{3}{2}x + \frac{15}{8}x^2 \mp \frac{35}{16}x^3 + \frac{315}{128}x^4 \mp \frac{693}{256}x^5 + \dots \tag{2.48}$$

$$(1 \pm x)^{-4} = 1 \mp 4x + 10x^2 \mp 20x^3 + 35x^4 \mp 56x^5 + \dots \tag{2.49}$$

Like the Taylor series, Eq. (2.41) will prove to be invaluable at various points of Chapters 3 and 8.

2.4.4 Fourier Series

A final type of series was developed by the renowned French mathematician Joseph B. Fourier (1768-1830), and it bears his name in modern practice, i.e. *Fourier Series*. While this type of series is not used in the derivation of diagenetic models, it plays an important role in the solution of linear PDEs and the stability of numerical methods. An example is included here only for comparison with the other series, i.e.

$$C(x) = \frac{a_o}{2} + \sum_{m=1}^{\infty}\left(a_m \cos\left(\frac{m\pi x}{\ell}\right) + b_m \sin\left(\frac{m\pi x}{\ell}\right)\right) \tag{2.50}$$

where a_o, a_m, and b_m are parameters, m is an index parameter, $\pi = 3.14159...$, and ℓ = length of the depth or space interval of interest. The constants a_o, a_m, and b_m must be calculated in a special way, and this is explained later in Chapter 6.

2.5 The Mean Value Theorems

The Mean Value Theorems for integrals are useful in a number of places in this book. The two of interest in our context state that

$$\int_a^b C(x)dx = \overline{C(x)}\,[b-a] \tag{2.51}$$

$$\int_a^b C_1(x)C_2(x)dx = \overline{C_1(x)}\int_a^b C_2(x)dx \tag{2.52}$$

where the overbar indicates the mean value of the concentrations C(x) and $C_1(x)$ in the interval from x = a to x = b. The only catch to these powerful equations is that we need to know the mean values *a priori* for these theorems to be useful in numerical calculations; nevertheless, this fact does not negate their utility in some theoretical derivations. These rules can be applied to integrals involving quantities other than concentration (see Chapter 4).

2.6 Leibnitz's Rule and Gauss' Theorem

It may sometimes be necessary to differentiate an integral that involves a time- and space-dependent quantity, such as a concentration or a flux; then, Leibnitz's Rule applied to such a quantity, say a concentration C(x,t), tells us that in one spatial dimension (Hildebrand, 1976, p. 365)

$$\frac{d}{dt}\int_{a(t)}^{b(t)} C(x,t)dx = \int_{a(t)}^{b(t)} \frac{\partial C}{\partial t}dx + \left[C(b,t)\frac{db(t)}{dt} - C(a,t)\frac{da(t)}{dt}\right] \tag{2.53}$$

where a(t) and b(t) are the lower and upper bounds (depths) of integration, which may be time dependent. If the bounds in Eq. (2.53) are not functions of time, that is to say constants, then the quantity in the square brackets disappears. While Eq. (2.53) may appear at first to be rather esoteric, in fact it says something quite commonsensical. It tells us that the time change in the total amount of some solute or solid species in the chosen depth interval will be the same as the sum (integral) of all the time changes in the concentration at each point in this interval, if the interval is not shifting its location with time. If the interval moves, we then need to correct for this movement.

The 3-D extension of Leibnitz's Rule states that for concentration (Hildebrand, 1976)

$$\frac{d}{dt}\iiint_V C\,dV = \iiint_V \frac{\partial C}{\partial t}dV + \iint_S C(v_s \cdot \mathit{n})dS \tag{2.54}$$

where V(t) is a (closed) volume of sediment that may be deforming its shape with time, S is the surface of that same volume, v_s is the velocity of each point on that surface as it moves, and n is the outwardly directed normal to the surface. Again, this theorem tells us that the change in the total amount of the species will be equal to the total amount of change in the concentration of the species at each point, plus the effects of having the volume change its shape.

Another theorem for vector quantities (like flux), which will prove its usefulness in deriving diagenetic models, is the divergence theorem or Gauss's theorem. For a 3-D case, it states that (Kaplan, 1973, p. 310)

$$\int_V \nabla \cdot \mathrm{F}\, dV = \int_S \mathrm{F} \cdot n\, dS \tag{2.55}$$

where F is a vector quantity (e.g. flux), $\nabla\cdot$ is the divergence operator (see Chapter 3), S is the geometrically closed surface bounding a chosen volume V, and n is the outwardly directed normal to that surface (i.e. a vector). The Cartesian divergence operator is

$$\nabla \cdot \equiv \left(\mathbf{i}\frac{\partial}{\partial x} + \mathbf{j}\frac{\partial}{\partial y} + \mathbf{k}\frac{\partial}{\partial z} \right). \tag{2.56}$$

where **i**, **j**, **k** are the unit vectors in the x, y, and z directions, respectively. If F is the flux, then Eq. (2.55) states that the total flux of a species across a surface bounding a volume is equal to the total (integral) divergence (spatial change) of the flux within the volume. In two dimensions, this same theorem reads as

$$\int_A \nabla \cdot \mathrm{F}\, dA = \int_\ell \mathrm{F} \cdot n_\ell\, d\ell \tag{2.57}$$

where A is a closed area, ℓ is the line that bounds that area, and n_ℓ is the normal to that line (also a vector).

2.7 Reynolds Decomposition and Averaging

In deriving transport-reaction models, one is sometimes faced with describing time-fluctuating concentrations. The classic example comes from fluid dynamics where turbulence causes velocity and concentration distributions (fields) to fluctuate randomly with time at a given point in the flow. Reynolds (1895) was the first physical scientist to consider how to model such fields; he noted that, even though velocity (and by inference concentration) appeared to fluctuate randomly when observed instantaneously, it usually had a well-defined time-averaged behavior. He then advanced a method to model the mean behavior, while taking into account the effects of the fluctuations.

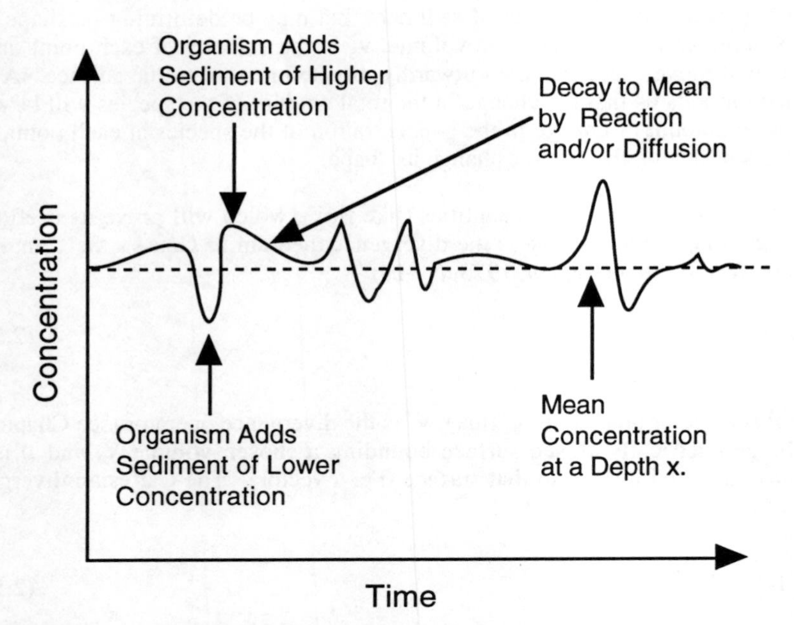

Figure 2.3. A hypothetical time-series record of the concentration of a solid or solute species at a chosen point in a bioturbated sediment. Bioturbators cause random fluctuations when they transport material (solids or porewater) from adjacent depths with higher or lower concentrations. This substance is allowed to decay to a mean value by the presence of reaction and/or diffusion, as discussed in the next chapter. This decay to a mean is not essential for a mean to exist.

In diagenetic modelling, a primary cause of concentration fluctuations is the activity of infaunal organisms in the upper decimeters of sediments. These organisms move through the sediment by displacing sediment grains, all the while disturbing the porewaters trapped between the grains. This activity is called *bioturbation*, and it is discussed more thoroughly in Chapters 3 and 4. What is important here is that bioturbation can cause seemingly random variations in the concentration of a chosen species (solid or porewater) at a given point, just like turbulence in fluid flows.

Figure 2.3 illustrates a hypothetical time-series record of a species concentration at a chosen depth, x, in a sediment undergoing bioturbation. Note that there exists a time-average mean in this record. Thus, the mean,

$$\overline{C}(x,t) = \frac{1}{\Delta t_o} \int_t^{t+\Delta t_o} C(x,t)\,dt \qquad (2.58)$$

is a well-defined quantity, where the averaging period Δt_o must be suitably chosen. Reynolds (1895) suggested, quite astutely, that one could write C(x,t) as a sum of a mean and a fluctuating quantity,

$$C(x,t) = \overline{C}(x,t) + C'(x,t) \tag{2.59}$$

and this decomposition bears his name. He then also observed that

$$\overline{C'}(x,t) = 0 \tag{2.60}$$

$$\overline{\overline{C}}(x,t) = \overline{C}(x,t) \tag{2.61}$$

For two fluctuating quantities, say a velocity, v, and a concentration, C,

$$\overline{v'\overline{C}} = \overline{v'}\,\overline{C} = 0 \tag{2.62}$$

$$\overline{\overline{v}C'} = \overline{v}\,\overline{C'} = 0 \tag{2.63}$$

Therefore,

$$\overline{vC} = \overline{(\overline{v} + v')(\overline{C} + C')} = \overline{v}\,\overline{C} + \overline{(v'C')} \tag{2.64}$$

and

$$\overline{\frac{\partial C'}{\partial t}} = \frac{\partial \overline{C'}}{\partial t} = 0 \tag{2.65}$$

As implied by Eq. (2.64), the average of the product of fluctuations is not in general zero because of correlations between the two fluctuating quantities. By introducing the Reynolds decomposition into an equation governing a random concentration field and then time averaging, it is possible to obtain an equation for the mean field, plus some terms to account for the effects of the fluctuations. The resolution of the fluctuating terms into quantities involving mean values is called the *closure*, and we shall discuss an example in the next chapter.

2.8 The Dirac Delta Function

A final topic is that of the Dirac Delta function, written as $\delta(x\text{-}a)$. The Dirac Delta is not a true function in the classical sense; however, it has the handy property that

$$\delta(x-a) = \begin{cases} 0 & x \neq a \\ \infty & x = a \end{cases} \tag{2.66}$$

If this function represents a concentration, it says that at some single depth, x=a, an infinite concentration exists at time t=0. This sort of function cannot really be graphed, but see the figure in Chapter 10 of Spanier and Oldham (1987). This function has several uses in geochemical modelling. In particular, it can be used to simulate the sudden introduction of material at a depth, x=a, e.g. dumping a very thin, but concentrated, layer of contaminant on the surface of a sediment, a=0.

There are some very important integral relations involving the Dirac Delta function. One of particular relevance to this book is

$$\int_{x_o}^{x_1} \delta(x-a)f(x)dx = \begin{cases} f(a) & x_o \leq a \leq x_1 \\ 0 & x_o > a \quad \text{or} \quad x_1 < a \end{cases} \tag{2.67}$$

which has as a special case,

$$\int_{x_o}^{x_1} \delta(x-a)\,M\,dx = \begin{cases} M & x_o \leq a \leq x_1 \\ 0 & x_o > a \quad \text{or} \quad x_1 < a \end{cases} \tag{2.68}$$

where M is a constant. This last equation tells us in the case of concentration that, while a Delta function input of a contaminant may produce an infinite concentration, the total amount of (mass) contaminant in the input is finite.

3 MODEL CONSTRUCTION

Humpty Dumpty looked doubtful.
"I'd rather see that done on paper," he said.

Alice couldn't help smiling as she took out her
memorandum-book, and worked the sum for him:

$$\begin{array}{r} 365 \\ \underline{+1} \\ 364 \end{array}$$

Humpty Dumpty took the book, and looked at it
carefully. "That seems to be done right -" he began.

"You're holding it *upside down*!" Alice interrupted.

(Lewis Carroll, *Through the Looking-Glass*, 1871)

3.1 The Standard Approach

Assuming that we have an appropriate question for diagenetic modelling, we now need to create a mathematical description of the processes responsible for the diagenesis in question. As a starting point, it is fair to ask what sediment properties the model will describe. Specifically, these should be the intensive continuum properties that characterize the sediment-porewater system. To identify them, let us start by defining these terms.

Intensive: a variable that does not depend on the absolute size of the system, as is a quantity normalized to unit mass or volume of porewater or sediment, e.g. concentration.

Continuum: at the grain scale, the sediment is heterogeneous (Fig. 3.1). Trying to capture this detail in a model is probably both impossible and meaningless. Nevertheless, as a result of this heterogeneity, the value of a chosen property, e.g. the concentration of a porewater or solid species, will change with scale, e.g. porosity as in Fig. 3.2. But, there is a spatial range over which the value of the property (variable) is relatively constant, also in Fig. 3.2.

To obtain a meaningful concentration, it is necessary to average over a volume of the size where the concentration has ceased to change because of variations in fine-scale properties. The characteristic dimension of such a volume is not fixed, but should be larger than several grain diameters and much smaller than the scale defined by the bulk (macroscopic) gradient in the concentration, which we wish to explain with the model. Given such a volume, then we can define

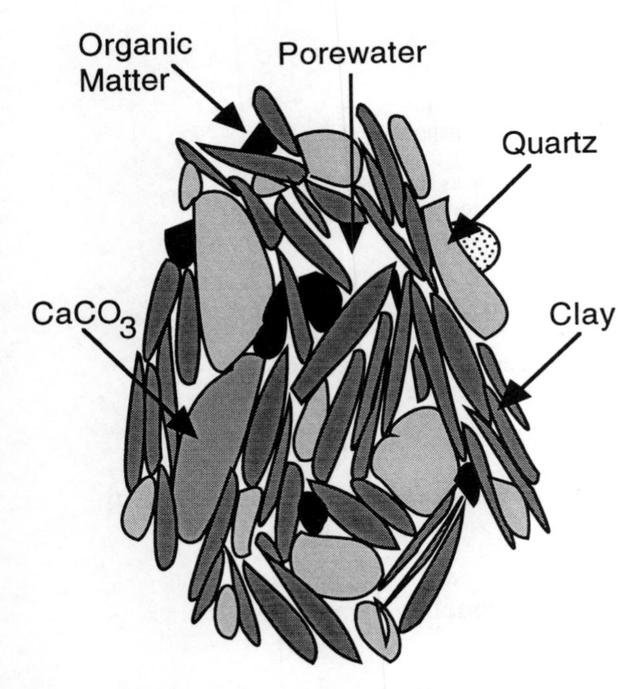

Figure 3.1. Typical heterogeneous mixture of interstitial waters and various types of solid particles that make up aquatic sediments when examined at a fine enough scale.

$$\hat{C}(x, y, z, t) \equiv \frac{1}{V_{avg}} \int_{V_{avg}} C(x, y, z, t) dV \tag{3.1}$$

where $\hat{C}$ ≡ average bulk concentration (Berner, 1980), C ≡ microscopic value of the concentration at a point in the porewater or solids, and V_{avg} ≡ averaging volume. Real sampling techniques measure the bulk average or a closely related quantity. Note that Eq. (3.1) does not preclude the existence of a finite number of real interfaces across which the bulk concentration may be discontinuous, e.g. the sediment-water interface.

The averaging in Eq. (3.1) can be applied to various sediment properties, e.g. porosity, solute concentrations, solid species concentrations, etc., to obtain a set of continuous sediment properties (fields) that overlap. This approach allows us to talk about both the porosity and the solute concentration at a same given depth, even though that precise point may be occupied by a particle of the solids in reality. I now advance the idea that we can create models for these continuous fields.

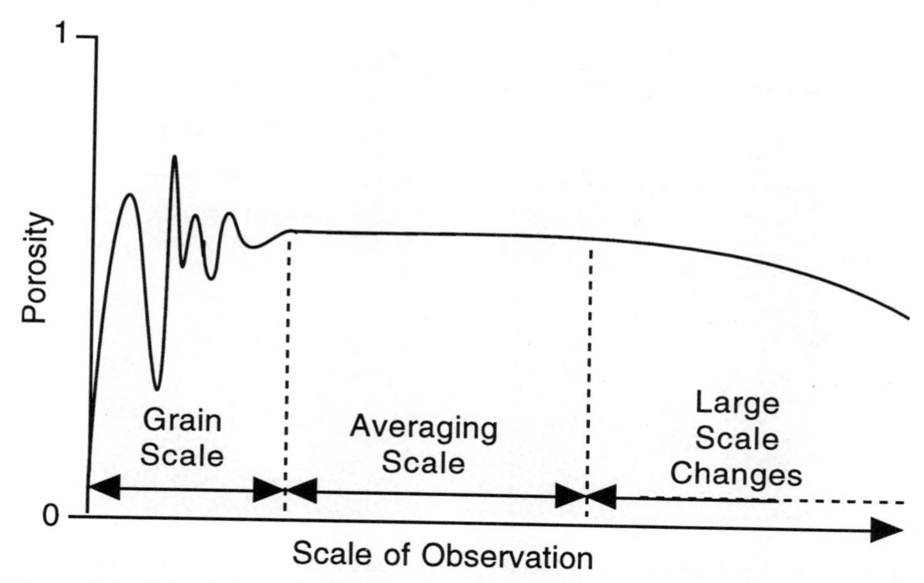

Figure 3.2. Schematic representation of the "value" of the porosity as the scale of observation (i.e. the size of the parcel of sediment being observed) goes from the scale of a grain to that of the large-scale (macroscopic) diagenetic changes, e.g. compaction. The porosity begins at zero as we arbitrarily start the process in a sediment grain. The initial value would be unity if we started in a pore. A meaningful continuous function for porosity exists only at the indicated averaging scale. Beyond that, diagenetic variations are included.

As a first step, a diagenetic model needs a well-defined *reference frame*. In our case, this means stating the location of the origin and orientation of the spatial axes. Normally, the Cartesian system is used; however, this still leaves an infinite number of possible origins, e.g. 1) at mean sea level, 2) at the basement-sediment interface, and 3) at the sediment-water interface.

Primary interest is directed at changes near the sediment-water interface; therefore, this interface is the choice of convenience, with x down into the sediment (Fig. 3.3). The consequence of this choice is that the reference frame moves with time if the sediment is accumulating (Berner, 1971). Moreover, Berner (1980) states that most diagenetic changes are vertical in sediments. In many cases, this allows us to consider only one spatial dimension, and this assumption is adopted unless otherwise stated.

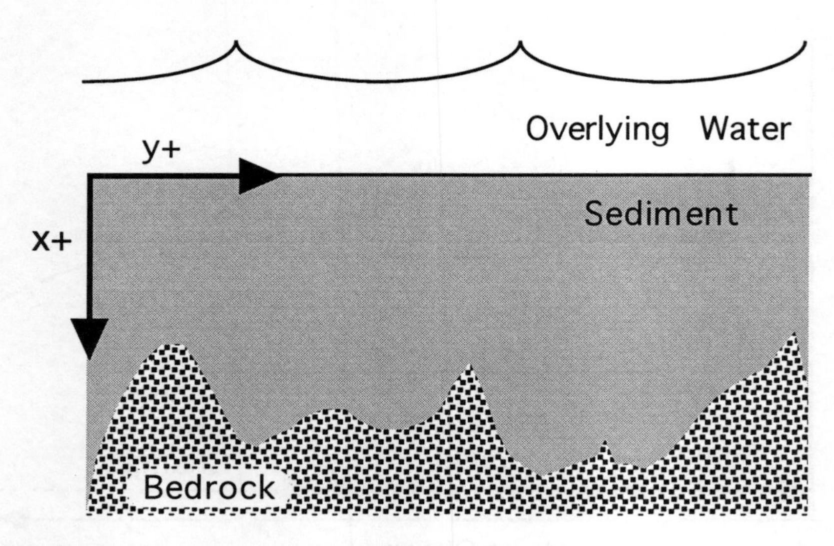

Figure 3.3. The Cartesian coordinate system used in modelling early diagenesis in aquatic sediments. (The z-axis is out of the plane of the diagram.)

3.2 Total Derivatives and Diagenesis

We can now ask, what is diagenesis in mathematical terms? Not every process affecting the distribution of a sediment/porewater species is diagenesis. Berner (1980) defines diagenesis as the total changes we would see following a selected parcel or layer of *solid* sediment as it is buried. From the mathematical review, this is the total or material derivative of the bulk concentration, Eq. (2.24) of Chapter 2,

$$\left(\frac{D\hat{C}}{Dt}\right)_{\substack{\text{fixed}\\\text{layer}}} = \left(\frac{\partial\hat{C}}{\partial t}\right)_{\substack{\text{fixed}\\\text{depth}}} + \frac{dx}{dt}\left(\frac{\partial\hat{C}}{\partial x}\right)_{\substack{\text{fixed}\\\text{time}}} \tag{3.2}$$

where x is the depth relative to the sediment-water interface. Berner (1980) uses slightly different notation:

$$\underbrace{\left(\frac{D\hat{C}}{Dt}\right)_{\substack{\text{fixed}\\\text{layer}}}}_{\text{notation used here}} = \underbrace{\left(\frac{d\hat{C}}{dt}\right)_{\substack{\text{fixed}\\\text{layer}}}}_{\text{Berner's notation}} \tag{3.3}$$

which has the possibility of being confused with an ordinary derivative.

Diagenesis is the left-hand side of Eq. (3.2), and it is balanced by the changes on the right-hand side. The term dx/dt is the velocity of the observed layer as perceived in the coordinate system anchored at the sediment-water interface; thus, dx/dt is the burial velocity of the solids, w, (see Chapter 2)

$$\frac{dx}{dt} = w \tag{3.4}$$

relative to the sediment-water interface. Using Eq. (3.4) and converting to simplified notation, we can identify diagenesis mathematically as:

$$\left(\frac{D\hat{C}}{Dt}\right)_{\substack{\text{fixed}\\\text{layer}}} = \left(\frac{\partial\hat{C}}{\partial t}\right)_{\substack{\text{fixed}\\\text{depth}}} + w\left(\frac{\partial\hat{C}}{\partial x}\right)_{\substack{\text{fixed}\\\text{time}}} \tag{3.5}$$

Berner (1980) identified some interesting limiting forms of Eq. (3.5). For simplicity, the subscripts can be dropped, as this should not lead to any confusion, i.e.

$$\frac{D\hat{C}}{Dt} = \frac{\partial\hat{C}}{\partial t} + w\frac{\partial\hat{C}}{\partial x} \tag{3.6}$$

Then, *no diagenesis* occurs if

$$\frac{D\hat{C}}{Dt} = 0 \tag{3.7}$$

so that

$$\frac{\partial\hat{C}}{\partial t} = -w\frac{\partial\hat{C}}{\partial x} \tag{3.8}$$

Thus, the changes with time one sees at a set depth are caused simply by the burial of a time-varying input (Fig. 3.4).

Steady-state diagenesis relative to the sediment-water interface, occurs if the concentration always has the same value at the same depth. This can only be true if there are no temporal changes in input; therefore,

$$\frac{\partial\hat{C}}{\partial t} = 0 \tag{3.9}$$

and, consequently,

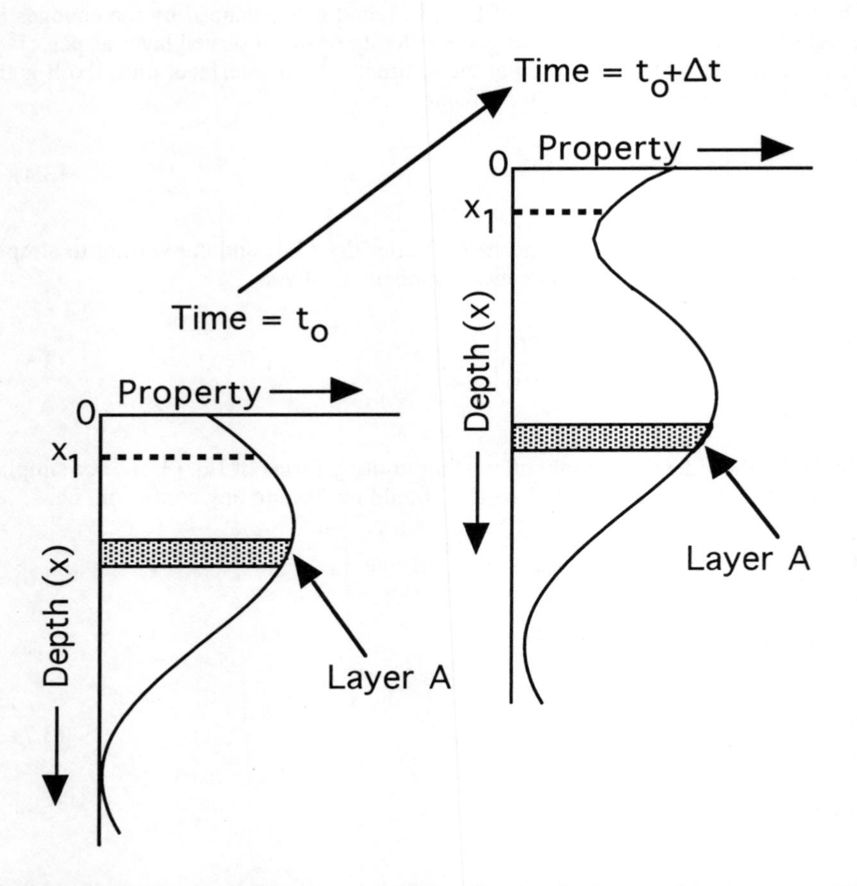

Figure 3.4. Illustration of non-steady-state diagenesis for a stable (non-reactive) species. The property changes with time at a depth x_1 that is fixed relative to the sediment-water interface. However, there are no changes with time if one follows a chosen mass of sediment, such as layer A. (Adopted from Berner, 1972; reproduced with the kind permission of the National Association of Geology Teachers, Inc.)

$$\frac{D\hat{C}}{Dt} = w\frac{\partial \hat{C}}{\partial x} \tag{3.10}$$

Equations (3.9) and (3.10) mean that there are indeed time changes in concentration during steady-state diagenesis if we follow a chosen layer of sediment, but there are no changes at a depth fixed relative to the interface. Steady state is accomplished by balancing the diagenetic changes with a changing burial flux with depth (Fig. 3.5).

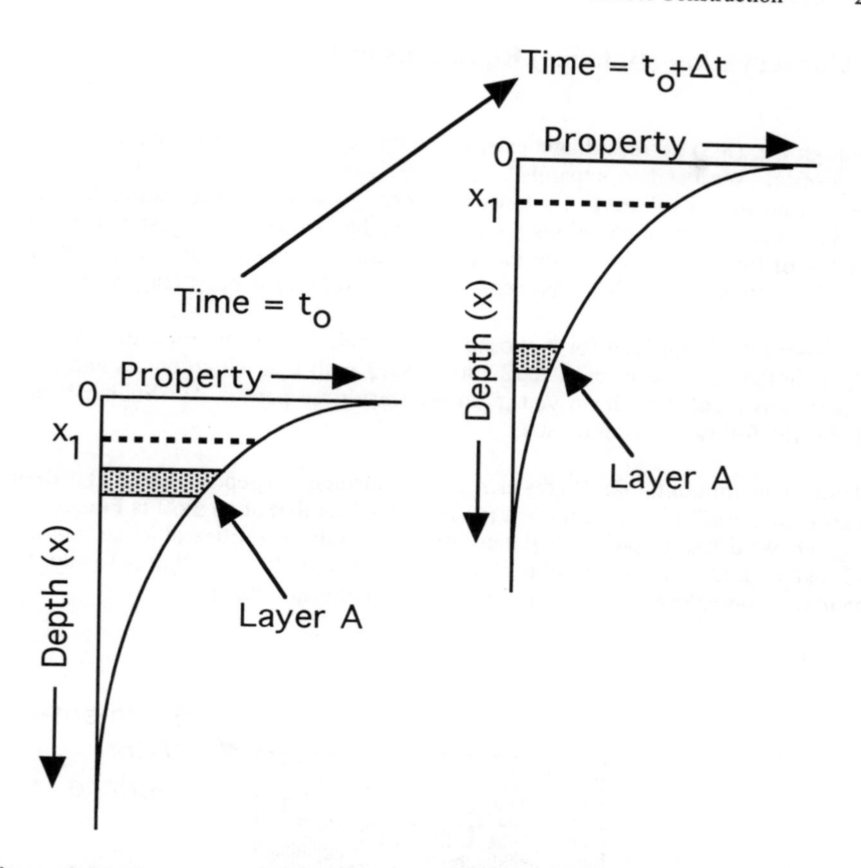

Figure 3.5. Illustration of steady-state diagenesis of a decaying sediment property, e.g. a radioisotope. In any given time interval, Δt, the value of the property at a depth x_1 that is fixed relative to the sediment-water interface does not change. However, if we follow a chosen sediment mass, layer A, the property changes with time. (Adopted from Berner, 1972; reproduced with the kind permission of the National Association of Geology Teachers, Inc.)

Before leaving the topic of the total derivative, let me state that Eq. (3.5) does not have a privileged position from a mathematical point of view. In particular, it is not the only material derivative that is definable in a sediment system (Bedford and Drumheller, 1983; Hassanizadeh, 1986). Only the traditional definition of diagenesis affords Eq. (3.5) its apparent special status. It is quite possible and, as shown later, quite useful to consider other material derivatives, for example that based on following the movement of a chosen mass of porewater,

$$\frac{D\hat{C}}{Dt} = \frac{\partial \hat{C}}{\partial t} + u\frac{\partial \hat{C}}{\partial x} \qquad (3.11)$$

3.3 Conservation (Balance) Equations in 1-D

Although Eq. (3.5) defines diagenesis, it cannot be solved to generate a concentration profile. We need to expand the total derivative into its component terms in order to find such a solution. To do this, we need to write a conservation or balance equation. Conservation equations are known as the "Diagenetic Equations" through the work of Berner (1980). To derive a conservation equation, consider an arbitrary slice of sediment of thickness, Δx, centered on an arbitrary depth x (Fig. 3.6).

The conservation equation for a chosen species balances what concentration accumulates in the layer between $x-\Delta x/2$ and $x+\Delta x/2$ with that which enters and leaves (fluxes) or is created or destroyed in an arbitrary time period, Δt. Such a balance includes the following components.

i) The mean amount that fluxes across the surface perpendicular to the depth-direction at $x-\Delta x/2$ per unit time is $F(x-\Delta x/2,t)$, while that at $x+\Delta x/2$ is $F(x+\Delta x/2,t)$. However, we'd like to write a balance involving only quantities at x, i.e. not at $x-\Delta x/2$ and $x+\Delta x/2$. This is possible if the processes are all *local*; that is to say, they depend only on values of concentrations and gradients at x itself.

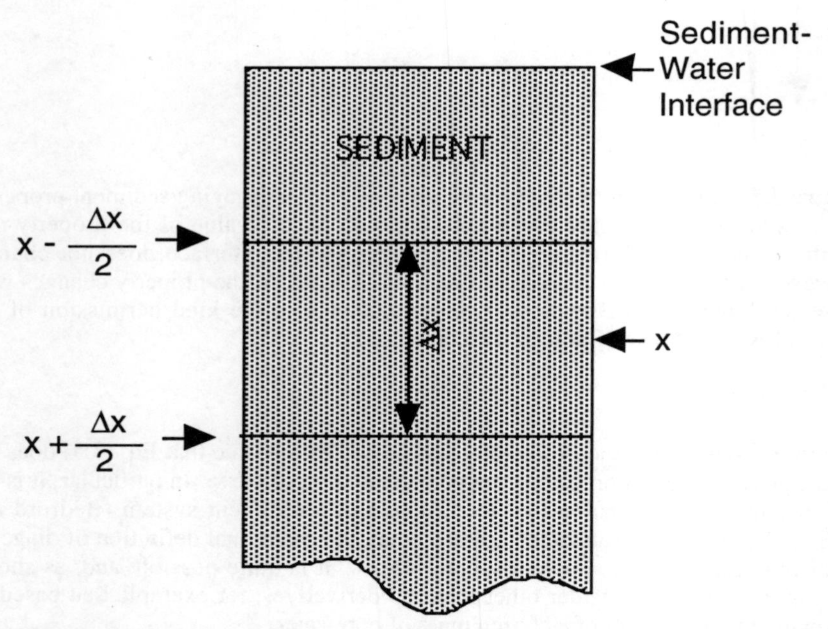

Figure 3.6. A small arbitrary layer of sediment used to create the 1-D conservation (diagenetic) equation.

If the fluxes are indeed local, then we can express both $F(x-\Delta x/2,t)$ and $F(x+\Delta x/2,t)$ in terms of the value at x, i.e. $F(x,t)$. This is done with Taylor series expansions (see Chapter 2),

$$F(x-\frac{\Delta x}{2},t)=F(x,t)-\frac{\partial F}{\partial x}\frac{\Delta x}{2}+\frac{\partial^2 F}{\partial x^2}\frac{\Delta x^2}{8}-.... \tag{3.12}$$

and

$$F(x+\frac{\Delta x}{2},t)=F(x,t)+\frac{\partial F}{\partial x}\frac{\Delta x}{2}+\frac{\partial^2 F}{\partial x^2}\frac{\Delta x^2}{8}+... \tag{3.13}$$

If Δx is kept sufficiently small, then only the first two terms on the right-hand side of Eqs (3.12) and (3.13) need be retained, i.e.

$$F(x-\frac{\Delta x}{2},t)\approx F(x,t)-\frac{\partial F}{\partial x}\frac{\Delta x}{2} \tag{3.14}$$

and

$$F(x+\frac{\Delta x}{2},t)\approx F(x,t)+\frac{\partial F}{\partial x}\frac{\Delta x}{2} \tag{3.15}$$

ii) The total amount of species consumed or produced in the layer during Δt by reactions is $\Sigma R\Delta t\Delta x$, where ΣR is the net time-averaged source or sink caused by all the reactions that affect the bulk concentration.

iii) Given that the accumulated mass, ΔM, in the layer in time Δt is

$$\Delta M = [\hat{C}(x,t+\Delta t) - \hat{C}(x,t)]\,\Delta x \tag{3.16}$$

then, mass balance requires that

$$\left[\hat{C}(x,t+\Delta t)-\hat{C}(x,t)\right]\Delta x\approx\left(F(x-\frac{\Delta x}{2})-F(x+\frac{\Delta x}{2})\right)\Delta t+\sum R\,\Delta t\Delta x \tag{3.17}$$

or, with (3.14) and (3.15),

$$\left[\hat{C}(x,t+\Delta t)-\hat{C}(x,t)\right]\Delta x\approx-\frac{\partial F}{\partial x}\Delta x\Delta t+\sum R\,\Delta t\Delta x \tag{3.18}$$

Finally, divide by $\Delta x \Delta t$ and let Δx and $\Delta t \to 0$,

$$\underbrace{\frac{\partial \hat{C}}{\partial t}}_{\text{accumulation}} = -\underbrace{\frac{\partial F}{\partial x}}_{\substack{\text{divergence}\\ \text{of the flux}}} + \underbrace{\sum R}_{\substack{\text{production or}\\ \text{consumption}}} \tag{3.19}$$

where F and ΣR are now the instantaneous values in this limit. This equation is the *conservation or balance equation* for a generic species in one dimension (depth), which we can recognize as Berner's diagenetic equation. It is important to realize that Eq. (3.19) balances the change with depth of the flux, F, against reaction and accumulation, and not the flux itself.

Equation (3.19), or its higher dimensional analogs as given below, must be solved in order to calculate a concentration profile. However, that is not possible without identifying this bulk concentration and specifying the mathematical forms of the flux, F, and reaction, ΣR. The first of these points is the easiest to address. The concentration may refer to a solute in the porewaters, in which case,

$$\hat{C} = \varphi C \tag{3.20}$$

where C is now the amount or mass per unit volume of porewater, and φ is the porosity, which is defined as

$$\varphi \equiv \frac{\text{volume of porewater}}{\text{volume of solid sediment} + \text{porewater volume}} \tag{3.21}$$

Conservation of total fluid is obtained by setting $C = \rho_f$, which is the density of the pore fluid. If a species is part of the solids, then

$$\hat{C} = (1-\varphi)\, B \tag{3.22}$$

where B is the amount or mass of the species per unit volume of solids. To obtain conservation of total solids, set $B = \rho_s$, which is the intrinsic density of these solids.

3.4 Types of Fluxes in 1-D

The specification of flux components within F involves the mathematical representation of certain transport processes. However, let me caution that there are transport processes, for both solutes and solids, that need not appear in F, but actually form part of ΣR (see nonlocal processes, described below).

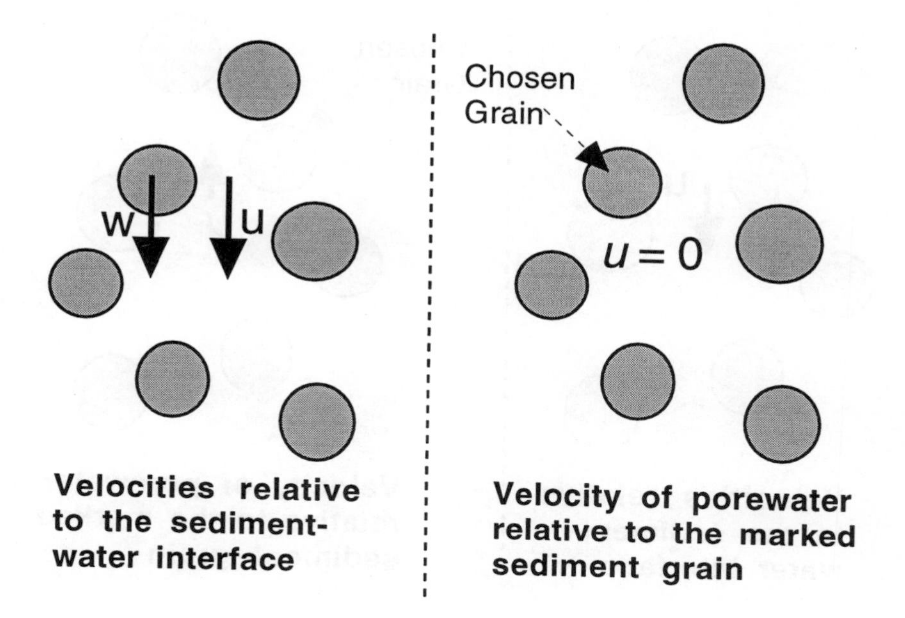

Figure 3.7. A magnified region of a hypothetical sediment with no compaction. The left-hand diagram shows the velocity vectors of the porewater, u, and solid sediment, w, at a chosen depth relative to the sediment-water interface. In the right-hand diagram is the porewater velocity relative to a reference frame fixed to the chosen sediment grain, i.e. *u*. The direction of the arrows indicates the direction of flow, and their length the magnitude. (No externally impressed flow - see below.)

3.4.1 Advection

Advection is the bulk flow of an environmental medium, e.g. water or sediment, regardless of its cause. The advective flux, F_A (mass or moles moving through a unit area per unit time), for a solute in a 1-D model is simply

$$F_A = \varphi u C \tag{3.23}$$

where u is the velocity of the porewater flow in a chosen reference frame. In a 1-D model, this advection can be due to burial, compaction, and/or externally impressed flow (hydrological flow).

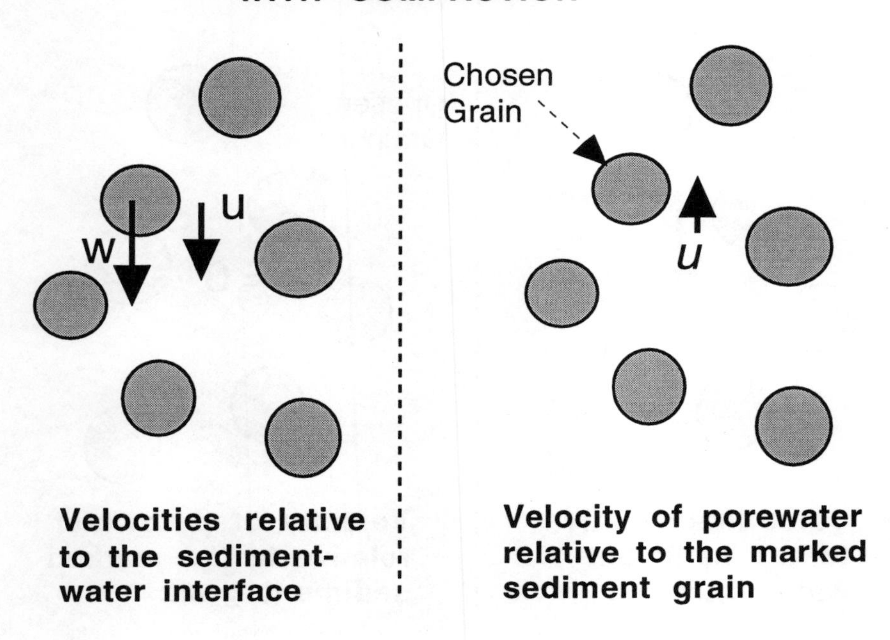

Figure 3.8. Same magnified region as in Fig. 3.7, but this time with compaction. Notice that the porewater advection, u, is smaller than the solid sediment advection, w, when viewed from the sediment-water interface. The pore fluid advection relative to a sediment grain, *u*, is upward. (No external flow - see below.)

Burial and Compaction. The *burial* component of porewater advection results from the sediment-water interface moving away from porewater as sediment accumulates. *Compaction* is the closer packing of sediment particles, caused by the weight of the overlying solid sediment column with a concomitant expulsion of pore fluids. This advection of porewater is relative to the sediment grains (Figs 3.7 and 3.8), and it does not necessarily result in net porewater movement out through the sediment-water interface (Berner, 1980).

During compaction, porewater moves up toward the sediment-water interface when viewed from a sediment grain (Fig. 3.8 right), but in an accumulating sediment column, it usually still moves downward in a reference frame anchored to that same interface (Fig. 3.8 left). In the latter case, the porewater velocity downward appears to be smaller than that of the solids.

Externally Impressed Flow. This type of flow results from gradients in the effective pressure, p′, or thermal gradients (Imboden, 1975; Berner, 1980; Phillips, 1991).

Such flows are not the rule in muddy aquatic sediments because of their low permeabilities, but sandy sediments offer much greater opportunities.

Externally impressed flow sometimes occurs in lacustrine and near-shore coastal sediments, where aquifers come into contact with bottom sediments (Fig. 3.9). In addition, sediments overlying cooling basalt in the deep ocean can be the locus of thermally induced convection (Fig. 3.10). Porewater flows can also results from the pressure variations caused by gravity waves (see later in this chapter).

Pressure-driven flow is governed by *Darcy's Equation*, which has the 1-D form (Phillips, 1991)

$$u_x = -\frac{\mathbf{k}}{\varphi\mu}\frac{\partial p'}{\partial x} \tag{3.24}$$

where $\mathbf{k}$ is the permeability, μ is the dynamic viscosity of the porewater, and p' is the reduced pressure (the difference between the actual and average hydrostatic pressures) - see Chapter 4.

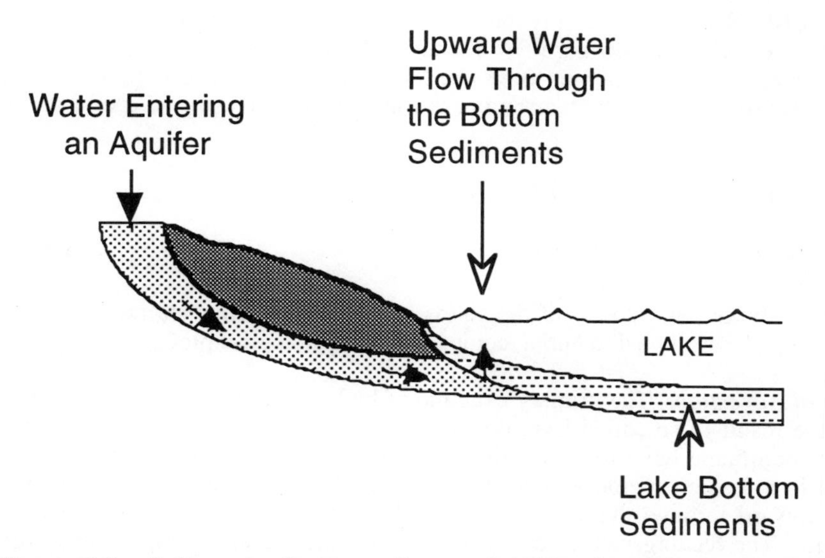

Figure 3.9. Schematic diagram of upward infiltration (flow) of groundwater through lake-bottom sediments caused by aquifer discharge. In near-shore marine areas, seawater may also seep into aquifers, leading to a downward flow (advection) through the sediment. Note that the scale of lateral variations in the sediment flow is often so large that the resulting advection is essentially vertical.

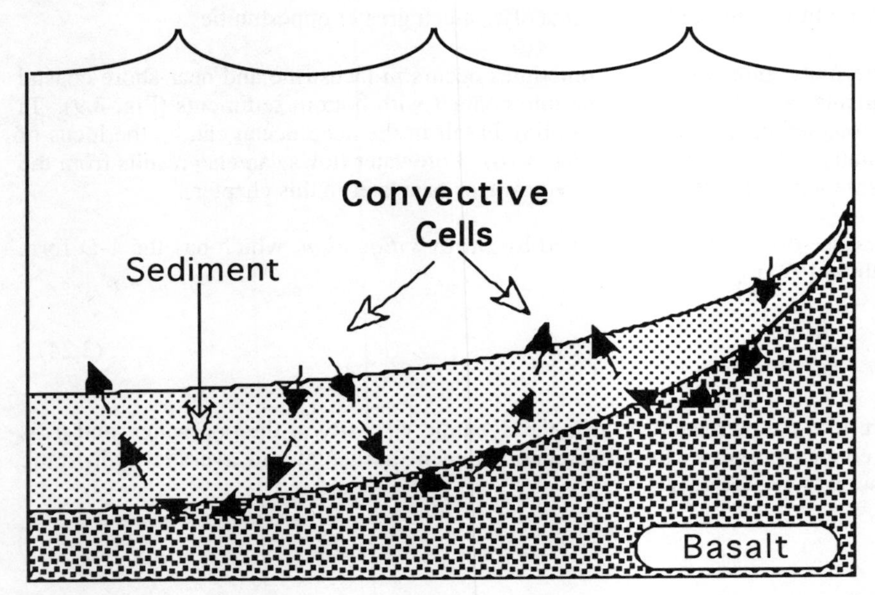

Figure 3.10. Schematic representation of hydrothermal flow through sediment overlying cooling oceanic basalt. Because of the large spatial scale of the convective cells in hydrothermal flow, the local porewater motions are, to a good degree of approximation, either upward or downward, and a 1-D model can be used.

For a species of the solids, the advective flux, F_A, due to burial in 1-D is simply

$$F_A = (1 - \varphi)wB \tag{3.25}$$

where w is the burial velocity of the solids in the chosen reference frame. Solid advection can be attributed to burial, compaction, and biological processes.

Biological Advection. The primary example of biologically mediated solid advection is the result of so-called head-down deposit feeders (Fig. 3.11). These burrowing macrofauna have their mouths at depth in the sediment and ingest solid material in a feeding region around their mouths. The ingested sediment is then carried upward through the animal's gut to the surface, where it is defecated or deposited. The feeding void at depth is filled by collapse and compaction of the sediment, which engenders a downward advection from all depths between the void and the surface.

A collection of such organisms living in a sediment will create a time-averaged transport of sediment up to the surface through their guts, and a downward advection of sediment to fill the feeding voids. The upward motion is sometimes called a biological "conveyor-belt", and it is modelled as a "*nonlocal*" transport, as described

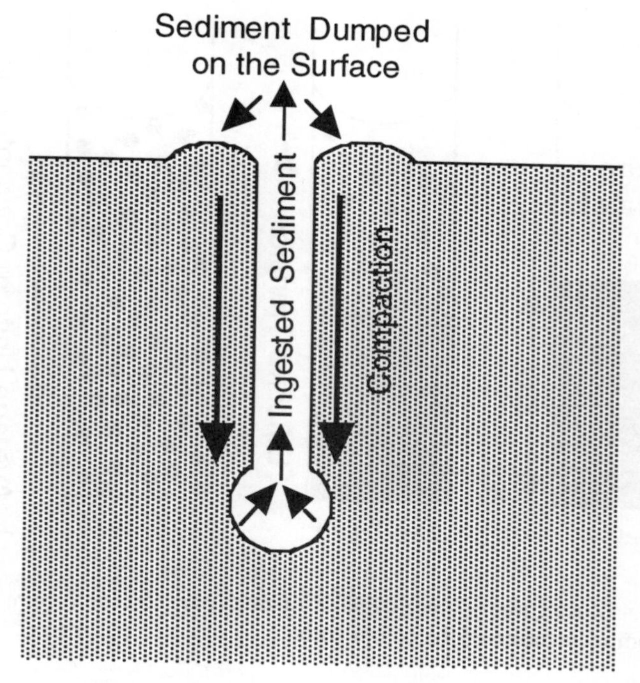

Figure 3.11. Schematic representation of sediment motion in the advective loop caused by a *head-down* deposit feeder. Only the downward motion is modelled as an advection. Transport to the surface is a *nonlocal* process (see later in this chapter).

later. It is only the downward component of this motion due to compaction that is treated as an advection.

3.4.2 Diffusive Fluxes

Molecular and Ionic Diffusion. Diffusion of solutes "... is the process by which matter is transported from one point of a system to another as a result of random molecular motions" (Crank, 1975, p. 1). If there are more ions/molecules of a dissolved substance (or even colloids) in one region of a fluid, random molecular motions due to collisions between solutes and water molecules will eventually lead to a homogenization (Fig. 3.12).

Therefore, diffusion acts to eliminate gradients of concentration (or chemical potential) by moving material down gradient. The 1-D flux, F_D, resulting from molecular/ionic diffusion in the porewaters is (see Chapter 4 for a derivation):

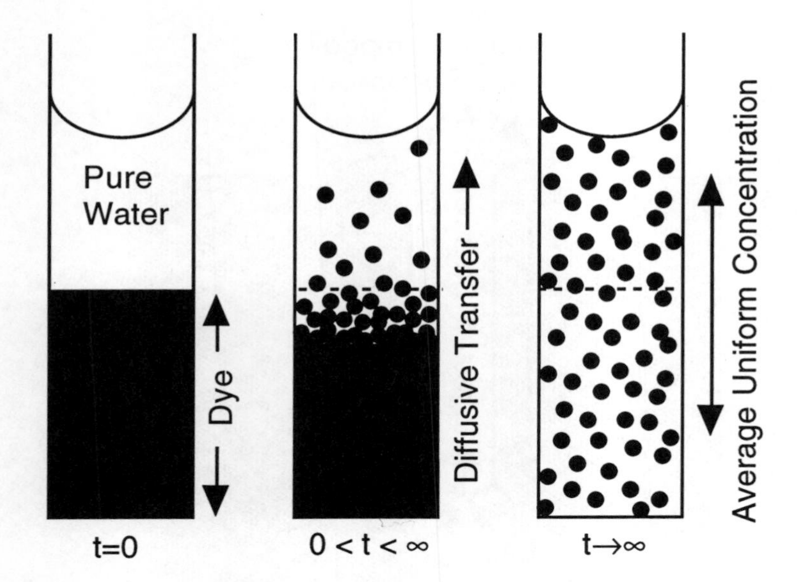

Figure 3.12. Illustration of the diffusion of dye (particles) into pure water. The result is a homogenization of the composition.

$$F_D = -\varphi D' \frac{\partial C}{\partial x} \tag{3.26}$$

where D′ is the effective diffusion coefficient of the solute (colloid) in the pores (see below). The negative indicates the flux in a direction opposite to the gradient. As the flux is a vector quantity, it has both a magnitude and a direction. If total fluid is being considered, then there is no diffusive flux of total fluid from molecular/ionic diffusion, $F_D = 0$. (Derivation of Eq. (3.26) is delayed until Chapter 4.)

Equation (3.26) is known as *Fick's First Law* for a porous medium. It is an example of a *constitutive relation* (see the next chapter). Constitutive relations are equations or definitions that must be added to the conservation equation(s) in order to resolve unknown terms, e.g. the form of F_D. They are separate statements, derived from experiments and/or independent theory.

The effective diffusivity, D′, that appears in these porous media equations is related to the value of the molecular/ionic diffusion coefficient in free solution, D_o, but they are not equal because the path followed by ions/molecules/colloids during diffusion in sediments is tortuous (Fig. 3.13).

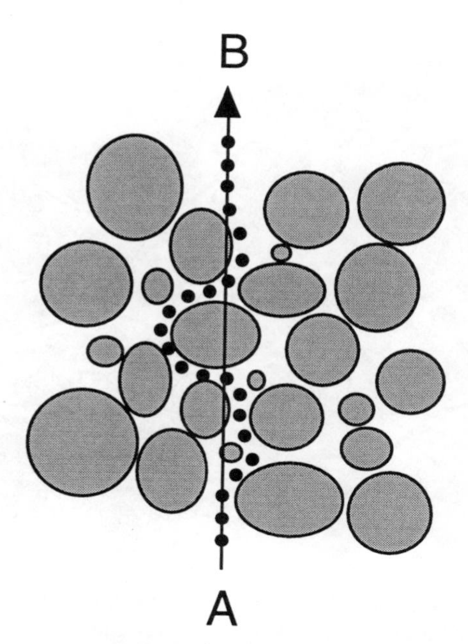

Figure 3.13. A solute trying to diffuse from point A to point B cannot pass through solid sediment particles (filled shapes). The solute must travel around these particles and so traverses the *tortuous* dotted path, rather than the direct solid-line path.

In one dimension, the molecular/ionic/colloid diffusive flux is given then by

$$F_D = -\frac{\varphi D_o}{\theta^2}\frac{\partial C}{\partial x} \tag{3.27}$$

where θ is called the *tortuosity*. The tortuosity is always greater than one, and reflects the longer path that diffusing molecules/ions must follow (see Chapter 4).

Hydrodynamic Dispersion. Dispersion is a type of mixing that accompanies hydrodynamic flows (see above) through more permeable sediments, such as sands. At the pore level, not all the paths available for porewater flow are the same. Some are more constricted and tortuous than others and, therefore, harder to follow. The flow concentrates in the more open channels, and the fluid in these preferred paths moves faster that the average bulk velocity would suggest. Preferred paths just as commonly diverge from each other, causing a solute mass to spread out (Fig. 3.14).

The convergence and divergence of micro-flows cause mixing of the flowing porewater and, consequently, mixing of any solutes being carried. A tracer released at a point will spread out laterally to create a "cloud". This mixing is usually modelled as a type of diffusion,

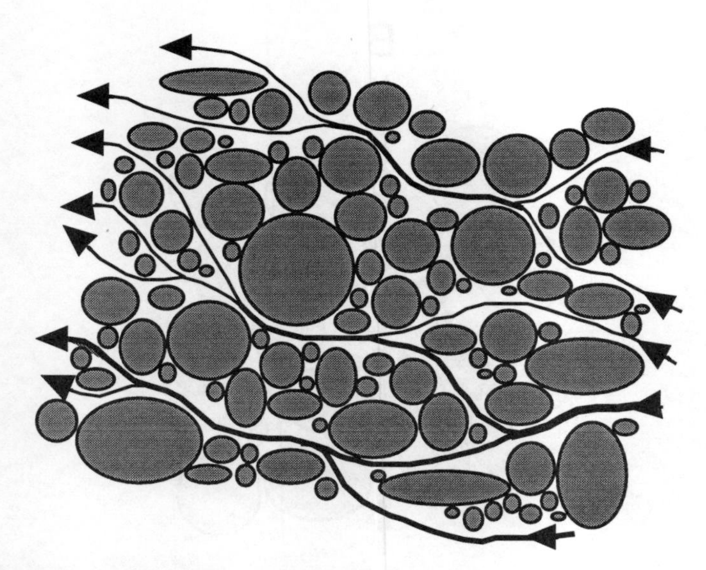

Figure 3.14. Possible flow paths in a random porous sediment. The presence of the haphazardly placed grains causes flow paths (stream lines) of flowing porewater to both converge and diverge. This leads to hydrological dispersion of solutes.

$$\left(F_{dispersion}\right)_i = -\varphi D^* \frac{\partial C}{\partial x} \qquad (3.28)$$

where D^* is the effective vertical dispersion coefficient. (Derivation of this last equation can be found in Chapter 4.)

Another source of advection and dispersion in permeable (e.g. sandy) sediments is the movement of gravity waves at the surface of the overlying water. The pressure changes induced by these waves are transmitted into the seabed (Putnam, 1949; Sleath 1984), and these pressure variations generate Darcian flows of porewater (Fig. 3.15).

For a flat sediment-water interface subject to symmetrical gravity waves of relatively high frequency, as illustrated schematically in Fig. 3.15, the trajectories of the porewater parcels will be circles, which diminish in radius with depth, if the sediment column is very much thicker than the water column height. If the sediment column is of the same order of thickness as the water column or thinner, then the paths are ellipses that flatten to lines at the sediment base (Webster and Taylor, 1992). In either case, the paths are closed, and the porewater parcels return to their original positions. When these flows are averaged over a wave length/period, there is no net advective movement of interstitial waters.

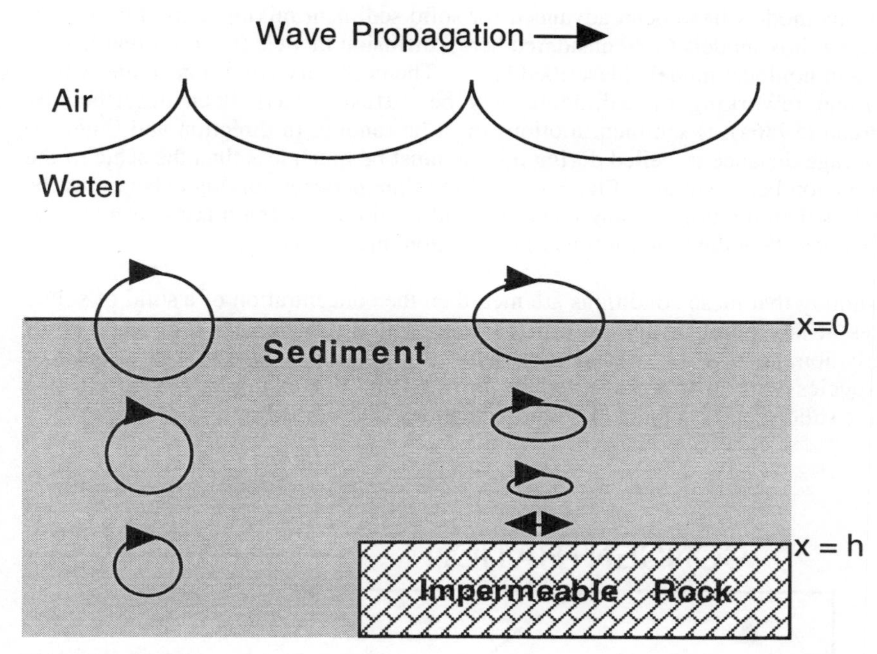

Figure 3.15. Porewater flow paths due to gravity waves over a permeable sediment with a flat sediment-water interface. The circular paths occur when the sediment is much thicker that the water column depth, while the ellipsoidal paths occur when the sediment thickness is of the same order of magnitude or thinner.

While the porewater motions due to symmetric gravity waves on a flat bed appear to cancel out when averaged, they nevertheless generate non-zero dispersion of porewater and solutes. In general, this dispersion is a function of the magnitude of the porewater velocity and the grain size (Rutgers van der Loeff, 1981; see Chapter 4). The germane consequence of this fact is that Eq. (3.28) will contribute to the 1-D porewater transport if there exists a vertical gradient of a solute.

Bioturbation. We need to consider next fluxes that result from biological mixing. As intimated earlier, bioturbation comprises all kinds of displacements within unconsolidated sediments and soils produced by the activity of organisms (paraphrased from Richter, 1952). These activities include burrow and tube excavation and their ultimate collapse or infilling, ingestion and excretion of sediment, plowing through the surface sediment, and building of mounds and digging of craters. During these activities, both solids and pore fluids are moved.

Bioturbation is to sediment movement what turbulence is to flowing fluids, but unlike the theory for turbulence, the study of bioturbation is relatively new and its theory is both less developed and less well known. Consequently, it is necessary to spend a little time here delving into the mathematical description of bioturbation.

Various models have been advanced for solid sediment mixing caused by bioturbation, i.e. box models (not considered here), diffusion models (i.e. ones that lead to flux), and nonlocal models (described later). The necessary conditions under which biological reworking of sediments may be diffusive have been described by Boudreau (1986a): i) sediment motions must be random in direction and time; ii) the average distance travelled during mixing must be much less than the scale of the phenomenon being studied (Fig. 3.16); iii) the time between mixing events must be much less than the time for any reactions; and iv) time between mixing events must be much less than the time for tracer introduction into the sediments.

Assuming that these conditions are met, then the concentration of a solid or solute species at any point in the bioturbated zone will fluctuate with time because of bioturbation; i.e. it will be a random variable. Instantaneous conservation balance of that species will still be governed by Eq. (3.19). For example, the burial of a reactive solid species subject to these fluctuations is described by

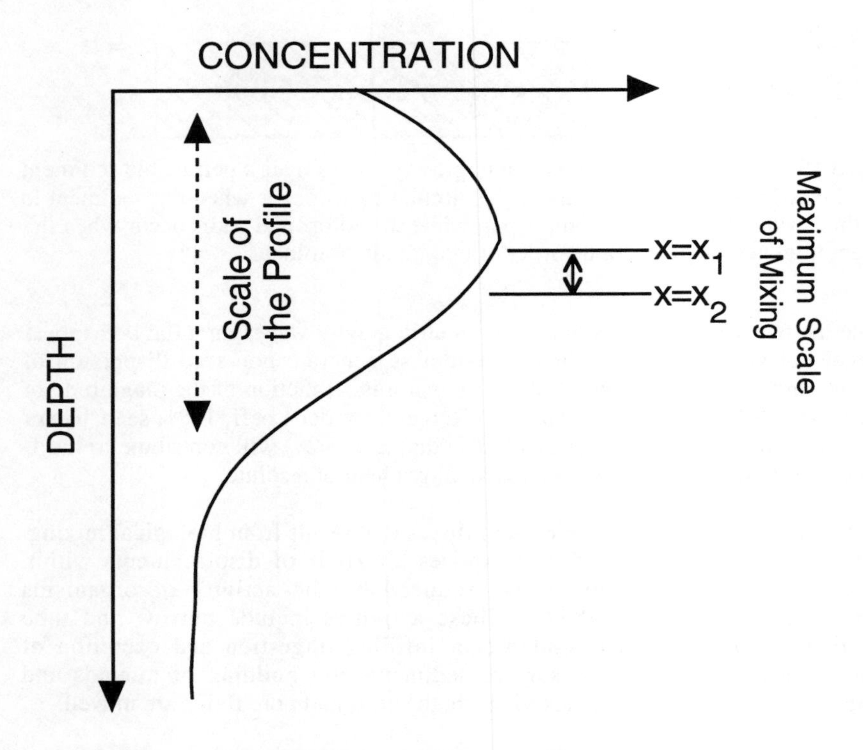

Figure 3.16. Schematic illustration of (small scale) local mixing. The interval $x=x_1$ to $x = x_2$ represents the maximum distance sediment is displaced during bioturbation. The concentration profile under consideration has a much larger scale for change.

$$\frac{\partial \tilde{\varphi}_S \tilde{B}}{\partial t} = -\frac{\partial \tilde{\varphi}_S \tilde{w} \tilde{B}}{\partial x} + \sum \tilde{R}_B \tag{3.29}$$

where φ_S is the solid volume fraction of the sediment (1-φ), and the tildes in this last equation indicate the random (fluctuating) nature of these variables due to bioturbational motions. Notice that the flux is purely advective on the time scale of the fluctuations. No diffusion appears in this instantaneous balance because there is no equivalent to molecular diffusion for solid species.

However, we wish to model a time-averaged (non-fluctuating) concentration profile, i.e. one that does not contain the random variations. To do this, we employ the Reynolds decomposition and averaging scheme given in Chapter 2. To this effect we can write each variable as the sum of a mean value (with an overbar) and a fluctuation about this mean (with a prime):

$$\tilde{\varphi}_S = \overline{\varphi}_S + \varphi'_S \tag{3.30}$$

$$\tilde{B} = \overline{B} + B' \tag{3.31}$$

$$\tilde{w} = \overline{w} + w' \tag{3.32}$$

These mean values may still have long-term time-dependencies, but they no longer contain random noise.

Substituting these new quantities into Eq. (3.29) produces:

$$\begin{aligned}\frac{\partial \overline{\varphi}_S \overline{B}}{\partial t} + \frac{\partial \overline{\varphi}_S B'}{\partial t} + \frac{\partial \varphi'_S \overline{B}}{\partial t} + \frac{\partial \varphi'_S B'}{\partial t} = -\frac{\partial}{\partial x}\Big(&\overline{\varphi}_S \overline{w} \overline{B} + \overline{\varphi}_S \overline{w} B' + \overline{\varphi}_S w' \overline{B} \\ &+ \varphi'_S \overline{w} \overline{B} + \overline{\varphi}_S w' B' + \varphi'_S \overline{w} B' + \varphi'_S w' \overline{B} + \varphi'_S w' B'\Big) + \sum \overline{R}_B + \sum R'_B\end{aligned} \tag{3.33}$$

where the second reaction term contains the effects of fluctuating components on the reactions, which may be nonlinear.

Next, each term in Eq. (3.33) is time-averaged over a period encompassing many fluctuations, Δt_{av}, but much shorter than the time scale for long-term transient behavior. In doing this, one can take advantage of the averaging properties given by Eqs (2.59) through (2.65). In addition, the assumptions regarding the fluctuations imply that they are small compared with the mean values. If this were not the case, then mixing would be occurring on a scale comparable to the scale of the species profiles (Fig. 3.16), which is expressly forbidden (i.e. the mixing would be nonlocal, see below). Consequently, the fluctuating reaction term is small compared with the mean reaction term, and it can be neglected. The result of averaging Eq. (3.33) is

$$\frac{\partial \overline{\varphi}_S \overline{B}}{\partial t} + \frac{\partial \overline{\varphi'_S B'}}{\partial t} = -\frac{\partial}{\partial x}\left(\overline{\varphi}_S \overline{w} \overline{B} + \overline{\varphi}_S \overline{w' B'} + \overline{w}\,\overline{\varphi'_S B'} + \overline{B}\,\overline{\varphi'_S w'} + \overline{\varphi'_S w' B'}\right) + \sum \overline{R}_B \tag{3.34}$$

The terms $\overline{\varphi}_S \overline{w'B'}$ and $\overline{B}\,\overline{\varphi_S' w'}$ are of particular interest to us. What is their effect? Consider Fig. 3.17A, which illustrates what occurs during a bioturbational fluctuation that displaces material within a concentration gradient that decreases on average. A displacement moves sediment (or porewater) a (local) distance, ℓ, which is also a random variable. When w' is positive (in the x-coordinate system), then the material transported downward has a higher concentration than the material it mixes with; i.e. B' is a positive. If the displacement had been upward, then w' would be negative and so would B'. In either case, the time average of product $w'B'$ is a positive quantity; given that it is a flux, its net effect must be to transport material from high concentration to low concentration, in a direction opposite to that of the mean gradient.

In Fig. 3.17B, the mean concentration gradient is reversed compared with that in Fig. 3.17A. An upward displacement still means a negative w', but the upward bound material has a higher concentration than that which it encounters; thus, B' is positive in this case, and the mean flux caused by the fluctuations is a negative flux. A downward displacement reverses the sign of both w' and B', so that the flux remains negative. A negative flux is again opposite to the direction of the mean gradient. In both cases the bioturbational fluctuations act to eliminate the gradient, as in a diffusion.

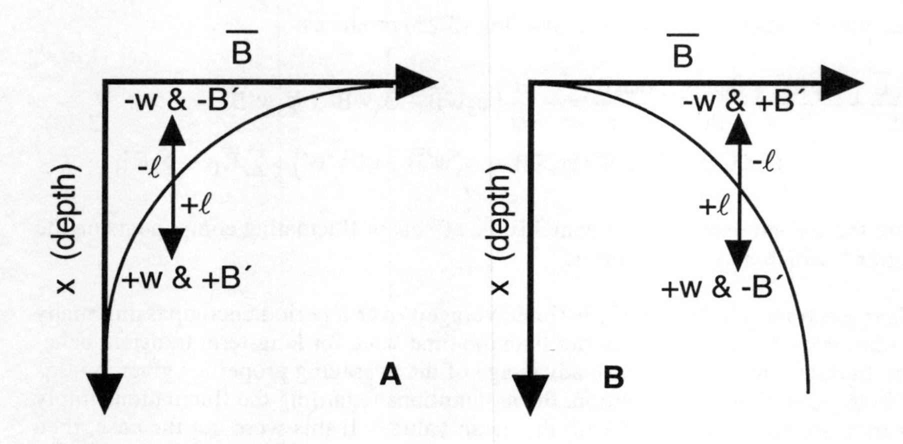

Figure 3.17. Effect of bioturbational fluctuations of two hypothetical sediment species distributions. Diagram A illustrates the case of a negative gradient, and Diagram B contains the case of a positive gradient. (These gradients are exaggerated for the purposes of illustration only.)

These fluxes can be expressed in terms of mean values, rather than fluctuations. First, consider an upward motion in Fig. 3.17A. Here B´ is B(x-ℓ) - B(x). If B(x-ℓ) is expressed in terms of a Taylor series in the mean gradient at x, then

$$B' \approx \ell \frac{\partial \overline{B}}{\partial x} \tag{3.35}$$

Secondly, w´ can be expressed as -ℓ divided by the total time between fluctuations, τ, which includes actual movement and rest times (Wheatcroft et al., 1990),

$$w' \approx -\frac{\ell}{\tau} \tag{3.36}$$

The term $\overline{w'B'}$ can be expressed, consequently, as

$$\overline{w'B'} \approx -\overline{\left(\frac{\ell^2}{\tau}\right)} \frac{\partial \overline{B}}{\partial x} \tag{3.37}$$

If the fluctuation in question is the downward one in Fig. 3.17A, then

$$B' \approx -\ell \frac{\partial \overline{B}}{\partial x} \tag{3.38}$$

and it is possible to write w´ as

$$w' \approx \frac{\ell}{\tau} \tag{3.39}$$

so that $\overline{w'B'}$ is still given by Eq. (3.37). In fact, the two remaining cases in Fig. 3.17B also lead to Eq. (3.37). Similarly, it is easily shown that

$$\overline{\varphi_s' w'} \approx -\overline{\left(\frac{\ell^2}{\tau}\right)} \frac{\partial \overline{\varphi}_s}{\partial x} \tag{3.40}$$

$$\overline{\varphi_s' B'} \approx \overline{\ell^2} \frac{\partial \overline{\varphi}_s}{\partial x} \frac{\partial \overline{B}}{\partial x} \tag{3.41}$$

$$\frac{\partial \overline{\varphi_s' B'}}{\partial t} \approx \overline{\left(\ell^2\right)} \frac{\partial}{\partial t} \left(\frac{\partial \overline{\varphi}_s}{\partial x} \frac{\partial \overline{B}}{\partial x} \right) \tag{3.42}$$

The three-part correlation $\overline{\varphi_s' w' B'}$ is proportional to $\overline{\ell}^3/\tau$.

Now take the limit as both $\overline{\ell^2} \to 0$ and $\tau \to 0$, but in such a way that the ratio $\overline{\ell^2}/\tau$ remains a non-zero value; i.e. define

$$D_B \equiv D_B(x,t) \equiv \overline{\left(\frac{\ell^2}{\tau}\right)} \tag{3.43}$$

which may be different at each depth and also a function of time. The parameter D_B is known as the *biodiffusion coefficient.* In addition, the terms defined by Eqs (3.41) and (3.42) and the three term correlation involving $\varphi_s' w' B'$ reduce to zero. Therefore, Eq. (3.34) now reads as

$$\frac{\partial \overline{\varphi}_s \overline{B}}{\partial t} = \frac{\partial}{\partial x}\left\{D_B\left(\overline{\varphi}_s \frac{\partial \overline{B}}{\partial x} + \overline{B}\frac{\partial \overline{\varphi}_s}{\partial x}\right)\right\} - \frac{\partial(\overline{\varphi}_s \overline{w} \overline{B})}{\partial x} + \sum \overline{R}_B \tag{3.44}$$

This is the conservation equation for the time-averaged concentration, in which local bioturbation is present as a diffusive process.

Now, bioturbation may act in two different ways with respect to the mixing of total solids and total porewater fluids. One type of behavior is to mix solids and fluid together to remove porosity gradients, which I call *interphase mixing*. In this case, both diffusive flux terms in Eq. (3.44) should be kept. The bioturbational flux for a solid species with interphase mixing is then

$$(F_B)_{solid} = -D_B \frac{\partial \varphi_s B}{\partial x} \tag{3.45}$$

Because fluid is being mixed simultaneously in interphase mixing, this implies that there is an additional flux for a dissolved species caused this process,

$$(F_B)_{solute} = -D_B \frac{\partial \varphi C}{\partial x} \tag{3.46}$$

This flux does not represent irrigation of porewaters (see below), and should not be confused with such a process (but it can include the effects of meiofauna - Aller and Aller, 1992). Equation (3.46) accounts only for small-scale mixing caused by the small displacements as a result of infaunal activity.

The second end-member case occurs if bioturbation does not mix solids with fluid to eliminate porosity gradients, which I call *intraphase mixing.* The flux for a solid species is then given by

$$(F_B)_{solid} = -\varphi_s D_B \frac{\partial B}{\partial x} \tag{3.47}$$

There may still be a solute flux in this case, and this flux is

$$\left(F_B\right)_{solute} = -\varphi D_B \frac{\partial C}{\partial x} \tag{3.48}$$

At this time there is no unequivocal evidence favoring interphase or intraphase mixing in natural sediments. Furthermore, it is likely that mixing is some combination of these two extremes.

The D_B in Eqs (3.45) and (3.46) and in Eqs (3.47) and (3.48) has the same value within each pair. Measurements of D_B for solids (see Chapter 4) indicate that prevalently $D_B << D_o/\theta^2$, and the biodiffusive term for solutes can often be ignored, except when dealing with strongly adsorbing species (Schink and Guinasso, 1978; Berner, 1980). Biological activity significantly affects porewater concentrations more commonly via irrigation, but that process is not a diffusion in a 1-D model. Physical mixing of sediments by resuspension has also been represented as a diffusive transport, but it is not obvious that this is a correct description. In particular, the typical mixing scale during a physical disturbance may or may not be local.

3.4.3 The Standard (Local) Diagenetic Equations

The stated forms of F can now be substituted to get the canonical forms of the conservation equations for solutes and solids in 1-D (see also Berner, 1980). Conservation of a porewater species with intraphase mixing is

$$\underbrace{\frac{\partial \varphi C}{\partial t}}_{\text{Accumulation}} = \underbrace{D_o \frac{\partial}{\partial x}\left(\frac{\varphi}{\theta^2}\frac{\partial C}{\partial x}\right)}_{\substack{\text{Divergence of}\\ \text{Molecular Diffusive}\\ \text{Flux}}} + \underbrace{\frac{\partial}{\partial x}\left(\varphi D_B \frac{\partial C}{\partial x}\right)}_{\substack{\text{Divergence of}\\ \text{Intraphase Mixing}\\ \text{Flux}}} - \underbrace{\frac{\partial \varphi u C}{\partial x}}_{\substack{\text{Divergence}\\ \text{of the Advective}\\ \text{Flux}}} + \underbrace{\sum R}_{\text{Reaction}} \tag{3.49}$$

and with interphase mixing,

$$\underbrace{\frac{\partial \varphi C}{\partial t}}_{\text{Accumulation}} = \underbrace{D_o \frac{\partial}{\partial x}\left(\frac{\varphi}{\theta^2}\frac{\partial C}{\partial x}\right)}_{\substack{\text{Divergence of}\\ \text{Molecular Diffusive}\\ \text{Flux}}} + \underbrace{\frac{\partial}{\partial x}\left(D_B \frac{\partial \varphi C}{\partial x}\right)}_{\substack{\text{Divergence of}\\ \text{Interphase Mixing}\\ \text{Flux}}} - \underbrace{\frac{\partial \varphi u C}{\partial x}}_{\substack{\text{Divergence}\\ \text{of the Advective}\\ \text{Flux}}} + \underbrace{\sum R}_{\text{Reaction}} \tag{3.50}$$

and a comparison with Eq. (3.5) shows that diagenesis of a solute is comprised of diffusion, bioturbation, *differential* compaction, and reaction.

One-dimensional conservation of a solid species with intraphase mixing is governed by:

$$\underbrace{\frac{\partial(1-\varphi)B}{\partial t}}_{\text{Accumulation}} = \underbrace{\frac{\partial}{\partial x}\left((1-\varphi)D_B\frac{\partial B}{\partial x}\right)}_{\substack{\text{Divergence of the}\\ \text{Intraphase Mixing}\\ \text{Flux}}} - \underbrace{\frac{\partial(1-\varphi)wB}{\partial x}}_{\substack{\text{Divergence of the}\\ \text{Burial Flux}}} + \underbrace{\sum R_B}_{\text{Reaction}} \tag{3.51}$$

or with interphase mixing:

$$\underbrace{\frac{\partial(1-\varphi)B}{\partial t}}_{\text{Accumulation}} = \underbrace{\frac{\partial}{\partial x}\left(D_B\frac{\partial(1-\varphi)B}{\partial x}\right)}_{\substack{\text{Divergence of the}\\ \text{Interphase Mixing}\\ \text{Flux}}} - \underbrace{\frac{\partial(1-\varphi)wB}{\partial x}}_{\substack{\text{Divergence of the}\\ \text{Burial Flux}}} + \underbrace{\sum R_B}_{\text{Reaction}} \tag{3.52}$$

respectively; therefore, diagenesis is mixing, differential compaction and reaction. A widely recognized special case of Eqs (3.51) and (3.52) occurs when both φ and D_B are constants and ΣR is a first-order decay, i.e. $-(1-\varphi)\lambda B$,

$$\frac{\partial B}{\partial t} = D_B\frac{\partial^2 B}{\partial x^2} - w\frac{\partial B}{\partial x} - \lambda B \tag{3.53}$$

This is the classic *bioturbation equation* for a solid radioactive tracer.

3.5 Special Equations for Total Solids and Total Fluid

Berner (1971,1980) examines a number of special cases for total solids and fluid that deserve repeating. Let us begin by setting $C = \rho_f$ and $B = \rho_s$. To a high order of approximation, both these variables are constant in time and space in the vast majority of diagenetic situations, and they are treated as such here.

Next we assume that there is only intraphase mixing or no mixing at all, there are no reactions that significantly affect the volume fraction of total fluid or total solids, and that there is no externally impressed flow of porewater. As a consequence of these assumptions, we obtain for the pore fluid,

$$\frac{\partial\varphi}{\partial t} = -\frac{\partial\varphi u}{\partial x} \tag{3.54}$$

and, for the solids (φ_s=1-φ),

$$\frac{\partial(1-\varphi)}{\partial t} = -\frac{\partial(1-\varphi)w}{\partial x} \tag{3.55}$$

or

$$\frac{\partial \varphi}{\partial t} = \frac{\partial (1-\varphi)w}{\partial x} = -w\frac{\partial \varphi}{\partial x} + (1-\varphi)\frac{\partial w}{\partial x} \tag{3.56}$$

and subtracting Eq. (3.56) from Eq. (3.54),

$$\frac{\partial[\varphi u + (1-\varphi)w]}{\partial x} = 0 \tag{3.57}$$

No Compaction. Consider a situation in which the material derivatives for the porewater and the solid volume fractions are both zero (at constant density); i.e.

$$\frac{D\varphi}{Dt} = \frac{\partial \varphi}{\partial t} + u\frac{\partial \varphi}{\partial x} = 0 \tag{3.58}$$

and

$$\frac{D(1-\varphi)}{Dt} = \frac{\partial (1-\varphi)}{\partial t} + w\frac{\partial (1-\varphi)}{\partial x} = 0 \tag{3.59}$$

or, equivalently,

$$\frac{D\varphi}{Dt} = \frac{\partial \varphi}{\partial t} + w\frac{\partial \varphi}{\partial x} = 0 \tag{3.60}$$

Substitution of Eq. (3.60) into Eq. (3.56) gives that

$$\frac{\partial w}{\partial x} = 0 \tag{3.61}$$

which says that the burial velocity cannot change as a function of depth; however, burial velocity of the entire (solid) sediment column may still change as a function of time (Berner, 1980). Substitution of Eq. (3.58) into Eq. (3.54) similarly tells us that

$$\frac{\partial u}{\partial x} = 0 \tag{3.62}$$

and, like the burial velocity of the solids, the porewater velocity is not a function of depth. Finally, using Eqs (3.61) and (3.62) in Eq. (3.57), there results

$$(u-w)\frac{\partial \varphi}{\partial x} = 0 \tag{3.63}$$

For this equation to be identically zero at all depths and at all times, either the porosity is a constant with depth (which is trivial) or $u = w$; i.e. the porewater velocity and the solids' burial velocity are equal. In Fig. 3.7 we argued heuristically that, when these two velocities are the same, there is *no compaction.* Therefore, Eqs (3.58) and (3.59) define the conditions for no compaction; i.e. both total derivatives need to be zero. Berner (1980) does not discuss the particular requirement set by Eq. (3.58). There is no restriction on φ, and it may change arbitrarily with space and time (Berner, 1980).

Steady-State Compaction. In Eqs (3.54) and (3.55), let the partial derivatives in time be zero. This is the steady state, and we obtain the two equations:

$$\frac{d\varphi u}{dx} = 0 \tag{3.64}$$

$$\frac{d(1-\varphi)w}{dx} = 0 \tag{3.65}$$

or upon integration,

$$\varphi u = \varphi_x u_x \tag{3.66}$$

$$(1\text{-}\varphi)w = (1\text{-}\varphi_x)w_x \tag{3.67}$$

respectively, where the subscript x indicates known values at some depth x. The usual treatment of these equations assumes that this point is some great depth or infinity, but it could very well be any depth. If, in addition, compaction ceases at this depth x, then $u_x = w_x$, and we further obtain the useful relationship (Berner, 1980)

$$\varphi u = \varphi_x\, w_x \tag{3.68}$$

3.6 Static Conservation of Volume

It is sometimes necessary to consider the total volume of solids plus fluid, e.g. Schink and Guinasso (1980) and Boudreau (1990a,b) in their models for opal diagenesis. Total volume is not a transportable quantity, nor is it affected by reactions. A statement for the conservation of total volume is simply

$$\varphi + \sum_{i=1}^{n} \varphi_i = 1 \tag{3.69}$$

where φ_i is the volume fraction of the i-th solid phase/mineral of the sediment (organic matter, opal, calcite, clay mineral, etc.), and n is the total number of mineral/phases in the solid sediment mixture. It is also possible to speak of the static conservation of total solid volume, i.e.

$$\sum_{i=1}^{n} \frac{B_i}{\rho_i} = 1 \tag{3.70}$$

Here ρ_i is the intrinsic density of i-th phase/mineral, and B_i is its concentration.

As an example of the use of these equations, let us consider the diagenesis of dissolvable solid, e.g. opal, in a sediment made of this solid plus an inert clay fraction. Conservation of the dissolvable solid at steady state and with constant porosity is governed by (Schink and Guinasso, 1980; Boudreau, 1990a)

$$D_B \frac{d^2B}{dx^2} - \frac{dwB}{dx} - k_{app}(C_{eq} - C)\frac{B}{\rho_B} = 0 \tag{3.71}$$

and the clay fraction by

$$D_B \frac{d^2M}{dx^2} - \frac{dwM}{dx} = 0 \tag{3.72}$$

where B, C, and M are the concentrations of the opal, dissolved silica, and clay, respectively, ρ_B and ρ_M are the intrinsic densities of pure opal and clay, respectively, k_{app} is the apparent rate constant for dissolution, and C_{eq} is the apparent equilibrium concentration for opal in water. In addition to these two equations, we have the boundary conditions (see Chapter 5)

$$F_B = \varphi_s\left(-D_B \left.\frac{dB}{dx}\right|_{x=0} + w(0)B(0)\right) \tag{3.73}$$

$$F_M = \varphi_s\left(-D_B \left.\frac{dM}{dx}\right|_{x=0} + w(0)M(0)\right) \tag{3.74}$$

$$\left.\frac{dB}{dx}\right|_{x\to\infty} = \left.\frac{dM}{dx}\right|_{x\to\infty} = 0 \tag{3.75}$$

where F_B and F_M are prescribed (known) fluxes of dissolvable solid (opal) and clay, respectively.

The important aspect of this example is that w is a function of the opal content of the sediment, not only in the broad sense that F_B influences w, but also because w will change with depth if dissolution removes opal with depth. This burial velocity can be written as a function of B and M; to do so, start by integrating Eq. (3.72),

$$D_B \left.\frac{dM}{dx}\right|_{x} - D_B \left.\frac{dM}{dx}\right|_{x=0} - w(x)M(x) + w(0)M(0) = 0 \tag{3.76}$$

However, boundary condition (3.74) immediately allows us to write this equation as:

$$\frac{F_M}{\varphi_s} = -D_B \left.\frac{dM}{dx}\right|_x + w(x)M(x) \tag{3.77}$$

Now we can make use of the appropriate form of Eq. (3.70), the static conservation of volume, i.e.

$$\frac{B}{\rho_B} + \frac{M}{\rho_M} = 1 \tag{3.78}$$

Substituting this into Eq. (3.77) for M gives

$$w(x) = \frac{\dfrac{F_M}{\rho_M} - \dfrac{\varphi_s D_B}{\rho_B}\dfrac{dB}{dx}}{\varphi_s\left(1 - \dfrac{B}{\rho_B}\right)} \tag{3.79}$$

This is the desired functionality of w with B, and its derivation was possible through the use of the static conservation equation (alternatively, see Keir, 1982).

3.7 Integral Conservation Balances in 1-D

It is also possible to derive equations that account for the mass balance of species over whole depth intervals, rather than at a specific depth. Calculation of concentration profiles from such equations is not possible, but these equations have very important roles to play in verifying experimental results and numerical calculations, as will be discussed later.

To derive these equations, it is necessary to perform a single definite integration on any of Eqs (3.49) through (3.52). (This is not the normal procedure to solve these equations, simply the method to obtain the integral balances.) For example, with Eq. (3.49), if the limits of integration are x=0 to x=L, then

$$\underbrace{\frac{\partial M}{\partial t}}_{\substack{\text{Total}\\\text{Accumulation}}} = \underbrace{D_o\left[\frac{\varphi}{\theta^2}\frac{\partial C}{\partial x}\right]_L}_{\substack{\text{Molecular Diffusive}\\\text{Flux at Depth } x = L}} + \underbrace{\left[\varphi D_B \frac{\partial C}{\partial x}\right]_L}_{\substack{\text{Intraphase Mixing}\\\text{Flux at } x = L}} - \underbrace{[\varphi u C]_L}_{\substack{\text{Advective Flux}\\\text{at } x = L}}$$

$$
-\underbrace{D_o\left[\frac{\varphi}{\theta^2}\frac{\partial C}{\partial x}\right]_0}_{\substack{\text{Molecular Diffusive}\\\text{Flux at the Sediment-}\\\text{Water Interface}}} - \underbrace{\left[\varphi D_B \frac{\partial C}{\partial x}\right]_0}_{\substack{\text{Intraphase Mixing}\\\text{Flux at the Sediment-}\\\text{Water Inteface}}} + \underbrace{[\varphi u C]_0}_{\substack{\text{Advective Flux}\\\text{at the Sediment-}\\\text{Water Interface}}} + \underbrace{\sum \int_0^L R\,dx}_{\substack{\text{Total Reaction}\\\text{through the}\\\text{Depth Interval}}} \tag{3.80}
$$

where M is the total amount (mass) of the solute in the interval. This equation says that the change in concentration in the total amount of solute is due to the difference between the input (by diffusion and advection at the sediment-water interface), the loss across the surface at x = L, and the total integrated amount of reaction that has occurred in the interval. (Close examination of this result also shows that this represents the technique for deriving box models.)

We are often interested only in the steady-state balance ($\partial M/\partial t = 0$), and the lower boundary is usually infinity, where gradients disappear. Porewater bioturbation and advection at the surface are also normally negligible; thus, Eq. (3.80) simplifies to

$$
\underbrace{D_o\left[\frac{\varphi}{\theta^2}\frac{\partial C}{\partial x}\right]_0}_{\substack{\text{Molecular Diffusive}\\\text{Flux at the Sediment-}\\\text{Water Interface}}} + \underbrace{[\varphi u C]_\infty}_{\substack{\text{Loss out the}\\\text{Bottom of the}\\\text{Model}}} = \underbrace{\sum \int_0^L R\,dx}_{\substack{\text{Total Reaction}\\\text{through the}\\\text{Depth Interval}}} \tag{3.81}
$$

This is an important result for diagenetic studies, both theoretical and applied. If an experimentalist measures the flux of some species at the sediment-water interface (e.g. with a chamber), and if the rate of reaction with depth of that species is obtained by some independent experimental technique, then Eq. (3.81) allows us to verify these results simply by depth integrating the rates and, thereafter, plugging the result into this equation. In numerical work, Eqs (3.80) and (3.81) also permit the modeller to check whether total mass balance prevails in a model calculation.

If the same operation is carried out on the 1-D conservation for a solid species, an interesting result is obtained if we know the total flux at the sediment-water interface, F_B,

$$
\underbrace{F_B}_{\substack{\text{Total Mixing + Advective}\\\text{Flux at the Sediment-}\\\text{Water Interface}}} - \underbrace{[\varphi_s w B]_\infty}_{\substack{\text{Advective Flux}\\\text{at the Base of}\\\text{the Model}}} = -\underbrace{\sum \int_0^L R_B\,dx}_{\substack{\text{Total Reaction}\\\text{through the}\\\text{Depth Interval}}} \tag{3.82}
$$

Again this can verify mass balance in experimental data or from modelling results. If the reaction terms in Eqs (3.81) and (3.82) can be equated (with the inclusion of some factor to account for unit differences), then a third integral balance between solutes and solid species can be obtained.

3.8 Well-Mixed Zones in 1-D

Mixing can be sufficiently fast, when compared with other processes, that the gradient in some variable(s) may disappear, to all intents and purposes. A classic example is a ^{14}C-distribution in the bioturbated region of sediments, but there are many others. How can well-mixed zones be treated within the context of the conservation equations derived above? This approach relies on averaging the appropriate conservation equation before taking the well-mixed limit. In this way, the derivation resembles that for an integral mass-balance.

Well-mixed zones are more typically a feature of solid species, so I will illustrate using Eq. (3.51). The particular choice of Eq. (3.51) is arbitrary, and Eq. (3.52) would do as well. The well-mixed zone is restricted to the depth range $0 \leq x \leq L$. Integrating Eq. (3.51) from 0 to L one obtains

$$\underbrace{\frac{\partial}{\partial t}\left(\int_0^L \varphi_s B\, dx\right)}_{\text{Total Accumulation}} = \underbrace{\left[\varphi_s D_B \frac{\partial B}{\partial x}\right]_L}_{\text{Intraphase Mixing Flux at x =L}} - \underbrace{\left[\varphi_s w B\right]_L}_{\text{Advective Flux at x = L}} - \underbrace{\left[\varphi_s D_B \frac{\partial B}{\partial x}\right]_0}_{\text{Intraphase Mixing Flux at the Sediment-Water Inteface}} + \underbrace{\left[\varphi_s w B\right]_0}_{\text{Advective Flux at the Sediment-Water Interface}} + \underbrace{\sum \int_0^L R\, dx}_{\text{Total Reaction through the Depth Interval}} \tag{3.83}$$

Designating the total flux at x = 0 by F_o and the total flux at x = L by F_L, Eq. (3.83) now reads

$$\frac{\partial}{\partial t}\left(\int_0^L \varphi_s B\, dx\right) = F_o - F_L + \sum \int_0^L R\, dx \tag{3.84}$$

Now take the limit, $D_B \rightarrow \infty$ in $0 \leq x \leq L$, such that everywhere

$$\frac{\partial B}{\partial x} = 0 \tag{3.85}$$

In other words, B and φ_s become independent of space, and Eq. (3.84) simplifies to:

$$\frac{d\varphi_s B}{dt} = \frac{1}{L}\left(F_o - F_L + \sum \int_0^L R\, dx\right) \tag{3.86}$$

which is the desired conservation equation for a well-mixed zone, i.e. a box model. (The flux F_L can be calculated from the expression for the flux at the top of the less-mixed layer below x=L.)

3.9 Multi-Dimensional Conservation Equations

3.9.1 Derivation

Some diagenetic phenomena must be described by multi-dimensional conservation equations. The results of the 1-D case can be generalized to two or three dimensions, but not by a simple extension of the derivation given above, as is aptly argued by Swan (1974). To appreciate this argument, consider a 2-D rectangular region ABB´A´ of sediment, as in Fig. 3.18.

In general, there are non-zero fluxes across all four sides of the sediment region ABB´A´. Let us designate the fluxes across the interfaces as F_{AB}, $F_{BB'}$, $F_{A'B'}$, and $F_{AA'}$, where the subscript indicates the corresponding edge. The flux components at the center of the square (x,y) are F_x and F_y, for the x and y directions, respectively. In order to express F_{AB} and $F_{A'B'}$ in terms of F_x through a Taylor Series, as in Eqs (3.12) and (3.13), F_{AB} cannot change with y along side AB. Similarly, $F_{A'B'}$ cannot change along A´B´. Conversely, in order to express $F_{AA'}$ and $F_{BB'}$ in terms of F_y at the center, these fluxes cannot vary in the x-direction. This creates a fundamental contradiction. These requirements are met only if there are no fluxes, which contravenes the original assumption that there were fluxes; therefore, the previous methodology will not work.

Conservation equations in two and three dimensions can be obtained only via integral mass balances. Integral mass balances can be formulated in a number of ways, and a simple approach is outlined below. An alternative methodology is contained in the later section on Conservation Equations via Volume Averaging.

The derivation is illustrated using the 2-D Cartesian area of sediment in Fig 3.18. The 3-D case in the Cartesian or any other coordinate system is a simple generalization, even if it involves a great deal more accounting of terms. The cumulative temporal change in bulk concentration over some arbitrary time interval, t to t+Δt, in ABB´A´ is

$$\int_t^{t+\Delta t} \int_A \frac{\partial \hat{C}}{\partial t} dA dt \tag{3.87}$$

where A is the area of ABB´A´. Likewise, the total production/consumption of the modelled species by chemical reactions in this area over the same time interval is

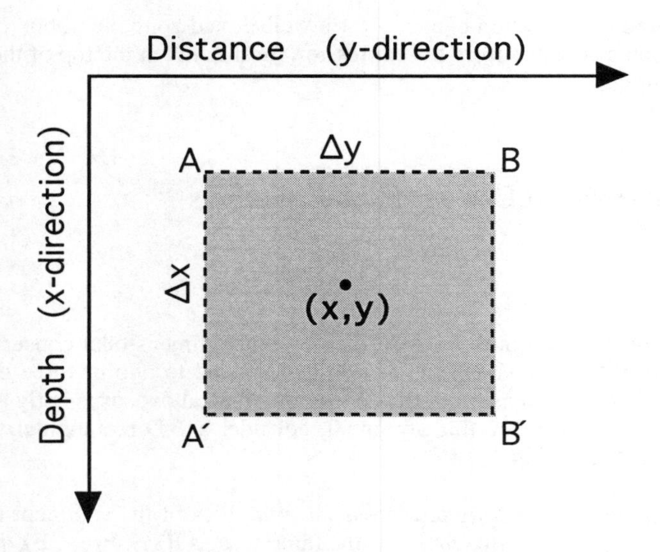

Figure 3.18. Arbitrary rectangular region of a sediment of size Δx in the depth-direction, Δy, in the lateral-direction and centered about a chosen point (x,y).

$$\int_t^{t+\Delta t} \int_A \sum R \, dA dt \tag{3.88}$$

The last component of the mass balance is the net effect of the fluxes across the sides. The total loss/gain by these fluxes is

$$\int_t^{t+\Delta t} \left(\int_{AB} F_{AB} \, dy + \int_{BB'} F_{BB'} \, dx + \int_{B'A'} F_{A'B'} \, dy + \int_{A'A} F_{AA'} \, dx \right) dt = - \int_t^{t+\Delta t} \int_{ABB'A'} F \cdot n \, d\ell \, dt \tag{3.89}$$

where ℓ is a unit of length around the perimeter and n is a unit vector that is directed outward (normal) to the sides of the surface. (Ignore the ambiguity at the corners).

The complete balance equates each of these changes, i.e. resulting from net accumulation, Eq. (3.87), to the net effects of reaction, Eq. (3.88), and the fluxes across the surface, Eq. (3.89); thus,

$$\int_t^{t+\Delta t} \int_A \frac{\partial \hat{C}}{\partial t} dA \, dt = - \int_t^{t+\Delta t} \int_{ABB'A'} F \cdot n \, d\ell \, dt + \int_t^{t+\Delta t} \int_A \sum R \, dA \, dt \tag{3.90}$$

With the use of the divergence theorem, Eq. (2.57), this last equation becomes

$$\int_t^{t+\Delta t}\int_A\left(\frac{\partial \hat{C}}{\partial t}+\nabla\cdot F-\sum R\right)dA\,dt=0 \tag{3.91}$$

The remarkable thing about the balance expressed as Eq. (3.91) is that it is valid for any area in the sediment and for any time interval because the position of ABB´A´ is quite arbitrary, as is the time interval. This integral can only be equal to zero for an arbitrary area and time interval if the integrand itself (what appears within the integral) is zero for all chosen areas and time intervals, i.e.

$$\frac{\partial \hat{C}}{\partial t}+\nabla\cdot F-\sum R=0 \tag{3.92}$$

This final result is also remarkable because it is true regardless of the chosen coordinate system or the number of dimensions being considered. We need only supply the correct form for the divergence operator $\nabla\cdot$, and the flux vector, F, to write an appropriate conservation equation. The conservation equations for each of the three main coordinate systems, i.e. Cartesian, Cylindrical, and Spherical, are presented below.

3.9.2 The Cartesian Coordinate System

With the explicit form for the divergence operator in a 3-D Cartesian system (e.g. Bird et al., 1960), the conservation equation reads

$$\frac{\partial \hat{C}}{\partial t}=-\frac{\partial F_x}{\partial x}-\frac{\partial F_y}{\partial y}-\frac{\partial F_z}{\partial z}+\sum R \tag{3.93}$$

where F_x, F_y, and F_z are the components of the (total) flux in the x, y, and z directions, respectively.

For a solute, Fick's First Law of diffusion (molecular or otherwise) in a 3-D Cartesian coordinate system has as its components:

$$F_{D,x}=-\varphi D'_x\frac{\partial C}{\partial x} \tag{3.94a}$$

$$F_{D,y}=-\varphi D'_y\frac{\partial C}{\partial y} \tag{3.94b}$$

$$F_{D,z}=-\varphi D'_z\frac{\partial C}{\partial z} \tag{3.94c}$$

where D_x', D_y', and D_z' are the x, y, and z-direction diffusivities, respectively.

For the moment, we will ignore the bioturbational flux. The advective flux components are simply, $\varphi u_x C$, $\varphi u_y C$,and $\varphi u_z C$, where the subscript indicates the direction of flow. Putting the advective and diffusive fluxes into Eq. (3.93), we obtain

$$\frac{\partial \varphi C}{\partial t} = \frac{\partial}{\partial x}\left(\varphi D'_x \frac{\partial C}{\partial x} - \varphi u_x C\right) + \frac{\partial}{\partial y}\left(\varphi D'_y \frac{\partial C}{\partial y} - \varphi u_y C\right) + \frac{\partial}{\partial z}\left(\varphi D'_z \frac{\partial C}{\partial z} - \varphi u_z C\right) + \sum R \tag{3.95}$$

A slightly modified version of this equation is useful and is derived by writing Eq. (3.95) for the conservation of total porewater, i.e. $C = \rho_w$. If the changes in porewater density are small (and they normally are) and if reactions don't remove or produce significant amounts of water and solutes, then there results the *continuity equation* for the porewater,

$$\frac{\partial \varphi}{\partial t} + \frac{\partial}{\partial x}(\varphi u_x) + \frac{\partial}{\partial y}(\varphi u_y) + \frac{\partial}{\partial z}(\varphi u_z) = 0 \tag{3.96}$$

Now if Eq. (3.96) is multiplied by C and subtracted from Eq. (3.95), there results the reduced solute conservation equation:

$$\varphi \frac{\partial C}{\partial t} = \frac{\partial}{\partial x}\left(\varphi D'_x \frac{\partial C}{\partial x}\right) + \frac{\partial}{\partial y}\left(\varphi D'_y \frac{\partial C}{\partial y}\right) + \frac{\partial}{\partial z}\left(\varphi D'_z \frac{\partial C}{\partial z}\right) - \varphi u_x \frac{\partial C}{\partial x} - \varphi u_y \frac{\partial C}{\partial y} - \varphi u_z \frac{\partial C}{\partial z} + \sum R \tag{3.97}$$

There are innumerable diagenetic phenomena that could be described with an equation similar to Eq. (3.95) or (3.97); however, one of the more interesting is pressure-induced advection of porewater related to bedforms (ripple, mount, etc.) at the sediment-water interface of more permeable sediments such as sands, e.g. Thibodeaux and Boyle (1987), Savant et al. (1987), and Shum (1992). This type of advection is not likely to occur in any sediment that contains an appreciable amount of clay, say above 10%; however when it does occur, the diffusion coefficients in Eq. (3.97) would not then simply represent molecular diffusion. There would also be dispersion, and these coefficients must account for the effects of both classical molecular diffusion and this physical dispersion.

The principle behind this flow is quite simple, and may be deduced qualitatively by considering a symmetric ripple (Fig. 3.19) and Bernoulli's principle. This principle states that regions of higher velocity in a flow have lower pressure, while zones of low velocity correspond to higher pressure regions. Thus, because a flow over a ripple must be faster at the crest than at the base to conserve total mass, there is an area of high pressure at the base (both front and back), and low pressure at the crest. Darcy's equations tell us that there will result a flow in the sediment because of this pressure gradient, directed into the sediment at the base, and out of the sediment at the crest of the ripple. (Of course, this assumes that the sediment is permeable.)

Bed-form asymmetry and turbulence will pervert this simple picture, as is shown by Thibodeaux and Boyle (1987), Savant et al. (1987), and Rutherford et al. (1995). Shum (1992,1993) considered the additional interactions between gravity waves and a wavy sediment-water interface. In all these cases it is necessary to solve an equation similar to Eq. (3.97) to obtain a concentration distribution.

3.9.3 The Cylindrical Coordinate System

Equation (3.93) can also be expressed in the cylindrical system as

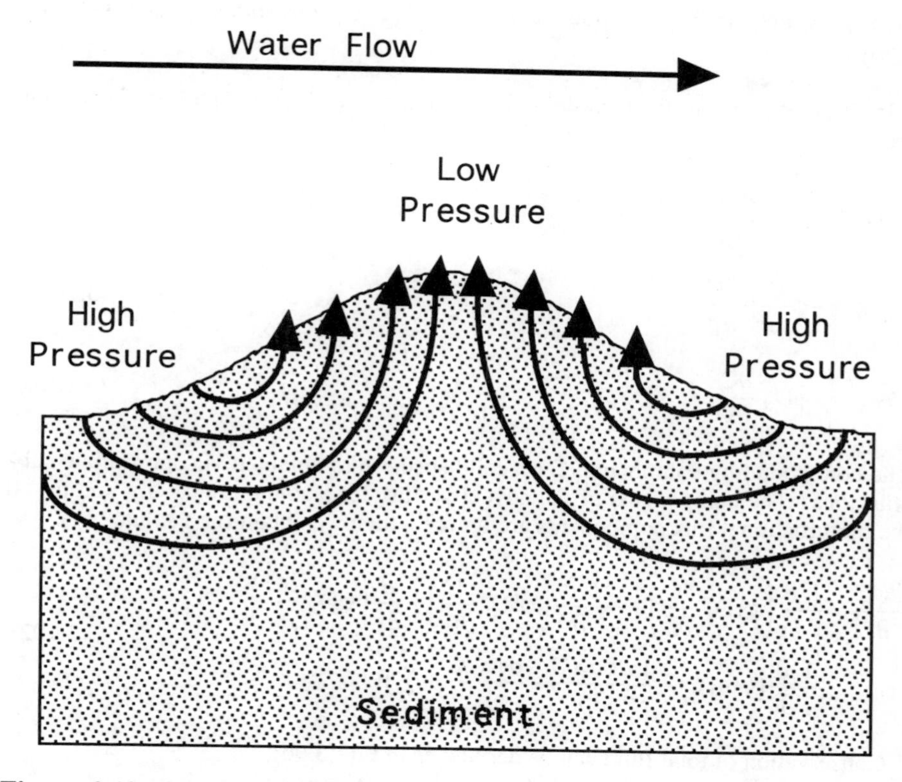

Figure 3.19. Highly idealized flow through a symmetric sand ripple caused by (laminar) flow over this bed-form. The arrows on the lines in the ripple indicate the direction of the flow. In a turbulent flow or with an asymmetric ripple (or both), the flow lines in the ripple become asymmetric, and the outward flow focuses on the leeward side of the bed-form (see Thibodeaux and Boyle, 1987).

$$\frac{\partial \hat{C}}{\partial t} = -\frac{1}{r}\frac{\partial}{\partial r}(r \cdot F_r) - \frac{1}{r}\frac{\partial F_\vartheta}{\partial \vartheta} - \frac{\partial F_x}{\partial x} + \sum R \tag{3.98}$$

where r is the radial coordinate at right-angle to the x (depth) coordinate,

$$r = \sqrt{y^2 + z^2} \tag{3.99}$$

and ϑ is the angle in the plane perpendicular to the x-axis,

$$\vartheta = \mathit{arctan}\left(\frac{y}{z}\right) \tag{3.100}$$

F_r, F_ϑ, F_x are the flux components in the r, ϑ, and x directions, respectively (Fig. 3.20).

In a cylindrical coordinate system, the components of Fick's First Law for a solute are:

$$F_{D,r} = -\varphi D'_r \frac{\partial C}{\partial r} \tag{3.101a}$$

$$F_{D,\vartheta} = -\varphi \frac{D'_\vartheta}{r}\frac{\partial C}{\partial \vartheta} \tag{3.101b}$$

$$F_{D,x} = -\varphi D'_x \frac{\partial C}{\partial x} \tag{3.101c}$$

The corresponding components for a solid species are obtained via appropriate substitutions. The components of the advective flux for a solute are u_r, u_ϑ, and u_z. If the bioturbational flux components are ignored, then we obtain for a solute:

$$\begin{aligned} \frac{\partial \varphi C}{\partial t} = & \frac{1}{r}\frac{\partial}{\partial r}\left(\varphi r D'_r \frac{\partial C}{\partial r}\right) + \frac{1}{r^2}\frac{\partial}{\partial \vartheta}\left(\varphi D'_\vartheta \frac{\partial C}{\partial \vartheta}\right) + \frac{\partial}{\partial x}\left(\varphi D'_x \frac{\partial C}{\partial x}\right) \\ & - \frac{1}{r}\frac{\partial}{\partial r}(\varphi r u_r C) - \frac{1}{r}\frac{\partial \varphi u_\vartheta C}{\partial \vartheta} - \frac{\partial \varphi u_x C}{\partial x} + \sum R \end{aligned} \tag{3.102}$$

Conservation of total fluid can be derived as in Eq. (3.96),

$$\frac{\partial \varphi}{\partial t} + \frac{1}{r}\frac{\partial}{\partial r}(\varphi r u_r) + \frac{1}{r}\frac{\partial \varphi u_\vartheta}{\partial \vartheta} + \frac{\partial \varphi u_x}{\partial x} = 0 \tag{3.103}$$

where again reactions are assumed to have little effect on total fluid. Multiplying Eq. (3.103) by C and subtracting from Eq. (3.102) leads to

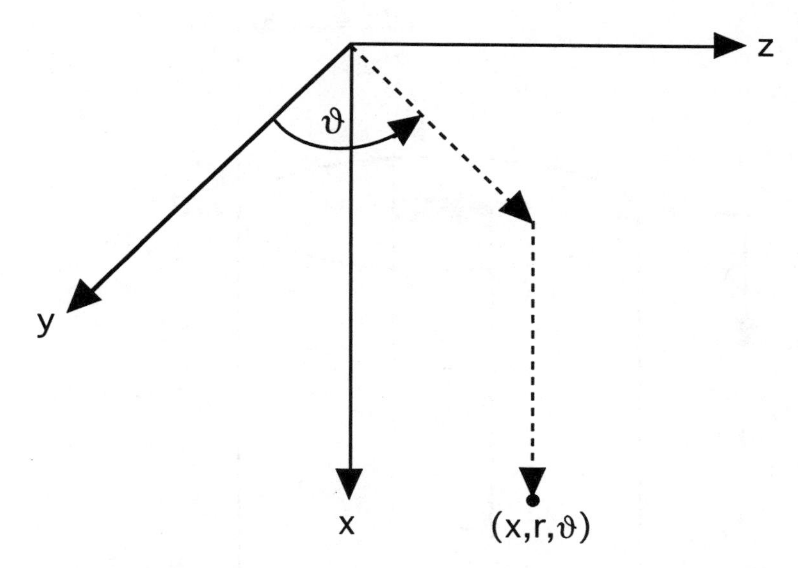

Figure 3.20. Specification of a point in the cylindrical coordinate system.

$$\varphi\frac{\partial C}{\partial t}=\frac{1}{r}\frac{\partial}{\partial r}\left(\varphi r D_r'\frac{\partial C}{\partial r}\right)+\frac{1}{r^2}\frac{\partial}{\partial \vartheta}\left(\varphi D_\vartheta'\frac{\partial C}{\partial \vartheta}\right)+\frac{\partial}{\partial x}\left(\varphi D_x'\frac{\partial C}{\partial x}\right)$$
$$-\varphi u_r\frac{\partial C}{\partial r}-\frac{\varphi u_\vartheta}{r}\frac{\partial C}{\partial \vartheta}-\varphi u_x\frac{\partial C}{\partial x}+\sum R \qquad (3.104)$$

which appears to be the canonical form for a porous medium.

The most widely recognized use of the cylindrical form of the diagenetic equation is the description of biological irrigation of porewaters. *Irrigation* is the pumping activity by burrow-dwelling animals which exchanges burrow water for overlying water (Aller, 1977). The model for this process is closely associated with the name of Robert Aller, who in a series of tour-de-force papers developed this remarkable and insightful mathematical description (Aller, 1980, 1982, 1988, and many others).

In a normal fine-grained sediment, organisms build vertical tubes as habitation. During respiration, such organisms can excrete into the burrow-waters metabolic by-products that may be self-poisoning. (Sanitation is a problem faced by most animals, not only man.) In addition, because of microbial organic-matter decay in the surrounding sediment, similar noxious solutes are released into the adjacent porewaters and can diffuse into the tube.

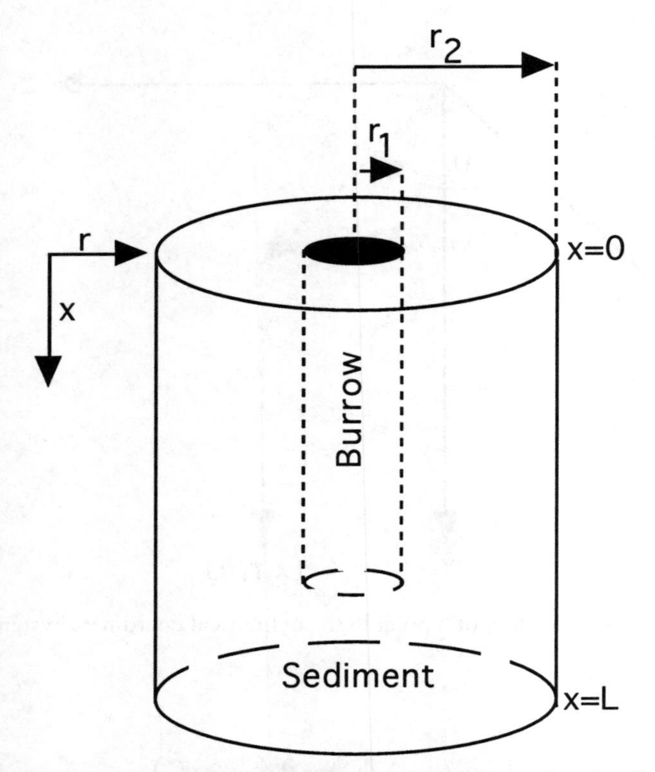

Figure 3.21. Sketch of a cylindrical burrow/tube and associated sediment idealized as a hollow annulus. The burrow/tube contains overlying water in Aller's original model. (Adapted from Aller, 1980; reproduced with the kind permission of Elsevier Science Ltd.)

To prevent the build up of these chemicals to dangerous levels, tube-dwelling organisms have developed a variety of strategies for exchanging their contaminated burrow-water with overlying water. As a result, overlying waters are brought into contact along the tube wall with porewaters relatively deep in the sediment (typically up to 15-20 cm, but sometimes much deeper), which have a very different composition. Normally diffusive exchange between these deep porewaters and the overlying water must occur through the intervening sediment column; however, the presence of the irrigated tube allows some degree of nearly direct exchange.

Natural tubes tend to be somewhat random in their spacings. Aller (1980) proposed the clever idea of an equivalent "microenvironment" (Fig. 3.21). The sediment subject to irrigation is idealized as a collection of equally spaced annuli. Each annulus is a cylinder of sediment of radius r_2 containing at its center a hollowed-out smaller cylinder of radius r_1 that contains only overlying water. This smaller cylinder simulates the animal tube or burrow.

Both reaction and diffusion can occur within the porewater of each annulus, but the water in the hollow central cylinder is continuously maintained at the composition of the overlying water, at least in Aller's model. A solute field within the outer zone of an annulus is best described using the cylindrical coordinate system. Porewater advection is generally ignored and symmetry in the azimuthal direction is assumed. The resulting conservation equation derived from Eq. (3.104) is

$$\varphi \frac{\partial C}{\partial t} = \underbrace{\frac{\partial}{\partial x}\left(\varphi D' \frac{\partial C}{\partial x}\right)}_{\text{Vertical Diffusion}} + \underbrace{\frac{1}{r}\frac{\partial}{\partial r}\left(r \varphi D' \frac{\partial C}{\partial r}\right)}_{\text{Radial Diffusion}} + \sum R(C, x, r, t) \quad (3.105)$$

Here the radial diffusion term is an essential aspect of the model that cannot be ignored (although, as shown later, it may be represented differently). It accounts for the molecular transport to the tube, while the more familiar vertical term accounts for transport to/from the sediment-water interface. Aller proved that the extra radial diffusion can have a significant effect on the depth profiles of dissolved species such as silica and ammonia. (Wave-induced water movement though empty U-shaped burrows can also lead to significant irrigation of sediment, e.g. Webster, 1992.)

3.9.4 The Spherical Coordinate System

In the spherical system, the analogous conservation equation is

$$\frac{\partial \hat{C}}{\partial t} = -\frac{1}{r^2}\frac{\partial}{\partial r}\left(r^2 \cdot F_r\right) - \frac{1}{r \cdot \sin\vartheta}\frac{\partial}{\partial \vartheta}\left(F_\vartheta \cdot \sin\vartheta\right) - \frac{1}{r \cdot \sin\vartheta}\frac{\partial F_\phi}{\partial \phi} + \sum R \quad (3.106)$$

where, as illustrated in Fig. 3.22,

$$r = \sqrt{x^2 + y^2 + z^2} \quad (3.107)$$

$$\vartheta = \arctan\left(\frac{\sqrt{y^2 + z^2}}{x}\right) \quad (3.108)$$

$$\phi = \arctan\left(\frac{y}{z}\right) \quad (3.109)$$

In the spherical coordinate system, the components of Fick's First Law are:

$$F_{D,r} = -\varphi D'_r \frac{\partial C}{\partial r} \quad (3.110a)$$

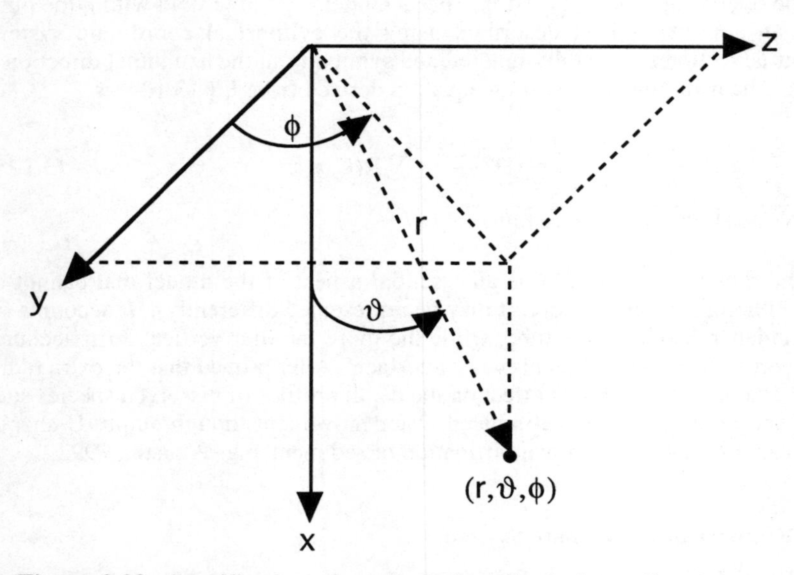

Figure 3.22. Specification of a point in the spherical coordinate system.

$$F_{D,\vartheta} = -\varphi \frac{D'_\vartheta}{r} \frac{\partial C}{\partial \vartheta} \tag{3.110b}$$

$$F_{D,\phi} = -\varphi \frac{D'_\phi}{r \cdot sin\vartheta} \frac{\partial C}{\partial \phi} \tag{3.110c}$$

and the velocity components are u_r, u_ϑ, and u_ϕ. Substituted into Eq. (3.106), there results

$$\frac{\partial \varphi C}{\partial t} = \frac{1}{r^2}\frac{\partial}{\partial r}\left(r^2 \varphi D'_r \frac{\partial C}{\partial r}\right) + \frac{1}{r^2 sin\vartheta}\frac{\partial}{\partial \vartheta}\left(sin\vartheta\, \varphi\, D'_\vartheta \frac{\partial C}{\partial \vartheta}\right) + \frac{1}{r^2 sin^2\vartheta}\frac{\partial}{\partial \phi}\left(\varphi D'_\phi \frac{\partial C}{\partial \phi}\right)$$
$$-\frac{1}{r^2}\frac{\partial}{\partial r}\left(r^2 \varphi\, u_r\, C\right) - \frac{1}{r \cdot sin\vartheta}\frac{\partial}{\partial \vartheta}\left(\varphi\, u_\vartheta C\, sin\vartheta\right) - \frac{1}{r \cdot sin\vartheta}\frac{\partial \varphi u_\phi C}{\partial \phi} + \sum R \tag{3.111}$$

If C is replaced with the fluid density, then the porewater continuity equation is

$$\frac{\partial \varphi}{\partial t} + \frac{1}{r^2}\frac{\partial}{\partial r}\left(r^2 \varphi u_r\right) + \frac{1}{r \cdot sin\vartheta}\frac{\partial}{\partial \vartheta}\left(\varphi u_\vartheta sin\vartheta\right) + \frac{1}{r \cdot sin\vartheta}\frac{\partial \varphi u_\phi}{\partial \phi} = 0 \tag{3.112}$$

Multiplying Eq. (3.112) by C and adding to Eq. (3.111) gives the alternative form for a solute,

$$\varphi\frac{\partial C}{\partial t}=\frac{1}{r^2}\frac{\partial}{\partial r}\left(r^2\varphi D'_r\frac{\partial C}{\partial r}\right)+\frac{1}{r^2 sin\vartheta}\frac{\partial}{\partial\vartheta}\left(sin\vartheta\,\varphi D'_\vartheta\frac{\partial C}{\partial\vartheta}\right)$$
$$+\frac{1}{r^2 sin^2\vartheta}\frac{\partial}{\partial\phi}\left(\varphi D'_\phi\frac{\partial C}{\partial\phi}\right)-\varphi u_r\frac{\partial C}{\partial r}-\frac{\varphi u_\vartheta}{r}\frac{\partial C}{\partial\vartheta}-\frac{\varphi u_\phi}{r\cdot sin\vartheta}\frac{\partial C}{\partial\phi}+\sum R \qquad (3.113)$$

The oft-quoted classic illustration for the use of the spherical coordinate system is Berner's (1968) model for concretion growth; however, another example is transport and reaction within spherical "microenvironments" (Jahnke, 1985) or "micro-niches" (Jørgensen, 1977). Organic matter arriving at the sea floor is in large part in the form of fecal pellets and other relatively large droppings. In addition, infaunal organisms that process sediments to extract organic matter may themselves create bundles of organic-rich pellets. The result is that the sediment is not a compositionally homogeneous medium, but one that contains small bodies of organic-rich substances, in addition to more disseminated organic materials. These (pelletized) micro-environments are speculated to be the locus of intense diagenetic reactions involving the reduction of electron acceptors, such as O_2, NO_3^-, and $SO_4^=$, and the oxidation of organic matter to CO_2 (Fig. 3.23).

It is possible to idealize a micro-environment/organic-matter particle in the sediment as a spherical body, symmetric in the ϑ and ϕ directions. These particles are usually sufficiently small that the outside porewater concentration can be taken as a constant over the surface of the sphere; thus, the problem of describing the micro-environment reduces to one of diffusion and reaction in a porous sphere. Under normal circumstances, there should be no advection in this example. The distribution of each solute, e.g. an oxidant, within the sphere will be described by the appropriate reduced form of Eq. (3.113), e.g.

$$\varphi\frac{\partial C}{\partial t}=\frac{1}{r^2}\frac{\partial}{\partial r}\left(r^2\varphi D'_r\frac{\partial C}{\partial r}\right)+\sum R \qquad (3.114)$$

or, more commonly,

$$\frac{\partial C}{\partial t}=D'_r\frac{\partial^2 C}{\partial r^2}+\frac{2D'_r}{r}\frac{\partial C}{\partial r}+\sum\frac{R}{\varphi} \qquad (3.115)$$

when the porosity of the micro-environment and the diffusion coefficient are reasonably constant.

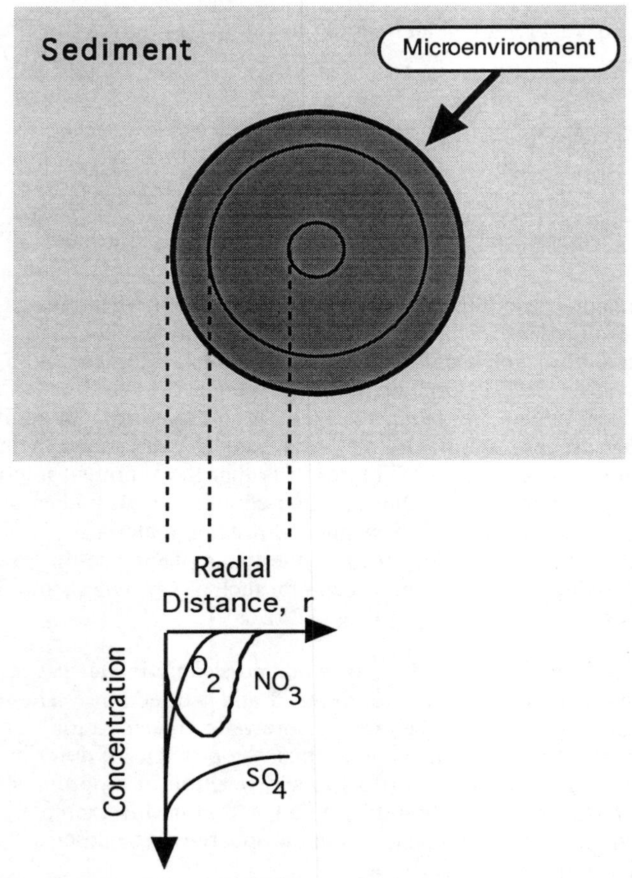

Figure 3.23. Schematic representation of an organic-rich spherical micro-environment that is the site of O_2, NO_3^-, and $SO_4^=$ reduction, as indicated by the radial concentration profiles on the left. (Adapted from Jahnke, 1985; reproduced with the kind permission of ASLO and Allen Press.)

3.10 Nonlocal Transport Processes

As alluded to earlier, some transport processes don't fit into the Fickian (local) flux formalism in that material is transported between points that are not adjacent to each other (Fig. 3.16). Such transport processes are then said to be *nonlocal*. It's important to take this nonlocality into account when:

i) sediment motions are not random, but directed (e.g. conveyor-belt mixing); and/or

ii) average distance traveled during mixing events is essentially of the same order as the full mixed-layer depth (Fig. 3.16); and/or

iii) a truly 2 or 3-D process must somehow be incorporated into a strictly 1-D model.

We consider four such processes below and how they appear in 1-D models.

3.10.1 Nonlocal Mixing within Sediments

Consider the sediment column illustrated in Fig. 3.24 (Boudreau and Imboden, 1987). This figure idealizes the sediment as a four-layer system, where the upper three layers are biologically active and the lowest is the non-mixed historical layer (extending to infinity). Sediment is added to the top at the sediment-water interface. An apparent velocity, w, caused by burial, moves sediment successively from the first, to the second, to the third, and finally into the historic layer, as the sediment column grows.

If organisms are sufficiently large, they may ingest material in layer 1 and excrete these solids in layers 2 and/or 3. Similarly, ingestion in layer 3 may be followed by defecation in layers 1 and/or 2. Consequently, material can be transferred to non-adjacent layers, layer 1 to 3 or layer 3 to 1 as indicated by the double arrows in Fig. 3.24. Burrowing activity may produce the same transfer between non-adjacent layers.

To arrive at a mathematical description of this process, let E_{ij} be the rate constant for nonlocal exchange of material from layer i to layer j. At the same time, let M_i be the mass of some component of interest in layer i. Then the component of the total derivative due to the biological transport is

$$\left[\frac{DM_i}{Dt}\right]_{\substack{\text{biological}\\ \text{transport}}} = \underbrace{\sum_{j=1}^{3} E_{ji} M_j}_{\substack{\text{transport}\\ \text{into layer i}}} - \underbrace{M_i \sum_{j=1}^{3} E_{ij}}_{\substack{\text{transport}\\ \text{out of layer i}}} \tag{3.116}$$

If we increase the number of layers in the mixed zone (e.g. as in Fig. 3.24) to some larger number n, then Eq. (3.116) simply becomes a sum over all n layers,

$$\left[\frac{DM_i}{Dt}\right]_{\substack{\text{biological}\\ \text{transport}}} = \underbrace{\sum_{j=1}^{n} E_{ji} M_j}_{\substack{\text{transport}\\ \text{into layer i}}} - \underbrace{M_i \sum_{j=1}^{n} E_{ij}}_{\substack{\text{transport}\\ \text{out of layer i}}} \tag{3.117}$$

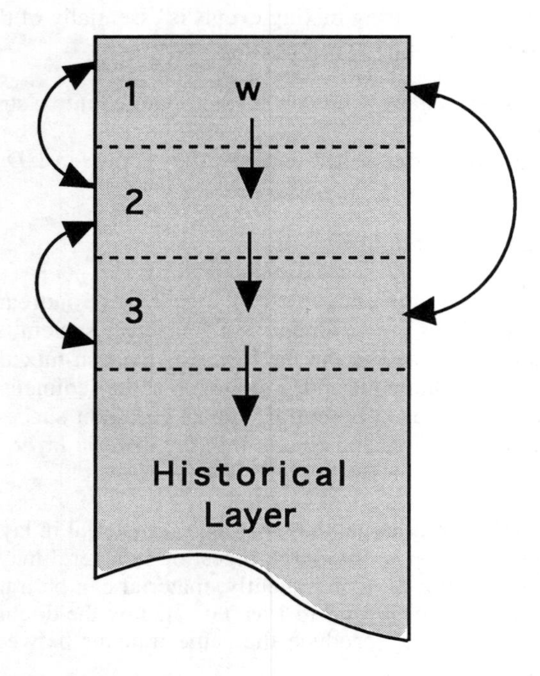

Figure 3.24. Schematic diagram of the exchanges possible in a n-layer sediment. In a 3-layer sediment, there can be local exchanges between adjacent layers 1 and 2 and layers 2 and 3, but also nonlocal exchange between 1 and 3. Local transport by burial is indicated by w. (After Boudreau and Imboden, 1987; reproduced with the kind permission of the American Journal of Science.)

Now, divide both M_i and E_{ij} (or E_{ji}) by Δx, which is the thickness of the layers, to define the bulk concentration in layer i, i.e. $(1-\varphi_i)B_i \equiv M_i/\Delta x$, and $K_{ij} \equiv E_{ij}/\Delta x$, which is the exchange constant from layer i to layer j (per unit time per unit length). Equation (3.117) becomes

$$\left[\frac{D(1-\varphi)_i B_i}{Dt}\right]_{\substack{\text{biological}\\ \text{transport}}} = \underbrace{\sum_{j=1}^{n} K_{ji}(1-\varphi_j)B_j\Delta x}_{\substack{\text{transport}\\ \text{into layer i}}} - \underbrace{(1-\varphi_i)B_i \sum_{j=1}^{n} K_{ij}\Delta x}_{\substack{\text{transport}\\ \text{out of layer i}}} \tag{3.118}$$

As the number of layers becomes very large ($\Delta x \rightarrow 0$), the following limits are obtained:

$$\operatorname*{limit}_{n\rightarrow\infty}\left(\sum_{j=1}^{n} K_{ji}(1-\varphi_j)B_j\Delta x\right) = \int_0^L K(x',x;t)(1-\varphi(x'))B(x')dx' \tag{3.119}$$

$$\underset{n\to\infty}{\text{limit}}\left((1-\varphi_i)B_i\sum_{j=1}^{n}K_{ij}\Delta x\right)=(1-\varphi(x))B(x)\int_0^L K(x,x';t)\,dx' \tag{3.120}$$

where L is the thickness of the mixed layer.

So we have transformed the layer-based function K_{ij} into the continuous exchange function K(x',x;t), which moves material from a chosen depth x to any other depth x', and likewise for K_{ji}. With these results, Eq. (3.118) becomes

$$\left[\frac{D(1-\varphi)B}{Dt}\right]_{\substack{\text{biological}\\ \text{transport}}} = \left\{\int_0^L K(x',x;t)(1-\varphi(x'))B(x')dx' - (1-\varphi(x))B(x)\int_0^L K(x,x';t)dx'\right\} \tag{3.121}$$

This term can be added to Eqs (3.51) or (3.52) to get a solids conservation equation that includes nonlocal internal mixing. If we assume that there is no separate biodiffusion term, the conservation equation reads:

$$\frac{\partial(1-\varphi)B}{\partial t}=-\frac{\partial(1-\varphi)wB}{\partial x}+\sum R_B + \int_0^L K(x',x;t)(1-\varphi(x'))B(x')dx' - (1-\varphi(x))B(x)\int_0^L K(x,x';t)dx' \tag{3.122}$$

Note that the nonlocal transport term in Eq. (3.122) is not within the $\partial/\partial x$ term; therefore, nonlocal exchange is not a flux in the sense of Fick's First Law. The first integral in this term is an *apparent source* and the second is an *apparent sink*, even though no chemical reactions are involved; there is just a juggling of material that appears as a combination of a source and a sink.

In the special case where K(x',x;t) = K(x,x';t), the same fraction of material is exchanged between x and x' per unit time. This is known as *symmetric nonlocal mixing*. If, in addition, K(x',x;t) and K(x,x';t) are equal to a constant K, then every point in the mixed layer exchanges the same fraction of sediment material with every other point. At constant porosity, Eq. (3.122) then reads

$$\frac{\partial B}{\partial t}=-w\frac{\partial B}{\partial x}+\sum R_B + K_{ex}\left\{\frac{1}{L}\int_0^L B(x)dx - B(x)\right\} \tag{3.123}$$

where K_{ex} = K·L, which has units of inverse time. Note that L must be a finite depth in this case. This last equation is the nonlocal analog to Eq. (3.53), i.e. constant local mixing.

In fact, a local model with D_B as a constant and a nonlocal model with constant K_{ex} can be thought of as the two extremes on a continuum of mixing functions based on the fraction of the total layer mixed in each exchange event (Fig. 3.25). Constant D_B produces effective mixing, even though the scale of exchange nears zero, because the intensity of that exchange is an infinite Delta function (see Boudreau and Imboden, 1987, for more mathematical details). We don't know very much about the occurrence of constant-K_{ex} nonlocal mixing in natural sediments. Currently Eq. (3.122) finds perhaps its greatest role as the basis for deriving models for the special case nonlocal mixing processes described below.

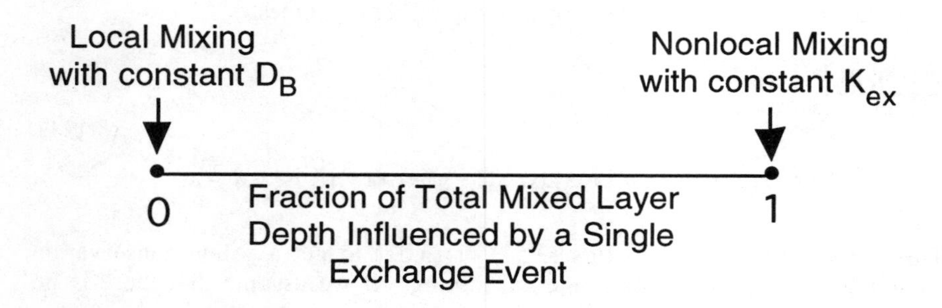

Figure 3.25. Conceptual diagram showing the relation between the end-member types of mixing, diffusion, and constant nonlocal exchange.

3.10.2 Conveyor-Belt Mixing (Head-Down Deposit-Feeding)

Conveyor-belt mixing is an important special case of nonlocal transport. Head-down deposit-feeding organisms are capable of creating a conveyor-belt type motion in the sediment (Fig. 3.11). These organisms ingest sediment at depth and defecate this material at the sediment-water interface. The void created at the feeding depth is eliminated by the downward compaction (advection) of the sediment. Shifting populations of such organisms of different sizes cause a general nonlocal transport between the interface and depth (Aller, 1982; Rice, 1986).

A mathematical model for this conveyor-belt transport can be derived from the nonlocal mixing model, Eq. (3.122). The removed material is dumped on the sediment-water interface, so that the removal function, $K(x,x';t)$, will be zero except for $x'=0$, i.e. $K(x,x';t) = \delta(x')K_R(x,t)$, where $\delta(x')$ is a Dirac Delta function (see Chapter 2). Removal to the sediment-water interface is technically dumping just on the negative side of $x=0$, so $K(x',x;t) = 0$ for all x'. Under these conditions, Eq. (3.122) for total solids, neglecting reaction, reduces to

$$\frac{\partial(1-\varphi)}{\partial t} = -\frac{\partial(1-\varphi)w}{\partial x} - (1-\varphi)K_R(x;t) \tag{3.124}$$

where $K_R(x;t)$ is the removal rate constant (per unit time). Note that Eq. (3.124) implies that w is, in general, a function of depth for this type of mixing. This is exactly the expected behavior, as continuous filling of the feeding void by compaction engenders an increased advection in the biologically active zone. For a solid species, the analogous conservation equation is

$$\frac{\partial(1-\varphi)B}{\partial t} = -\frac{\partial(1-\varphi)wB}{\partial x} + \sum R_B - (1-\varphi)K_R(x;t)B \tag{3.125}$$

Neither Eq. (3.124) or Eq. (3.125) contains a term for the return of excavated material to the sediments. The material dumped on the interface re-enters the model via a flux *boundary condition*; i.e. for total solids

$$\underbrace{[(1-\varphi)w]_0}_{\substack{\text{Burial of the Solids}\\ \text{at the Sediment-Water}\\ \text{Interface (x=0)}}} = \underbrace{F(t)}_{\substack{\text{Input of New}\\ \text{Solids to the}\\ \text{Interface}}} + \underbrace{\int_0^L (1-\varphi)K_R(x;t)\,dx}_{\substack{\text{Solids Cycled back to}\\ \text{the Sediment-Water Interface}\\ \text{by Head-Down Deposit Feeders}}} \tag{3.126}$$

and for the solid species

$$\underbrace{[(1-\varphi)wB]_0}_{\substack{\text{Burial of the Solid}\\ \text{Species at the}\\ \text{Sediment-Water}\\ \text{Interface}}} = \underbrace{F_0(t)}_{\substack{\text{Input of New}\\ \text{Amounts of the}\\ \text{Solid Species to}\\ \text{the Interface}}} + \underbrace{\int_0^L (1-\varphi)K_R(x;t)B(x)\,dx}_{\substack{\text{Solid Species Cycled to}\\ \text{the Sediment-Water Interface}\\ \text{by Head-Down Deposit Feeders}}} \tag{3.127}$$

The derivation of these boundary conditions is delayed until Chapter 5. Equations (3.124) through (3.127) constitute a model for conveyor-belt mixing as presented by Fisher et al. (1980), Boudreau (1986b), Robbins (1986), and Rice (1986).

3.10.3 Burrow-and-Fill Mixing

This type of mixing bears some similarity to head-down deposit-feeding in that sediment is excavated and dumped to the surface, but this time to create a burrow inhabited by an infaunal organism. Unlike head-down deposit feeding, once the burrow is formed, no more sediment is brought to the surface, and infilling occurs by slumping of surface material into the burrow once the organism abandons it (Fig. 3.26).

The situation being described mathematically is the spatially and time-integrated effect of the formation and filling of a large number of such burrows on the sediment distribution of a solid species. Furthermore, let us restrict ourselves to the case where all the infilling material originates at the sediment-water interface, $x = 0$, and for convenience, there is constant porosity.

Equation (3.122) establishes the basis for a quantitative model of burrow-and-fill mixing. Because material from burrow excavation is dumped on the sediment-water interface, i.e. $K(x,x';t) = \delta(x')K_R(x,t)$, this material movement must be accounted for in the boundary condition at the sediment-water interface (x=0).

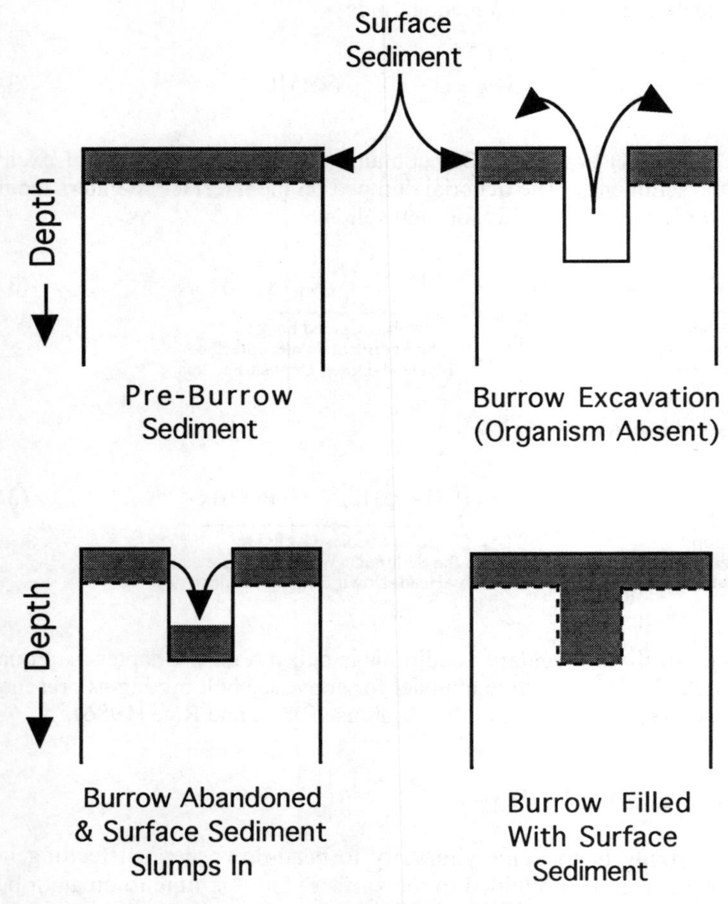

Figure 3.26. A schematic representation of the creation and filling of one burrow. The infilling material is the surface sediment, moved by currents and gravity. (After Boudreau and Imboden, 1987; reproduced with the kind permission of the American Journal of Science.)

In addition, filling of the burrows leads to a volumetric replacement of the excavated materials with surface sediment. This implies that there is a source of surface sediment equal to the removal rate at each depth, i.e. $K(x',x;t) = K(x,x';t)$; thus the conservation equation for a solid species in the zone of burrowing reads as

$$\frac{\partial B}{\partial t} = -w\frac{\partial B}{\partial x} + K_R(x;t)[B(0,t) - B(x,t)] + \sum R_B \tag{3.128}$$

subject to the boundary condition (see Chapter 5),

$$\underbrace{[(1-\varphi)wB]_0}_{\substack{\text{Advection from}\\ \text{the Interface}\\ \text{by Burial}}} + \underbrace{(1-\varphi(0))B(0,t)\int_0^L K_R(x,t)dx}_{\substack{\text{Surface Sediment}\\ \text{Removed to Depth by}\\ \text{Burrow Infilling}}}$$

$$= \underbrace{F_o(t)}_{\substack{\text{New}\\ \text{Input}}} + \underbrace{\int_0^L K_R(x,t)(1-\varphi(x))B(x,t)\,dx}_{\substack{\text{Sediment Cycled to}\\ \text{the Interface}\\ \text{by Burrowing}}} \tag{3.129}$$

which states that the input of material to the surface by new accumulation and excavation of burrows is balanced by burial and infilling of abandoned burrows. Equation (3.128) has been used, for example, to account for the effects of Fiddler Crab populations on ^{210}Pb distributions in marsh sediments (Gardner et al., 1987).

3.10.4 Resuspension-Redeposition

Resuspension is normally defined as the physical remobilization of previously deposited sediments resulting from the action of waves or turbulence (or both). Although a relatively common occurrence in near-shore sediments (even in muds), it appears to operate with less intensity as the water-column depth increases. Biological agents have also been implicated as contributors to resuspension. Resuspension events can be quite small scale, as occurs with a single turbulent "burst" (Fig. 3.27), or they can involve large areas of the sea floor, as happens during major storms over a shelf. In broad terms, solid sediment is removed to the water column by the initial resuspension event, creating a scallop or depression in the sediment bed, which is later refilled by redeposited sediment. On average, this process reconstitutes the original sediment-water interface. This infilling material may be partly some of the locally resuspended sediment, and just as likely, material resuspended elsewhere and transported by currents and waves. The infilling material is mixed while resuspended, so that gradients that existed originally in the sediment can be substantially modified or, conversely, there may be sorting of the sediment by size and/or density and consequent creation of gradients. In a fine-grained reasonably homogeneous sediment, resuspension should favor mixing.

One-dimensional mathematical models have been offered for the time-averaged effects over a great many such resuspension-redeposition events, and these have usually been of the diffusion variety (Vanderborght et al., 1977). Without casting aspersions on this approach, I offer another description of this phenomenon. When time-averaged, *in situ* resuspension-redeposition removes material from a given depth and replaces it with homogenized material from all depths affected by resuspension (Fig. 3.27). This is a nonlocal mixing process, and it can be modelled using Eq. (3.122).

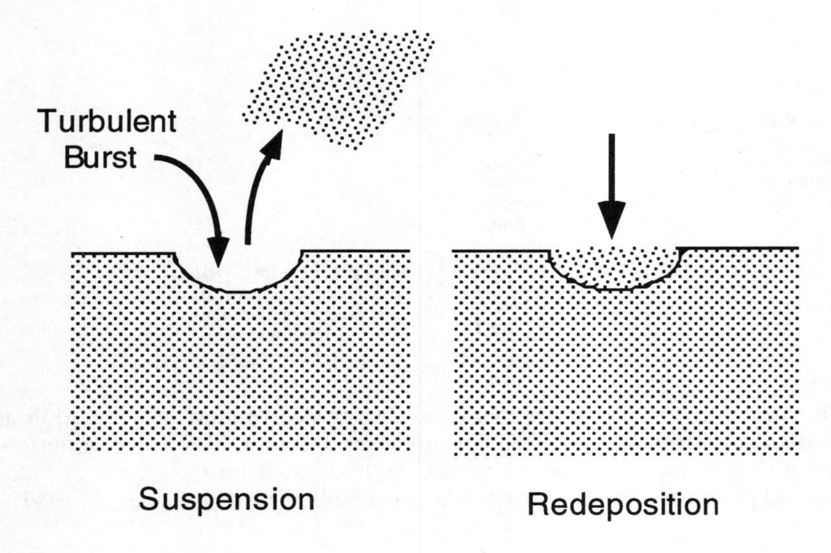

Figure 3.27. Schematic diagram of a resuspension-redeposition event in a muddy sediment, as caused by an occasional turbulent burst.

But what of the porewater during this process? Complete resuspension (not just mild fluidization) exchanges the porewater and its solutes at a given depth for overlying water. This also suggests a nonlocal exchange. To describe the distribution of a solute, start with Eq. (3.122) with a molecular diffusion term added and B and w replaced by C and u, respectively, i.e.

$$\frac{\partial \varphi C}{\partial t} = \frac{\partial}{\partial x}\left(\varphi D' \frac{\partial C}{\partial x} - \varphi u C\right) + \sum R + \int_0^L K(x', x; t)\varphi(x')C(x')dx' - \varphi(x)C(x)\int_0^L K(x, x'; t)dx' \tag{3.130}$$

The exchange in this case is between any depth x and the overlying waters. Let the concentration of the solute in the overlying waters be C_0. The porosity of the overlying waters will be essentially equal to unity as long as the suspension is typically dilute. The exchange functions are symmetric in this case, and they are non-zero only for one value of x', i.e. x'=0. Thus they can be written as:

$$K(x', x; t) = K(x, x'; t) = K(x, t)\cdot \delta(x') \tag{3.131}$$

where $\delta(x')$ is the Dirac Delta function, as explained in Chapter 2. Substituting this form into Eq. (3.130) and using the properties of the Dirac Delta function (Chapter 2), we arrive at

$$\frac{\partial \varphi C}{\partial t} = \frac{\partial}{\partial x}\left(\varphi D' \frac{\partial C}{\partial x} - \varphi w C\right) + \sum R + K(x,t)[C_o - \varphi(x)C(x)] \tag{3.132}$$

Mixing caused by resuspension does not appear in this last equation as an extra diffusion; instead it takes the form of an apparent sink that removes solute at depth x at a rate $K(x,t)C(x)$ and an apparent source that supplies new solute at the overlying water concentration at a rate $K(x,t)C_o$.

3.10.5 Porewater Irrigation as a 1-D Nonlocal Source/Sink

Aller's (1980) Tube-Model, as described previously, is a 3-D description of the irrigation phenomenon that employs only local transport processes (i.e. diffusion). However, there is a means to describe this phenomenon that retains the simplicity of a 1-D diagenetic equation. This 1-D approach leads to a representation of irrigation as a nonlocal transport process (Emerson et al., 1984; Boudreau, 1984). The derivation is given in some detail, as it illustrates the general principle that 2-D and 3-D processes may appear as nonlocal terms in 1-D models.

We start with Aller's steady-state axi-symmetric equation for a dissolved species in the porewaters surrounding an irrigated burrow (Fig. 3.21):

$$D' \frac{\partial^2 C}{\partial x^2} + \frac{D'}{r} \frac{\partial}{\partial r}\left(r \frac{\partial C}{\partial r}\right) + R = 0 \tag{3.133}$$

The presence or absence of the time derivative is of no consequence to this derivation, and so it is omitted. Equation (3.133) informs us that a dissolved species moves in the porewaters only by (hindered) molecular diffusion, and that it may be affected by a reaction of strength $R \equiv \Sigma R(C,x,r)$ at any arbitrary point. In addition, this equation is subject to the boundary conditions

$$C(0,r) = C_o \tag{3.134a}$$

$$C(x,r_1) = C_o \tag{3.134b}$$

$$\left.\frac{\partial C}{\partial r}\right|_{r=r_2} = 0 \tag{3.134c}$$

$$\left.\frac{\partial C}{\partial x}\right|_{x=L} = F \tag{3.134d}$$

where F is a prescribed gradient that can be set to zero without affecting the derivation. These boundary conditions specify the behavior of the concentration or its gradient along the borders of the modelled region. (Details of the derivation of boundary conditions are given in Chapter 5.) Conditions (3.134a,b) state that the concentration of the dissolved species is fixed at the value in the overlying waters,

C_o, at both the sediment-water and the tube-water interface, respectively. Boundary condition (3.134c) tells us that there is no gradient and, therefore, no diffusive flux at the outer boundary of the sediment cylinder, $r = r_2$.

The first step in this development is to define the average concentration in a horizontal slice at depth x, because a 1-D model resolves only in the x-direction:

$$\overline{C} = \frac{2}{\left(r_2^2 - r_1^2\right)} \int_{r_1}^{r_2} r\,C\,dr \tag{3.135}$$

Equation (3.133) can be expressed in terms of this radially averaged concentration by first dividing it by the denominator in Eq. (3.135) and multiplying by r, i.e.

$$D' \frac{\partial^2}{\partial x^2}\left(\frac{2rC}{r_2^2 - r_1^2}\right) + \frac{2D'}{r_2^2 - r_1^2} \frac{\partial}{\partial r}\left(r \frac{\partial C}{\partial r}\right) + \frac{2rR}{r_2^2 - r_1^2} = 0 \tag{3.136}$$

and finally integrating with respect to r from r_1 to r_2,

$$D' \frac{\partial^2 \overline{C}}{\partial x^2} + \frac{2D'}{r_2^2 - r_1^2}\left[r \frac{\partial C}{\partial r}\right]_{r_1}^{r_2} + \overline{R} = 0 \tag{3.137}$$

where $\overline{R}$ is the average rate of reaction in a radial slice of sediment. Now, condition (3.134c) gives that

$$\left[r \frac{\partial C}{\partial r}\right]_{r_1}^{r_2} = -r_1 \left.\frac{\partial C}{\partial r}\right|_{r=r_1} \tag{3.138}$$

There exists a radial distance $\bar{r}$ that has as its concentration the mean value, $\overline{C}$. The only catch is that this distance is not known *a priori*, but we do know that $r_1 \le \bar{r} \le r_2$. A Taylor series expansion for the mean concentration in terms of the concentration at $r=r_1$ then supplies the approximation

$$\left.\frac{\partial C}{\partial r}\right|_{r=r_1} \approx \frac{\overline{C} - C_o}{\bar{r} - r_1} \tag{3.139}$$

Therefore, Eq. (3.137) with (3.138) and (3.139) produces:

$$D' \frac{d^2 \overline{C}}{dx^2} - \alpha(x)\,(\overline{C} - C_o) + \overline{R} = 0 \tag{3.140}$$

where

$$\alpha(x) = \frac{2D' r_1}{(r_2^2 - r_1^2)(\bar{r} - r_1)} \tag{3.141}$$

The coefficient $\alpha(x)$ provides the rate of nonlocal exchange at depth x. A comparison of Eq. (3.140) with Eq. (3.133) shows that the 3-D radial-diffusion term has now become an apparent source-sink term in the 1-D model, and one that depends on the difference between the concentration at depth x and that of the overlying waters. This is the same kind of result as with resuspension, i.e. Eq. (3.132). Consequently, irrigation is not properly represented by an apparent increased diffusivity in 1-D diagenetic models, as is still occasionally done.

This last distinction is not trivial, as a nonlocal exchange term has a fundamentally different influence on concentration profiles than does local diffusion. Diffusion operates on the gradient, so that it is strongest where the gradient is steepest, which is usually near the sediment-water interface. Conversely, the nonlocal exchange is strongest where the difference in concentrations between C and C_o is greatest, which normally occurs at depth away from the sediment-water interface. (Irrigation has this same behavior in Aller's 3-D model.) This difference explains the sometimes anomalously high apparent diffusion coefficients sometimes needed to account for irrigation in near-shore sediments. (Values of $\alpha(x)$ are discussed in Chapter 4.)

3.11 Reaction Processes and Differential-Algebraic Equations

Little has been said so far about the reaction term, ΣR, present in the diagenetic equations derived above. A myriad of (algebraic) forms are possible for this term, related to the specific kinetics of the reaction(s) under consideration. The particular constitutive relations that define ΣR are called rate laws, and a few selected forms are discussed near the end of Chapter 4. These rate laws can be linear or nonlinear, and the latter impart a nonlinearity to the differential conservation equations that can make them difficult to solve (see Chapters 6 and 7). However, a detailed discussion of these forms is not warranted at this point. Instead, we need to examine the consequences of the classification of reactions into broad types: first, homogeneous vs. heterogeneous, and secondly, irreversible vs. reversible and, finally for reversible reactions, slow vs. fast (Berner, 1980; Lasaga, 1981; Rubin, 1983; Bahr and Rubin, 1987; Bahr, 1990; Weber et al., 1991).

3.11.1 Homogeneous and Heterogeneous Reactions

A reaction can take place wholly within the volume of a single phase, such as porewater, in which case it is known as a *homogeneous* reaction. Alternatively, it can occur at phase boundaries, as along the contact area between a sediment grain and porewater. The latter is called a *heterogeneous* reaction.

These two types of reactions are treated quite differently in diagenetic models. In principle, the ΣR term that appears in the conservation equations deals only with homogeneous reactions; heterogeneous reactions are normally described through boundary conditions (see Chapter 5). However, this dichotomy poses serious practical problems for modelling of heterogeneous reactions in sediments. Many of the reactions of interest in diagenesis, e.g. adsorption, dissolution, microbial organic matter decay, etc., are heterogeneous. At the same time, the number of reacting sediment grains in a macroscopic unit volume (say 1 cm^3) can be enormous, and each of these grains has a complex surface geometry. The specification of the location of this surface and the heterogeneous reaction(s) that occur there is an impossible task. Even for perfectly shaped grains with regular packing, this is a formidable problem.

A rather simple, if *ad hoc*, methodology has evolved to overcome this problem. If the reacting solid is reasonably well disseminated throughout each sediment volume, then it appears to the observer that the reaction is indeed occurring continuously through the volume. The reaction is said to be *pseudo-homogeneous*.

Formally, a given heterogeneous reaction is converted to a pseudo-homogeneous reaction via equations of the form

$$\left(\frac{DC}{Dt}\right)_{reaction} = \frac{A}{V} R_{het} \tag{3.142}$$

or

$$\left(\frac{DC}{Dt}\right)_{reaction} = \frac{A}{V} k_{het} f(C,...,x,t) \tag{3.143}$$

where A/V is the reactive area per unit volume of porewater, R_{het} is the rate of the heterogeneous reaction, k_{het} is the heterogeneous rate constant for the reaction in question, and f(C,...,x,t) is the rate's functional dependence on the concentration(s) and on x and t, if necessary. In practice, experimentalists often measure the product ($A\, k_{het}/V$) rather than these terms separately.

3.11.2 Reversible and Irreversible Reactions

This further classification of reactions can be illustrated by means of a hypothetical reaction of reactants A and B with the formation of products X and Y. An irreversible reaction is one in which only the forward direction is possible, as the backward reaction is inhibited, e.g.

$$a\mathrm{A} + b\mathrm{B} \rightarrow \chi\mathrm{X} + y\mathrm{Y} \tag{3.144}$$

where a, b, χ and y are appropriate stoichiometric coefficients for this reaction. ΣR will be the appropriate rate law for this reaction, e.g. if this is an elementary reaction, for species A (Lasaga, 1981),

$$\Sigma R = a(k_f[C_A]^{a}[C_B]^{b}) \quad (3.145)$$

where k_f is a forward rate constant.

If the reactions are slow, but reversible, then the rate expressions for both forward and backward reactions must be included in ΣR. For example, if

$$aA + bB \leftrightarrow xX + yY \quad (3.146)$$

and this is again elementary, then for species A,

$$\Sigma R = a(k_f [C_A]^{a}[C_B]^{b} - k_b [C_X]^{x}[C_Y]^{y}) \quad (3.147)$$

where k_b is the rate constant for the reverse or backward reaction. At all finite depths in the model, there is no chemical equilibrium, such that

$$K_{eq} \neq \frac{a_X^{x} a_Y^{y}}{a_A^{a} a_B^{b}} \quad (3.148)$$

except by chance at a given point. Here K_{eq} is the thermodynamic equilibrium constant and a designates the thermodynamic activity of the subscripted species.

Finally, if the reaction is reversible and the forward and backward reactions are fast, then equilibrium or near equilibrium will exist, i.e.

$$K_{eq} = \frac{a_X^{x} a_Y^{y}}{a_A^{a} a_B^{b}} \quad (3.149)$$

at all depths and times; however, the activities will remain functions of space and possibly time. While it is still possible to invoke a kinetic rate law of the form of Eq. (3.147) in this case, it is often much simpler to assume that Eq. (3.149) supplies an adequate description of the effects of this reaction; thus, if this last relationship holds, we need not specify the kinetics. In fact, adding such information would over-specify the problem.

A well-known example of this reversible-equilibrium type of reaction is linear equilibrium adsorption (as discussed by Berner, 1980, and Weber et al., 1991). For linear adsorption onto solid sediment, we have a reaction of the form

$$A \leftrightarrow \overline{A} \quad (3.150)$$

where A is the solute and $\overline{A}$ is the adsorbed species. For simplicity, assume that the porosity and solids density are constant and that the solute is subject to diffusion and advection, while the solids undergo biodiffusional mixing and burial. The conservation equation for the solute is

$$\frac{\partial C_A}{\partial t} = D'_A \frac{\partial^2 C_A}{\partial x^2} - u\frac{\partial C_A}{\partial x} + R_{ads} + \sum R_{others} \tag{3.151}$$

where C_A is the solute concentration per unit volume of porewater, D'_A is the tortuosity corrected diffusivity, u is the porewater advection velocity, R_{ads} is the net rate of adsorption onto the solids, and ΣR_{others} is the sum of all other slow reactions involving the solute.

For the adsorbed species (indicated by an overbar, following Berner, 1980), the conservation equation reads as

$$\frac{\partial \overline{C}_A}{\partial t} = D_B \frac{\partial^2 \overline{C}_A}{\partial x^2} - w\frac{\partial \overline{C}_A}{\partial x} + \overline{R}_{ads} + \sum \overline{R}_{others} \tag{3.152}$$

where $\overline{C}_A$ is the adsorbed concentration per unit volume of solids, D_B the biodiffusivity, w is the burial velocity, $\overline{R}_{ads}$ is the net rate of adsorption onto the solids, and $\Sigma \overline{R}_{others}$ is the sum of all other slow reactions involving the adsorbed species.

We will not make use of any kinetic information for the adsorption reaction, because reaction (3.150) is at equilibrium; thus, the two terms R_{ads} and $\overline{R}_{ads}$ are unknowns in these equations and must be eliminated. This is accomplished simply by multiplying Eq. (3.152) by $(1-\varphi)/\varphi$ and adding the result to Eq. (3.151). If we then recognize that mass conservation requires that $R_{ads} = -(1-\varphi)/\varphi\, \overline{R}_{ads}$, a single differential conservation is produced, i.e.

$$\begin{aligned}\frac{\partial\left(C_A + \frac{(1-\varphi)}{\varphi}\overline{C}_A\right)}{\partial t} &= D'_A \frac{\partial^2 C_A}{\partial x^2} + \frac{(1-\varphi)}{\varphi} D_B \frac{\partial^2 \overline{C}_A}{\partial x^2} \\ &\quad - u\frac{\partial C_A}{\partial x} - \frac{(1-\varphi)w}{\varphi}\frac{\partial \overline{C}_A}{\partial x} + \sum R_{others} + \frac{(1-\varphi)}{\varphi}\sum \overline{R}_{others}\end{aligned} \tag{3.153}$$

This last result is a single equation dealing with both unknown concentrations, C_A and $\overline{C}_A$. To resolve this indeterminacy we need to introduce a second equation. One of the original conservation equations, Eqs (3.151) or (3.152), cannot be employed because each contains an unknown kinetic term. Instead, the appropriate form of Eq. (3.149) provides the needed equation, i.e.

$$\overline{C}_A = K_{ads} C_A \tag{3.154}$$

Adoption of Eq. (3.154) is a specific example of the so-called *local equilibrium assumption* or LEA (Rubin, 1983; Jennings, 1987; Knapp, 1989). The modifier "local" indicates that the equilibrium expressed by Eq. (3.154) occurs at the local level, and that the porewater-sediment system as a whole is not in thermodynamic equilibrium, as is evidenced by the simultaneous existence of concentration gradients and non-thermodynamic processes, such as diffusion and other slow reactions (see also, Olander, 1960; Otto and Quinn, 1971; Chang and Rochelle, 1982; Meldon et al., 1982; Gallagher et al., 1986; Boudreau, 1987; DeCoursey and Thring, 1989; Goddard, 1990; Rein, 1992).

Together, Eqs (3.153) and (3.154) describe our example system, but one is a differential equation, while the other is an algebraic equation. Such combinations are known as *differential-algebraic equations* (DAEs). In general, DAEs can be a challenge to solve; however, in this example, it is possible to reduce these equations to a single ODE for which there may be an analytical solution, depending on the reaction terms. To see this, simply substitute Eq. (3.154) into Eq. (3.153) to obtain,

$$\frac{\partial C_A}{\partial t} = \frac{1}{1+K'_{ads}}\left((D'_A + K'_{ads}D_B)\frac{\partial^2 C_A}{\partial x^2} - (u + K'_{ads}w)\frac{\partial C_A}{\partial x} + \sum R_{others} + \frac{(1-\varphi)}{\varphi}\sum \overline{R}_{others}\right) \quad (3.155)$$

where $K'_{ads} \equiv (1-\varphi)K_{ads}/\varphi$. Most DAE systems are more complicated than this example, and the substitution of a nonlinear adsorption relation into the differential equation produces complicated ODEs that may or may not have analytical solutions. For example, examine Eq. (4.51) on p. 74 of Berner (1980).

Other types of reactions that lead to DAEs are acid-base reactions in porewaters (Boudreau, 1987); e.g. if HA is an acid,

$$HA \leftrightarrow H^+ + A^- \quad (3.156)$$

for which we have

$$K_{eq} = \frac{C_A C_H}{C_{HA}} \quad (3.157)$$

if the activity coefficients are equal to one. Since there are three dissolved species, we can write three diagenetic conservation equations,

$$\frac{\partial C_{HA}}{\partial t} = D'_{HA}\frac{\partial^2 C_{HA}}{\partial x^2} - u\frac{\partial C_{HA}}{\partial x} - R_{HA} + \sum R_{HA} \quad (3.158)$$

$$\frac{\partial C_A}{\partial t} = D'_A\frac{\partial^2 C_A}{\partial x^2} - u\frac{\partial C_A}{\partial x} - R_{HA} + \sum R_A \quad (3.159)$$

$$\frac{\partial C_H}{\partial t} = D'_H\frac{\partial^2 C_H}{\partial x^2} - u\frac{\partial C_H}{\partial x} + R_{HA} + \sum R_H \quad (3.160)$$

for HA, A^-, and H^+, respectively. Here R_{HA} is the net rate of dissociation of the acid HA into its constituent ions, and this is the unknown rate that is resolved by having Eq. (3.157). The DAE system to be solved is obtained by eliminating R_{HA}, first by adding Eqs (3.158) and (3.159),

$$\frac{\partial C_{HA}}{\partial t}+\frac{\partial C_A}{\partial t}=D'_{HA}\frac{\partial^2 C_{HA}}{\partial x^2}+D'_A\frac{\partial^2 C_A}{\partial x^2} - u\left(\frac{\partial C_{HA}}{\partial x}+\frac{\partial C_A}{\partial x}\right)+\sum R_{HA}+\sum R_A \tag{3.161}$$

and by subtracting Eq. (3.160) from Eq. (3.159), to get

$$\frac{\partial C_A}{\partial t}-\frac{\partial C_H}{\partial t}=D'_A\frac{\partial^2 C_A}{\partial x^2}-D'_H\frac{\partial^2 C_H}{\partial x^2}-u\left(\frac{\partial C_A}{\partial x}-\frac{\partial C_H}{\partial x}\right)+\sum R_A-\sum R_H \tag{3.162}$$

Equation (3.161) is a statement of conservation for total acid species, while Eq. (3.162) is a statement of conservation of charge, i.e. alkalinity. Together with Eq. (3.157), they constitute a system of three equations in three unknowns. Notice that Eqs (3.161) and (3.162) involve derivatives of more than one dependent species; in addition, the remaining reaction terms may involve various combinations of the dependent variables. Consequently, these equations are inherently *coupled*, and because of the linking through the derivatives, they are also called *implicit*. Solution methods are discussed in Chapters 6 and 8.

3.12 Conservation Equations via Volume Averaging

The previous derivation of conservation equations was based on the postulate that balances can be written for continuum properties defined by first averaging the sediment properties with Eq. (3.1). While geochemists have largely held such a belief as self-evident, this is not so for many hydrologists and engineers. They advance that the familiar conservation equations apply within a single phase only at the microscopic scale, but not at the sampling scale, V_{avg}, where the medium is heterogeneous. They further assert that conservation equations at the sampling scale should be derived by averaging the microscopic scale equations and not the properties themselves. Based on this differing view, they have erected a theory of volume-averaged conservation equations (e.g. Whitaker, 1967, 1985; Gray, 1975; Slattery, 1981; Marle, 1982; Hassanizadeh, 1986; Plumb and Whitaker, 1990; Gray et al., 1993), and it has been employed to describe some diagenetic problems (Domenico and Palciauskas, 1979; Nguyen et al., 1982). To complete the discussion of diagenetic model-building, here is a somewhat idiosyncratic version of this alternate theory.

Consider an averaging volume, V_{avg}, placed somewhere in a sediment (Fig. 3.28). This averaging volume is often called the *representative elementary volume* or REV. The volume V_{avg} is made up of two components, the volume of the porewater, V_f, and that of the sediment solids, V_s. If $C^\bullet$ is the actual concentration of a solute at a microscopic point in the porewaters (if we could measure such a quantity), then the average solute concentration, C, at the center of the averaging volume is given by

$$\left\langle C^\bullet \right\rangle \equiv \frac{1}{V_{avg}} \int_{V_{avg}} C^\bullet dV = \frac{\varphi}{V_f} \int_{V_f} C^\bullet dV = \varphi C \tag{3.163}$$

The conservation equation for a solute at any point in the porewater is

$$\frac{\partial C^\bullet}{\partial t} = -\nabla \cdot F + \sum R^\bullet \tag{3.164}$$

where F is the flux vector of the solute at this point and $\Sigma R^\bullet$ is the sum of all homogeneous reactions affecting the solute, again at this point. To get a conservation equation valid for V_{avg}, we take the average defined by Eq. (3.164),

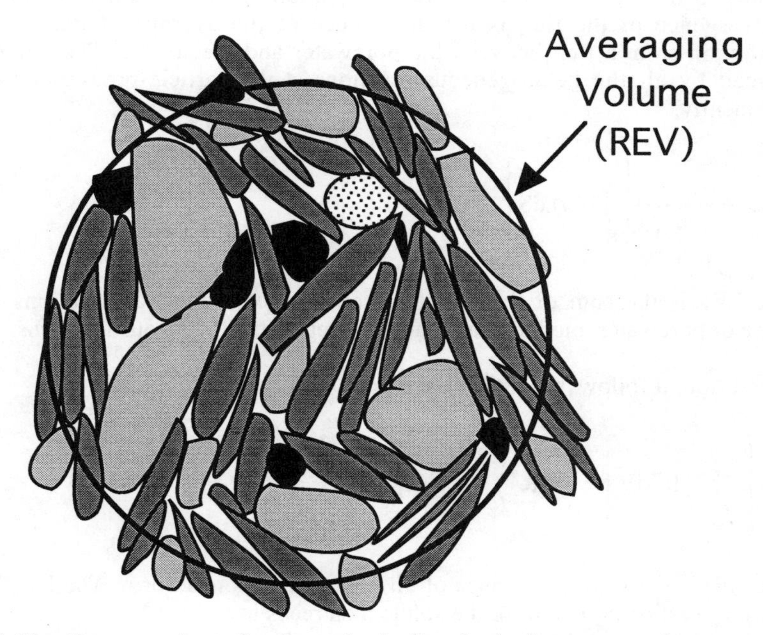

Figure 3.28. Cross-section of a (hypothetical) spherical representative averaging volume (REV) in an idealized aqueous sediment. The value of C(x,t), where x is at the center of the volume, is calculated by averaging the concentration over all points in this volume, as in Eq. (3.163).

$$\frac{1}{V_{avg}}\left[\int_{V_{avg}}\left(\frac{\partial C^{\bullet}}{\partial t}+\nabla\cdot F-\sum R^{\bullet}\right)dV\right]=0 \tag{3.165}$$

As the solid sediment particles are not detectably deformable, the average of the time-derivative term is simply

$$\frac{1}{V_{avg}}\left[\int_{V_{avg}}\frac{\partial C^{\bullet}}{\partial t}dV\right]=\frac{\partial\langle C^{\bullet}\rangle}{\partial t}=\frac{\partial\varphi C}{\partial t} \tag{3.166}$$

To average the divergence term, we utilize the Slattery-Whitaker averaging theorem (Slattery, 1981),

$$\langle\nabla\cdot F\rangle=\nabla\cdot\langle F\rangle+\frac{1}{V_{avg}}\left[\int_{S_s}F\cdot n\,dS\right] \tag{3.167}$$

where S_S is the contact surface between porewater and the solids in V_{avg} and n is the unit normal from porewater into solids. Equation (3.167) says that the average of the divergence of the flux is the divergence of the average of the flux plus a contribution from a flux between the porewater and the solids. The latter flux is produced by all the heterogeneous reactions at the particle-porewater interface; consequently,

$$\varphi\sum R_S\equiv-\frac{1}{V_{avg}}\left[\int_{S_s}F\cdot n\,dS\right] \tag{3.168}$$

where ΣR_S is the sum of the effects of all the heterogeneous reactions per unit volume of porewater, modelled as a sum of pseudo-homogeneous reactions.

In addition, it follows from the above that

$$\frac{1}{V_{avg}}\left[\int_{V_{avg}}\sum R^{\bullet}dV\right]=\left\langle\sum R^{\bullet}\right\rangle=\varphi\sum R_f \tag{3.169}$$

where ΣR_f is the sum of the rates of all homogeneous reactions. The locally averaged conservation equation for the solute then reads as

$$\frac{\partial\varphi C}{\partial t}=-\nabla\cdot\langle F\rangle+\varphi\sum R_f+\varphi\sum R_s \tag{3.170}$$

The last step is to expand the averaged flux. This flux is again composed of advective and molecular diffusive components,

$$F = u^{\bullet}C^{\bullet} - D_o\nabla C^{\bullet} \tag{3.171}$$

where $u^{\bullet}$ is the pointwise velocity vector; thus,

$$\langle F\rangle = \left\langle u^{\bullet}C^{\bullet}\right\rangle - \left\langle D_o\nabla C^{\bullet}\right\rangle \tag{3.172}$$

The molecular diffusion coefficient is very likely a constant over the averaging volume (i.e. the ionic strength and density are essentially constant); thus,

$$\left\langle D_o\nabla C^{\bullet}\right\rangle \approx D_o\left\langle \nabla C^{\bullet}\right\rangle \tag{3.173}$$

so that Eq. (3.170) now reads

$$\frac{\partial\varphi C}{\partial t} = \nabla\cdot\left(D_o\left\langle\nabla C^{\bullet}\right\rangle - \left\langle u^{\bullet}C^{\bullet}\right\rangle\right) + \varphi\sum R_f + \varphi\sum R_s \tag{3.174}$$

The final steps involve trying to express all concentrations and their gradients in terms of the mean concentration, C. This requires expanding the diffusion and advection terms. Starting with diffusion, application of the Slattery-Whitaker theorem (Gray, 1975, his Eq. (6)) gives

$$\left\langle\nabla C^{\bullet}\right\rangle = \nabla\left\langle C^{\bullet}\right\rangle + \frac{1}{V_{avg}}\int_{S_s} C^{\bullet}\, n\, dS \tag{3.175}$$

or, using Eq. (3.163),

$$\left\langle\nabla C^{\bullet}\right\rangle = \nabla\varphi C + \frac{1}{V_{avg}}\int_{S_s} C^{\bullet} n\, dS \tag{3.176}$$

At this point, Gray (1975) suggests representing $C^{\bullet}$ as the mean value C and a *spatial* deviation from this mean within the local averaging volume,

$$C^{\bullet} = C + \tilde{C} \tag{3.177}$$

where the tilde indicates this spatial deviation. This decomposition has the properties that

$$\langle C\rangle = C \tag{3.178}$$

and

$$\langle \tilde{C} \rangle = 0 \tag{3.179}$$

These are spatial equivalents to the Reynolds decomposition and averaging used above for bioturbation (see also Chapter 2). Substitution into Eq. (3.176) produces

$$\langle \nabla C^{\bullet} \rangle = \nabla \varphi C + \frac{1}{V_{avg}} \int_{S_s} C\, n\, dS + \frac{1}{V_{avg}} \int_{S_s} \tilde{C}\, n\, dS \tag{3.180}$$

Logically, there cannot be a diffusive flux if there is no porewater gradient in C, and porosity gradients cannot cause fluxes by molecular diffusion (Berner, 1980). Yet, Eq. (3.180) would lead to this conclusion, even if $\nabla C=0$, unless

$$\frac{1}{V_{avg}} \int_{S_s} C\, n\, dS = -C\, \nabla \varphi \tag{3.181}$$

Gray (1975) formally derives this requirement.

Secondly, the local deviation from the mean value is approximately equal to the first term in a Taylor series expansion about the mean, if the deviations are (presumably) small, i.e.

$$\tilde{C} \approx \delta \mathbf{r} \cdot \nabla C \tag{3.182}$$

where $\delta \mathbf{r}$ is a positional vector with its origin at the center of the REV. Thus, with Eqs (3.181) and (3.182), Eq. (3.180) becomes

$$\langle \nabla C^{\bullet} \rangle = \varphi \nabla C \left(\mathbf{I} + \frac{1}{V_f} \int_{S_s} \delta \mathbf{r} \cdot n\, dS \right) \tag{3.183}$$

where $\mathbf{I}$ is the unit vector, and where I have made use of the fact that $V_{avg} = V_f/\varphi$.

Next, the porewater velocity vector $u^{\bullet}$ can be written as a mean and a deviation, as was done for the concentration, so that

$$\langle u^{\bullet} C^{\bullet} \rangle = \varphi u C + \varphi \langle \tilde{u} \tilde{C} \rangle \tag{3.184}$$

Here, uC is the average advective flux in the porewaters, and the second term on the right-hand side is interpreted as the physical dispersion flux. Dispersion is not normally an important flux in diagenesis because flows are generally very small (i.e. < 1 cm yr^{-1}); therefore, this term is normally neglected and

$$\langle u^{\bullet} C^{\bullet} \rangle \approx \varphi u C \tag{3.185}$$

The results of retaining the dispersion term can be ascertained by consulting Whitaker (1967, 1985), Gray (1975), Slattery (1981), Hassanizadeh (1986), and Plumb and Whitaker (1990), and need not detain us here.

All these results are introduced into Eq. (3.174), to obtain

$$\frac{\partial \varphi C}{\partial t} = \nabla \cdot \left(\varphi D_o \nabla C \left(\mathbf{I} + \frac{1}{V_f} \int_{S_s} \delta \mathbf{r} \cdot n \, dS \right) - \varphi u C \right) + \varphi \sum R_f + \varphi \sum R_s \tag{3.186}$$

The term containing the surface integral causes a reduction in the apparent value of the diffusion coefficient, D_o, as a result of the irregular geometry of the diffusion paths in the sediment (Whitaker, 1967). Consequently, it can be identified as the tortuosity for this model. In one-dimension, Eq. (3.186) reads as

$$\frac{\partial \varphi C}{\partial t} = \frac{\partial}{\partial x} \left(\frac{\varphi D_o}{\theta^2} \frac{\partial C}{\partial x} - \varphi u C \right) + \varphi \sum R_f + \varphi \sum R_s \tag{3.187}$$

where the tortuosity is defined by a surface integral even in this 1-D form, i.e.

$$\frac{1}{\theta^2} \equiv 1 + \frac{1}{V_f} \int_{S_s} \delta \mathbf{r} \, n \, dS \tag{3.188}$$

in order to emphasize the fact that tortuosity is a result of the fine-scale 3-D properties of the sediment structure.

The apparent equivalency of Eq. (3.187) to previous diagenetic conservation equations should not come as a surprise. The same phenomena are being described in each case, and the difference is one of approach, not of fundamental interpretation. I would consider neither method as superior in the context of diagenetic modelling. Perhaps the surface integral representation of tortuosity in Eq. (3.188) conveys more information than the postulated θ, but that remains to be proven.

Yet another approach to the formulation of conservation equations is the theory-of-mixtures (e.g. Atkin and Craine, 1976; Drumheller, 1978; Bedford, 1978; Bedford and Drumheller, 1983; Drumheller and Bedford, 1980; Nunziato and Walsh, 1980; Demiray, 1981), and it has been applied to diagenesis by Kirwan and Kump (1987). This theory considers simultaneously the conservation of both mass and momentum of the species of interest. While the theory-of-mixtures does differ from the other theories developed above, these differences appear to be significant only on short time scales and when diffusion is relatively slow, e.g. bioturbation (Boudreau, 1989).

3.13 Modelling Continua

So far in this chapter, concentration has been treated as if it were only a function of the independent variables space and time. In most situations that is a perfectly accurate description, but this may not always be the case. For example, a process may depend on the "age" of the particles in the sediment, i.e. the elapsed time since a particular particle entered the sediment, in addition to time itself. As particles enter sediments continuously with time and the surface zone of sediments is usually bioturbated, any given layer will contain a continuous mixture of particles of different ages. Furthermore, a concentration that appears in a diagenetic equation is, in a very real sense, also a function of the parameters that this equation contains. Normally, we consider only the case where such parameters take on discrete values, but what if the parameter can have a *spectrum* of values in a given sediment? For example, what if the interfacial flux introduces organic matter particles characterized by a continuous distribution of different reactivities? These reflections lead to an examination of two specific continuum models.

3.13.1 Age-Dependent Processes

By its operational definition, time is what is measured by a clock starting from an arbitrary datum. Every observer in a diagenetic system and every particle "sees" the same time. (No relativistic effects here.) On the other hand, age is the time that has passed since a particle (or whatever) was introduced into the sediments. Each group of particles that initially entered the sediments together have the same age. Because particles are continuously accumulating in most sediments and most surface sediments are mixed, there results a spectrum of ages in any sediment horizon.

Particles may have age-dependent properties, which lead to altered responses to diagenetic processes. For example, the rate of mixing of organic matter particles may very well be related to their age. C.R. Smith et al. (1993) propose that as an organic matter particle ages it becomes less "food-rich" or labile. Thus, infaunal organisms preferentially ingest and mix newer/younger particles, in comparison with older particles. This theory of age-dependent mixing is based largely on the previous theory of age-dependent dispersal that has developed in biomathematical theory (see Okubo, 1980, for a readable summary of that theory). But particle mixing is not the only application of age-dependence. We also know that soluble mineral species, such as biogenic opal, appear to dissolve less readily as they age.

To take into account the effects of age, Okubo (1980) introduces a new type of concentration variable, b(x,a,t), which is the amount (number or mass) of the species per unit volume and per unit of age (a = age). Age is not normally a measurable quantity of mixed sediments because such sediments are a collection of particles with different ages. What we usually know is the total concentration, B(x,t). These two variables are related by

$$B(x,t) = \int_0^\infty b(x,a,t)da \qquad (3.189)$$

Thus b(x,a,t) is a population variable because it describes how much material there is between the ages a and a+da, where da is an infinitesimal increment.

Diagenetic processes act, nevertheless, at the b(x,a,t)-level and should be described as such. Okubo (1980) suggests that this can be done by treating age as a third independent variable; the total derivative now reads as (at constant porosity)

$$\frac{Db}{Dt} = \left(\frac{\partial b}{\partial t}\right)_{x,a} + \left(\frac{\partial b}{\partial a}\right)_{x,t} + v\left(\frac{\partial b}{\partial x}\right)_{a,t} \tag{3.190}$$

where v is a generic velocity. This last equation leads to the altered conservation equation for b(x,a,t)

$$\frac{\partial b}{\partial t} + \frac{\partial b}{\partial a} = \kappa_b \frac{\partial^2 b}{\partial x^2} - v\frac{\partial b}{\partial x} + R_b \tag{3.191}$$

where κ_b is a generic diffusivity and R_b is the net rate of reaction affecting b(x,a,t). With a first-order decay, Eq. (3.191) becomes

$$\frac{\partial b}{\partial t} + \frac{\partial b}{\partial a} = \kappa_b \frac{\partial^2 b}{\partial x^2} - v\frac{\partial b}{\partial x} - \lambda_b b \tag{3.192}$$

The parameters κ_b and λ_b are themselves functions of x, t, and a, in general ,which allows for the effects of age on these processes. By making either κ_b or λ_b decaying functions of age, one can simulate decreasing mixing intensity with age (e.g. Smith et al., 1993) or even the decreasing solubility or reactivity of a solid as it ages in a sediment, a phenomenon which has yet to be modelled.

3.13.2 Reactive Continua

In the final topic of this chapter, let us consider the results when reactive particles have not one or even many possible reactivities, but a spectrum or distribution of reactivities. Such a theory has been developed to describe organic matter decay in sediments (Boudreau and Ruddick, 1991) and iron oxy-hydroxide dissolution (Postma, 1993). (Bosatta and Ågren (1991, 1995) have independently developed related models for soil organic matter decomposition.)

The basic premise of this continuum theory is the existence a spectrum of reactive types that are characterized by a distribution function, g(k,t), which determines the concentration of species having rate constants between k and k+dk, where dk is an infinitesimal increment in k (see Aris and Gavalas, 1966; Aris 1968, 1989; Hutchinson and Luss, 1970; Ho and Aris, 1987). Let us assume that the members of the continuum are solid species and that each member decays according to a first-order reaction in its own concentration. Then, in the bioturbated zone of a sediment, with diffusive mixing and constant porosity, the conservation equation for a component of the continuum is then

$$\frac{\partial g}{\partial t} = D_B \frac{\partial^2 g}{\partial x^2} - w \frac{\partial g}{\partial x} - kg \tag{3.193}$$

and, in the historical zone underneath,

$$\frac{\partial g}{\partial t} = -w \frac{\partial g}{\partial x} - kg \tag{3.194}$$

Normally, the observable quantity in natural situations is again the total amount or concentration of the population, B(x,t). This quantity is calculated as the integral of the distribution function over the range of all possible values of k, which can be taken to be zero to infinity (Aris, 1968),

$$B(x,t) = \int_0^\infty g(k,x,t)\,dk \tag{3.195}$$

where g(k,x,t) is obtained as the solution of either Eq. (3.193) or Eq. (3.194), subject to appropriate boundary and initial conditions. The challenge in using this continuum model lies in specifying the initial distribution of reactive types, g(k,x,0), and the concentration or flux at x=0. The initial distribution can be thought of as a constitutive equation, and this is a topic revisited in Chapter 4.

Equation (3.194) also forms the basis for the interpretation of laboratory-measured decomposition rates. In the historical layer at steady state, there is a unique relation between time and depth (Berner, 1980), i.e. $x = t \cdot w$, which cannot exist in the mixed zone. If this relation is substituted into the steady-state form of Eq. (3.194), there results

$$\frac{dg}{dt} = -kg \tag{3.196}$$

This is the famed decay equation, but one that applies to each member of the continuum. As we shall see in Chapter 5, the solution to Eq. (3.196) is

$$g(k,t) = g(k,0)\,exp(-kt) \tag{3.197}$$

where $g(k,0) \equiv$ initial distribution. Integrating this equation for all k values, one gets

$$B(t) = \int_0^\infty g(k,0)\,exp(-kt)\,dk \tag{3.198}$$

and the problem is solved once g(k,0) is specified. If some appropriate form of g(k,0) is placed in this equation, then there results an equation for B as a function of time. This result can be fitted to laboratory-measured values of B(t) versus t to evaluate any free parameters in g(k,0).

4 CONSTITUTIVE EQUATIONS

"The time has come," the Walrus said,
"To talk of many things:
Of shoes - and ships - and sealing wax -
of cabbages - and kings ...
and why the sea is boiling hot -
and whether pigs have wings."

(Lewis Carroll, *Through the Looking-Glass*, 1871)

4.1 Introduction

Equations such as Fick's First Law are needed to resolve quantities like the flux vector, F, in the conservation equations. This required substitution reflects the fact that conservation equations are general statements, independent of the particular situation or environment. Constitutive equations specify the intrinsic response of the diagenetic transport or reaction processes to the forces that create them. Without constitutive equations, the specification of a diagenetic model is incomplete.

Constitutive equations are themselves conceptual models. Because they play such a fundamental role in diagenetic modelling, care must be exercised in adopting particular forms. Modellers should be fully aware of their meaning and limitations. This chapter deals with the origin and properties of the constitutive equations normally used in diagenetic conservation equations, and it examines how some of their parameters can be estimated.

Before doing this, however, there is a set of properties any well-formulated constitutive relation should possess, and these deserve repeating (Truesdell and Toupin, 1960; Bear, 1972). A constitutive relation should have:

i) *consistency*, in that it conserves mass;

ii) *coordinate invariance and material indifference*, in that the mathematical description of the process should not be dependent on the particular coordinate system one chooses, nor on the position of the observer;

iii) *dimensional invariance*, in that changing the units should not alter the effect of the process; and

iv) *just setting*, which simply means that substitution of the correct constitutive equations into a conservation equation does indeed lead to a mathematically well-defined problem that has a unique and physically meaningful solution.

Failure to possess any of these properties is a very good reason to reject a proposed constitutive relationship.

4.2 Fickian Transport Processes

The common mathematical relations for the fluxes of momentum, energy, and mass in environmental systems are all of the form, Flux = Constant X Force, where the force is represented as the spatial derivative of an appropriate variable, i.e. velocity, temperature, or concentration/chemical potential. All such "laws" can be called *Fickian* because of their resemblance to Fick's First Law. (However, they could just as easily be called Ohmian, Fourian, Newtonian, or even Darcian; the choice simply suits the present context.) The descriptions given below highlight the striking similarities amongst these relationships.

4.2.1 Fluid Viscosity

Viscosity is a measure of a fluid's resistance to flow under the action of a force. In particular, we are interested in the viscosity that resists shearing stresses in a fluid. The fluid goes into motion as a consequence of the transfer of momentum. This transfer is not instantaneous, but occurs gradually over time, spread by net differences in molecular motions. This sounds very much like diffusion of momentum, and indeed, it is. The constitutive equation governing this process is of the form

$$\sigma = -\mu \frac{\partial v_x}{\partial y} \tag{4.1}$$

where, in this case, x is the direction of flow, y is the direction normal to the flow, σ is the shear rate (flux), v_x is the velocity component parallel to the surface, $\partial v_x/\partial y$, is the velocity gradient (the stress), and μ is a proportionality constant called the *dynamic viscosity*. The dynamic viscosity thus characterizes the resistance to flow.

This molecular viscosity does not play a direct major role in most diagenetic models. However, it is a prominent parameter in various formulas for solute diffusion coefficients. Consequently, some knowledge of its origin is desirable, and I adopt a simple phenomenological derivation.

To this aim, consider two adjacent, arbitrary "layers" in a water mass in laminar flow (Fig. 4.1). One layer moves with velocity v and the other at v+Δv. Observations on momentum transfer show that momentum is transported between these two layers because of the difference in their velocities (Bird et al., 1960), i.e.

$$\sigma \propto \frac{\rho(v + \Delta v) - \rho v}{\Delta y} = \rho \frac{\Delta v}{\Delta y} \tag{4.2}$$

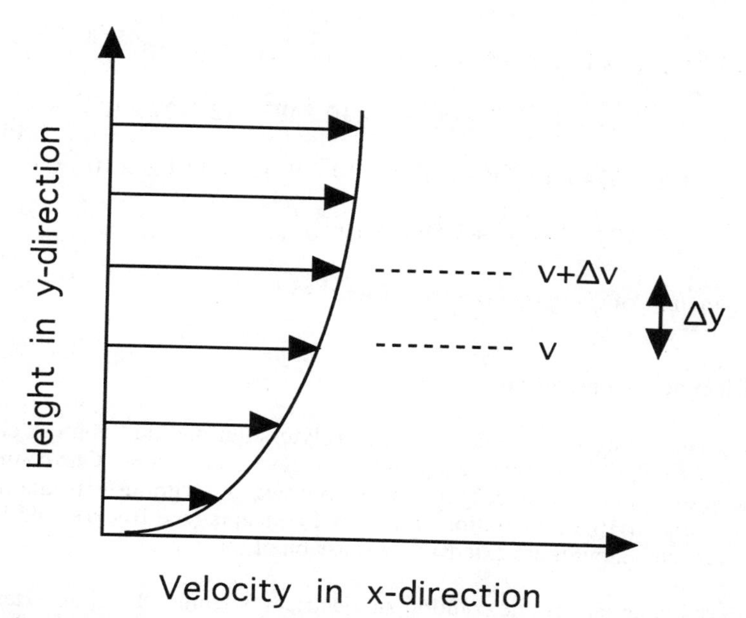

Figure 4.1. Vertical velocity profile in an arbitrary flow near the boundary with a solid.

where the operator $\propto$ indicates proportionality, ρ is the water density (assumed constant), and Δy is the distance between the layers.

To make Eq. (4.2) an equality, we must recognize first that momentum goes from a layer of high velocity to one of low velocity, i.e. in the direction negative to $\Delta v/\Delta y$. Secondly, there is a major difference in the units on either side of this proportionality. Specifically, σ has units of mass divided by the product of length and squared time. On the other hand, $\rho\Delta v/\Delta y$ has units of mass divided by the product of volume and time. There must be a constant of proportionality, ν, with units of length squared per unit time,

$$\sigma = -\nu\rho\frac{\Delta v}{\Delta y} \tag{4.3}$$

The constant ν is called the *kinematic viscosity*; however, ν and ρ can be combined into a single constant, $\mu \equiv \nu\rho$, which is the dynamic viscosity of Eq. (4.1). Finally, take the limit of $\Delta y \rightarrow 0$, and the ratio of differences becomes a derivative, and Eq. (4.1) is recovered.

Values of the dynamic viscosity can be calculated from an empirical equation developed by Matthaus (as quoted in Kukulka et al., 1987), which is claimed to be accurate to within 0.7% for the temperature, t, salinity, S, and pressure, P, ranges of $0 \leq t^{o}C \leq 30$, $0 \leq S \leq 36$, and 1 to 1000 bars, respectively,

$$\mu = 1.7910 - (6.144 \times 10^{-2})t + (1.4510 \times 10^{-3})t^2 - (1.6826 \times 10^{-5})t^3$$
$$- (1.5290 \times 10^{-4})P + (8.3885 \times 10^{-8})P^2 + (2.4727 \times 10^{-3})S$$
$$+ t(6.0574 \times 10^{-6}P - 2.6760 \times 10^{-9}P^2) + S(4.8429 \times 10^{-5}t$$
$$- 4.7172 \times 10^{-6}t^2 + 7.5986 \times 10^{-8}t^3) \quad (4.4)$$

where μ is in units of centipoise (i.e. 10^{-2} g cm^{-1} s^{-1}).

4.2.2 Electrical Conductivity

Porewaters (including freshwaters) are electrolyte solutions that contain charged ions as solutes and, consequently, can conduct an electrical current. One cannot talk about the movement (diffusion) of ions in porewaters without introducing the *electrical conductivity* of a solution and related concepts (see Bockris and Reddy, 1970, for a readable, but more extensive, discussion of this topic).

There exists in an electrolyte solution an electrical potential, Ψ. If an externally-imposed electrical current is absent or if the ions themselves do not move as a result of thermally induced motions, the potential would be constant throughout the solution. However, ions are in constant motion and currents are often applied by inquisitive experimentalists. In either case, a perturbation in the potential, i.e. a gradient that is either local to the moving ions or global from the applied current, generates a force on the ions. This force per unit charge, X, is called the electric field, and potential theory gives that

$$X = -\frac{\partial \Psi}{\partial x} \quad (4.5)$$

The negative sign indicates that this force acts on a positive charge in a direction opposite to that of the positive gradient in potential.

This force will engender a flux of ions, say $(F_e)_i$ for the i-th ion. Without loss of generality, we can assume that $(F_e)_i$ and X are related via a power series,

$$(F_e)_i = b_o + b_1 X + b_2 X^2 + \ldots \quad (4.6)$$

where b_o, b_1, b_2, etc., are constants to be determined. If X is sufficiently small, the terms in X^2 and higher can be ignored, so that

$$(F_e)_i \approx b_o + b_1 X \quad (4.7)$$

$$\left(F_e\right)_i = b_1 X \tag{4.8}$$

or, with Eq. (4.5),

$$\left(F_e\right)_i = -b_1 \frac{\partial \Psi}{\partial x} \tag{4.9}$$

Now $(F_e)_i$ represents the amount or number of ions of type i that cross a unit area per unit time, while X is the actual applied force (e.g. in dynes); therefore, b_1 must define the flux caused by the application of a unit force. The flux can be stated as the local concentration, C_i, multiplied by a velocity; therefore, if we write the velocity of these ions that is caused by the application of unit force as υ_i, then

$$b_1 = \upsilon_i C_i \tag{4.10}$$

The parameter υ_i is called the (absolute) mobility of the particular ion. As a solution approaches infinite dilution, $\Sigma C_i \rightarrow 0$, the value of υ_i approaches a constant, called the (absolute) limiting mobility, υ_i°. This limiting mobility plays a fundamental role in defining diffusion coefficients of ions in electrolyte solutions and will reappear later in this chapter. As a consequence of Eq. (4.10), the flux generated by the electrical potential can be written as

$$\left(F_e\right)_i = -C_i \upsilon_i \frac{\partial \Psi}{\partial x} \tag{4.11}$$

In addition to the flux of ions, there also results a flux of charge, $(\mathbf{i}_e)_i$, given by

$$\left(\mathbf{i}_e\right)_i = z_i \boldsymbol{F} C_i \upsilon_i \frac{\partial \Psi}{\partial x} \tag{4.12}$$

where $\boldsymbol{F}$ is Faraday's constant (96490 C mol^{-1})† and z_i is the charge of ion i. In this last equation it is convenient to define the (conventional) equivalent conductance or ionic conductivity as

$$\lambda_i \equiv z_i e_o \upsilon_i \boldsymbol{F} \tag{4.13}$$

where e_o is the charge on the electron; therefore,

$$\left(\mathbf{i}_e\right)_i = \frac{\lambda_i C_i}{e_o} \frac{\partial \Psi}{\partial x} \tag{4.14}$$

---

† Cussler (1984) makes an informative remark about Faraday's constant: it "is even more of a cultural artifact; it is a unit conversion factor explicitly included whenever this equation is written. The apparent suspicion is that no one can properly use electrostatic units without warning." - p. 149.

At infinite dilution,

$$\lim_{\sum C \to 0} \lambda_i \equiv \lambda_i^\circ \tag{4.15}$$

In principle, λ°_i is a measurable quantity, and as we shall see, it forms part of the definition of ionic diffusion coefficients. Tables containing values of λ°_i for various cations and anions in the temperature range of interest to diagenetic modellers are available in Nigrini (1970), Miller (1982), Dobos (1975), and CRC (1993) in units of $cm^2\ ohm^{-1}\ eqv^{-1}$. Nigrini (1970) provides the following equation to estimate the value of λ°_i at 25°C for cations not in those tables,

$$\lambda_i^\circ = 10.56 + 90.72 \log_{10}(z_i) + \frac{42.95\, r_i}{z_i} \tag{4.16}$$

where r_i is the crystal ionic radius, which may be found in the CRC (1993).

Dobos (1975) suggested that λ°_i is essentially a linear function of temperature in the range observed during early diagenesis. Linear regressions of λ°_i against temperature for the data found in Nigrini (1970), Miller (1982), Dobos (1975), and CRC (1993) are presented in Tables 4.1 and 4.2 for cations and anions, respectively. The fits are excellent, with R^2 values always greater than 0.95. Some two data-point linear equations are also included, and their slopes are generally consistent with the linear regressions. These relationships allow calculation of λ°_i at other temperatures.

4.2.3 Molecular/Ionic Diffusion Coefficients

Fickian diffusion is frequently a significant component of solute fluxes in porewaters. Diffusion is caused by thermally generated random motions of molecules/ions/colloids. The macroscopic result of this phenomenon is the transfer of solutes from areas of high concentration to areas of low concentration (Fig. 3.12). Adolf Fick (1829-1901) was the first researcher to advance an acceptable mathematical model for the flux of material that ensues from diffusion, i.e. Eq. (3.42),

$$(F_D)_i = -D_i^o \frac{\partial C_i}{\partial x} \tag{4.17}$$

a relationship that now bears his name, i.e. Fick's First Law (see Tyrrell, 1964, for an informative biographical sketch).

In order to utilize Eq. (4.17), it is necessary to possess appropriate values of the molecular and/or ionic diffusion coefficient for the species in question, D°_i. Before broaching this subject further, we can gain much insight by first examining theoretical justifications for Eq. (4.17).

Table 4.1. Linear Regressions of λ_i° Against Temperature for Cations. (cm^2 ohm^{-1} eqv^{-1})

+1 Cations	m_o	m_1	r^2	n†
H^+	225.8	4.93	0.99	6
Li^+	18.7	0.81	0.99	6
Na^+	25.5	1.00	0.99	6
K^+	40.0	1.35	0.99	6
Rb^+	43.0	1.40	0.99	6
Cs^+	43.2	1.37	0.99	6
Ag^+	33.2	1.16	0.99	3
NH_4^+	39.8	1.37	0.99	4
+2 Cations	m_o	m_1	r^2	n
Ba^{2+}	33.9	1.18	0.99	3
Ca^{2+}	30.0	1.20	0.99	5
Cd^{2+}	27.7	1.02	0.99	3
Co^{2+}	27.7	1.02	0.99	3
Cu^{2+}	28.4	1.06	0.98	3
Fe^{2+}	27.7	1.00	0.99	3
Mg^{2+}	28.7	0.95	0.99	3
Mn^{2+}	26.7	1.04	0.99	3
Ni^{2+}	28.0	0.86	*	
Sr^{2+}	30.9	1.13	0.99	3
Pb^{2+}	37.3	1.33	0.99	3
Ra^{2+}	32.9	1.34	0.99	3
Zn^{2+}	27.7	1.01	0.99	3
+3 Cations	m_o	m_1	r^2	n
Al^{3+}	29	1.36	*	
Ce^{3+}	37	1.31	*	
La^{3+}	34.9	1.38	0.99	3
Pu^{3+}	34	1.20	*	

$\lambda^\circ = m_o + m_1 t$ with t in °C
* two-point slope
† number of points in the regression

Table 4.2. Linear Regressions of λ_i° Against Temperature for Anions. ($cm^2\ ohm^{-1}\ eqv^{-1}$)

-1 Anions	m_o	m_1	r^2	n†
OH^-	108.0	3.65	0.99	5
F^-	26.5	1.16	*	
Cl^-	40.2	1.46	0.99	6
Br^-	41.8	1.47	0.99	6
I^-	41.1	1.44	0.99	6
ClO_4^-	36.9	1.22	1.00	3
$H_2PO_4^-$	7.4	1.14	*	
HS^-	43.3	0.85	0.99	6
HSO_3^-	27.0	0.92	*	
HSO_4^-	25.1	1.04	*	
NO_2^-	43.1	1.06	0.96	3
NO_3^-	9.6	1.29	0.99	4
Acetate	20.1	0.83	1.00	3

-2 Anions	m_o	m_1	r^2	n
CO_3^{2-}	36.1	1.34	0.99	3
SO_4^{2-}	40.9	1.55	0.99	3
$S_2O_4^{2-}$	34.0	1.26	*	

$\lambda^\circ = m_o + m_1 T$ with T in °C

* two-point slope

† number of points in the regression

Fick himself based his law of diffusion on an analogy to the empirical equation for heat conduction, i.e. Fourier's Law, and the agreement he obtained with experimental results using his assumed form (Fick, 1855). Equation (4.17) can also be derived theoretically, using either our modern understanding of molecular motions (a microscopic approach) or the modern theory of irreversible thermodynamics (a macroscopic/phenomenological approach). The material that follows illustrates two simplified (heuristic) derivations based on both these theories.

Molecular/Statistical Basis. Although diffusion is a random process, it does lead to a net flux in a direction opposite to a concentration gradient. To prove this point, consider a body of porewater with a concentration gradient of some (uncharged) solute in the x-direction. Place two closely spaced planes, A and B, perpendicular to the x-axis. Because of the gradient, there are N molecules of solute at plane A and $(1+\varepsilon)N$ molecules at plane B, where ε is a (small) difference (Fig. 4.2).

If the motions are truly random and N is sufficiently large, then at any given instant half of the particles at A, i.e. N/2, will move toward plane B and half in the opposite direction. At the same time, half of the molecules at B, i.e. $(1-\varepsilon)N/2$, will move toward plane A and the other half in the opposite direction. Given that the planes are adjacent, then the net exchange of molecules, $\aleph$, from plane B to plane A at the chosen instant is

$$\aleph = (1+\varepsilon)\, N/2 - N/2 = \varepsilon N/2 \tag{4.18}$$

While this outcome does not constitute a proof in a formal sense, it nonetheless demonstrates the intended result that random motions of molecules lead to a net transfer in a direction opposite to the gradient.

To derive Fick's First Law with this simple, purely heuristic molecular model, we must assume that the fluid is at rest, and that the diffusion process is the result of random discrete molecular jumps of mean distance, ℓ_D, and mean time, τ. Let the direction of the concentration gradient be the x-direction, and place three planes, L, T, and R, normal to this gradient at a distance $\ell_D/2$ apart (Fig. 4.3).

Moreover, let the concentration of the solute be C_L, C_T, and C_R at the planes L, T, and R, respectively. The net flux of molecules across the plane T, F_D, will be the difference between what moves from plane L to plane R and what moves in reverse, if ℓ_D is indeed small and reactions involving the solute are not instantaneous. Let $\hat{u}$ be the mean velocity of the molecules during their random motions; then

$$F_D = \frac{\hat{u}C_L}{2} - \frac{\hat{u}C_R}{2} \tag{4.19}$$

because only half of the molecules at planes L and R move towards T.

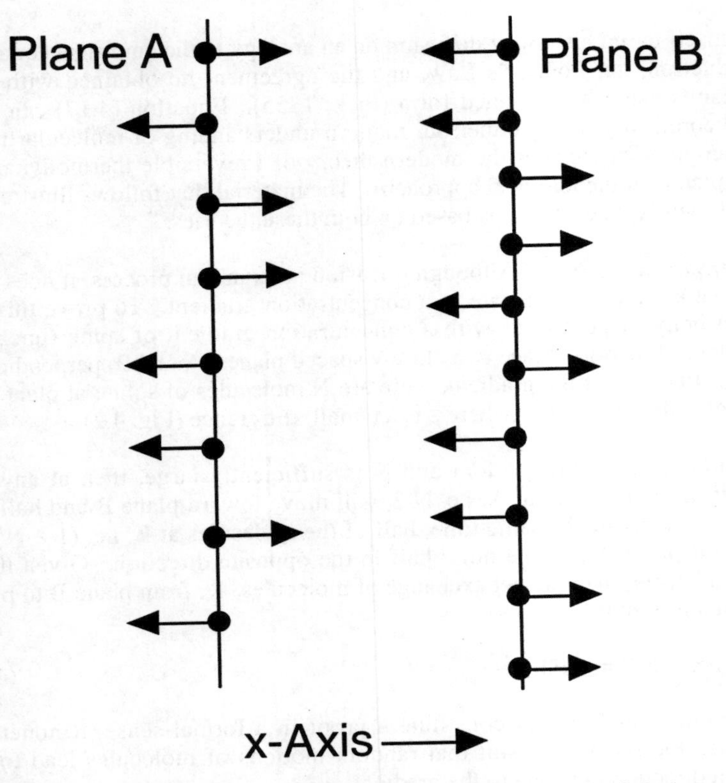

Figure 4.2. Hypothetical movement of a collection of particles as a result of diffusion in one dimension. The dots are molecules/ions/colloidal particles and the arrows indicate the direction of 1-D motion.

As long as the concentration gradient is not too strong, then the concentrations at the two planes, L and R, can be expressed in terms of the concentration at the center plane, T, via Maclaurin series (x-axis origin arbitrarily placed at T),

$$C_L \approx C_T - \frac{\ell_D}{2}\frac{\partial C}{\partial x} \tag{4.20}$$

and

$$C_R \approx C_T + \frac{\ell_D}{2}\frac{\partial C}{\partial x} \tag{4.21}$$

Substitution of Eqs (4.20) and (4.21) into Eq. (4.19) gives

$$F_D = -\frac{\hat{u}\ell_D}{2}\frac{\partial C}{\partial x} \tag{4.22}$$

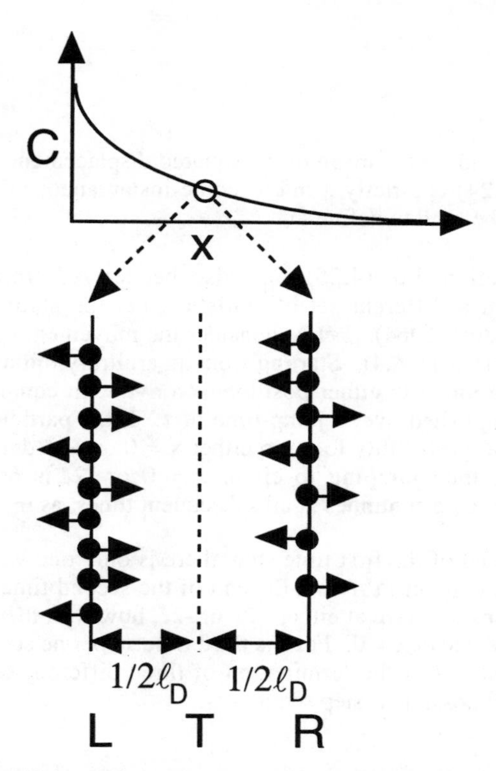

Figure 4.3. Three arbitrary planes, L, T, and R, placed normal to the direction of a diffusion gradient and at a distance $1/2\ell_D$ apart, where ℓ_D is the distance involved in a typical diffusive movement. The solid circles represent the diffusing particles on planes L and R; however, particles are also present at T and in the space between the planes, but they are not represented.

If we set

$$D_o \equiv \frac{\hat{u}\ell_D}{2} \tag{4.23}$$

then, Fick's First Law is recovered, and D_o is the diffusion coefficient. An alternative definition of D_o is obtained by writing $\hat{u}$ as the quotient of ℓ_D and a jump time, τ,

$$D_o \equiv \frac{\ell_D^2}{2\tau} \tag{4.24}$$

This latter definition of D_o should not be confused with the justly famous Einstein or Einstein-Smoluchowski relation, which relates the macroscopic displacement of a concentration impulse, χ, to the elapsed time, t, i.e.

$$\langle \chi^2 \rangle = 2\,D_o\,t \tag{4.25}$$

where the left-hand side is the mean of the squared displacement and t is the elapsed time. Equation (4.24) is strictly a microscopic instantaneous description, even if both equations can be used to define D_o.

The Einstein relation, Eq. (4.25), can also be derived from molecular level dynamics, but from a different set of statistics, i.e. the statistical properties of random walks (see Jost, 1964). Let us consider the movement of a single diffusing particle on a 1-D grid (Fig. 4.4). Starting from an arbitrary initial point at the top, a particle, i.e. ●, can move to either position $+\ell$ or $-\ell$ with equal probability. This movement is accomplished over a jump-time of τ. If the particle is now at $+\ell$, i.e. ⊘, then it has equal probability to go to either x = 0 or $+2\ell$ during the next jump-time. If it's at $-\ell$, then jumping to either x = 0 or -2ℓ is possible with equal probability. This process continues at all subsequent times, as in Fig. 4.4.

Note that, at the end of the first time step, there is only one way to be at position $+\ell$ and only one way to get to $-\ell$. At the end of the second time step, there is only one sequence of steps to arrive at either $+2\ell$ or -2ℓ; however, there are two different sequences that terminate at x = 0. For the third time step, one sequence leads to $+3\ell$ or -3ℓ, while $+1\ell$ or -1ℓ is the termination of three different sequences each. In general, there are 2^n possible n-step sequences.

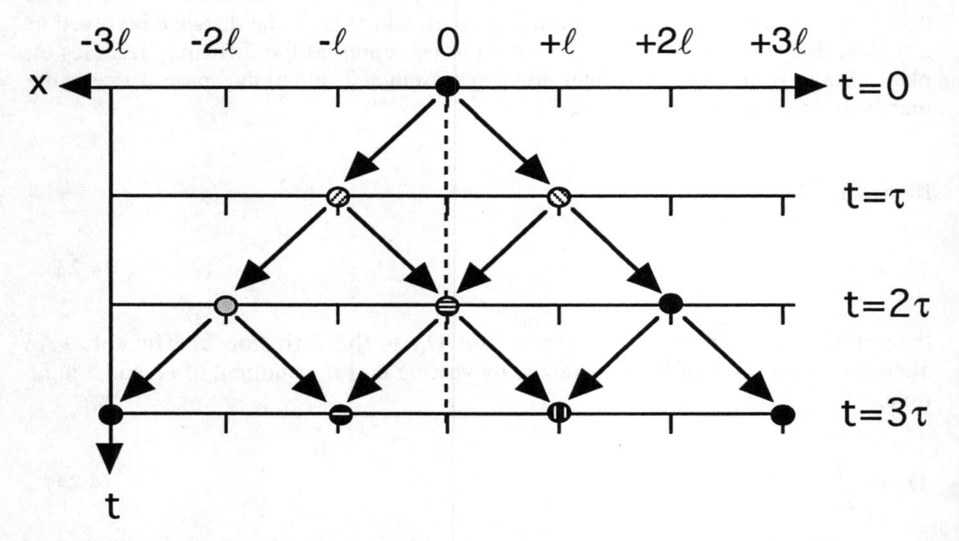

Figure 4.4. Possibilities allowed during the random walk of a particle initially at x = 0 on an equally spaced 1-D grid. (Adopted from Jost, 1964; reproduced with the kind permission of VCH Verlagsgesellschaft mbH.)

Let us sum the displacements, χ, for each possible sequence of spatial steps at each time step. For the first jump, $\Sigma\chi = \ell - \ell = 0$. At the second time step, $\Sigma\chi = 2\ell - 2\ell = 0$, and for the third, $\Sigma\chi = 3\ell + \ell + \ell + \ell - 3\ell - \ell - \ell - \ell = 0$, etc. Thus, there is no mean displacement when all the possibilities are considered. The lack of mean dis-placement originates with the symmetric nature of unbiased random walks, in which all possible movements to the right cancel out all possible movements to the left.

On the other hand, the square of the displacements will not sum to zero, as these are related to the total distance travelled from the origin, regardless of movement to the right or left. Thus, at the first time step, $\Sigma\chi^2 = \ell^2 + \ell^2 = 2\ell^2$; at the second, $\Sigma\chi^2 = 4\ell^2 + 4\ell^2 = 8\ell^2$; and at the third, $\Sigma\chi^2 = 9\ell^2 + \ell^2 + \ell^2 + \ell^2 + 9\ell^2 + \ell^2 + \ell^2 + \ell^2 = 24\ell^2$; etc. Generally, for n time steps, $\Sigma\chi^2 = n(2^n)\ell^2$. The average of the sum of the squared displacements for all step sequences is simply $\Sigma\chi^2$ divided by the number of sequences that can lead to this displacement,

$$\langle\chi^2\rangle = \frac{\Sigma\chi^2}{2^n} \tag{4.26}$$

or

$$\langle\chi^2\rangle = \frac{n2^n\ell^2}{2^n} = n\ell^2 \tag{4.27}$$

Moreover, the number of time steps, n, is simply the elapsed time, t, divided by the jump-time, τ,

$$n = \frac{t}{\tau} \tag{4.28}$$

so that

$$\langle\chi^2\rangle = \frac{t\ell^2}{\tau} \tag{4.29}$$

With the definition of the diffusion coefficient given by Eq. (4.24), Eq. (4.29) becomes Eq. (4.25).

Irreversible Thermodynamic Basis. From the perspective of this macroscopic theory, diffusion fluxes are generated by gradients in chemical potential (Bockris and Reddy, 1973; de Groot and Mazur, 1984). The chemical potential per mole of the i-th chemical species, ψ_i, is

$$\psi_i = \psi_i^o + RT\,ln(a_i) \tag{4.30}$$

where $\psi°_i$ is the standard chemical potential per mole, R is the gas constant, T is the absolute temperature, and a_i is the activity of solute i. Equation (4.30) can also be written as

$$\psi_i = \psi_i^o + RTln(\gamma_i C_i) \tag{4.31}$$

where γ_i is the activity coefficient for solute i and C_i is its concentration.

The force, f_i, that drives the flux, F_D, of solute i is then

$$f_i = -\frac{\partial \psi_i}{\partial x} \tag{4.32}$$

This force can be related, in a general sense, to F_D through a power series,

$$F_D = b_o + b_1 f_i + b_2 f_i^2 + b_3 f_i^3 + ... \tag{4.33}$$

where b_o, b_1, b_2, etc., are arbitrary constants. If the terms in $f_i{}^2$ and higher powers are small, then

$$F_D \approx b_o + b_1 f_i \tag{4.34}$$

This approximation is the weak point in this derivation, because it is difficult to judge *a priori* that the higher order terms are indeed negligible. Assuming that Eq. (4.34) is valid, then $b_o = 0$; otherwise, a flux would result even if the diffusive force were absent, i.e. $f_i = 0$. Therefore,

$$F_D \approx -b_1 \frac{\partial \psi_i}{\partial x} \tag{4.35}$$

Now, $\partial\psi_i/\partial x$ is the gradient of chemical potential per mole of solute i. The actual concentration is unlikely to be exactly one mole, so b_1 must include the actual concentration that produces F_D at x, i.e.

$$b_1 = b_1' C_i \tag{4.36}$$

Substituting Eqs (4.31) and (4.36) into (4.35) produces

$$F_D = -b_1' \frac{RT}{\gamma_i} \frac{\partial \gamma_i C_i}{\partial x} \tag{4.37}$$

or, expanding,

$$F_D = -b_1' RT \frac{\partial C_i}{\partial x} - \frac{b_1' RTC_i}{\gamma_i} \frac{\partial \gamma_i}{\partial C_i} \frac{\partial C_i}{\partial x} \tag{4.38}$$

and, using the definition of the derivative of a logarithm,

$$F_D = -b_i' RT\left(1+\frac{\partial ln\gamma_i}{\partial lnC_i}\right)\frac{\partial C_i}{\partial x} \tag{4.39}$$

From this last equation, we obtain the classic thermodynamic definition for the diffusion coefficient (Bockris and Reddy, 1973),

$$D_o \equiv b_i' RT\left(1+\frac{\partial ln\gamma_i}{\partial lnC_i}\right) \tag{4.40}$$

and Eq. (4.39) becomes Fick's First Law. The activity coefficient, γ_i, is largely a function of total salinity of the porewater. Diagenesis rarely creates an appreciable gradient in salinity, so that usually

$$\frac{\partial ln\gamma_i}{\partial lnC_i} << 1 \tag{4.41}$$

and

$$D_o = b_i' RT \tag{4.42}$$

The final step is to identify the constant b_i'. Examination of its units shows that it is the velocity of a molecule/ion of solute i when subject to a unit force. This velocity is then the mobility, υ_i, which appeared previously in the treatment of conductivity.

Infinite-Dilution Diffusion Coefficients. Various investigators have developed models (formulas) to predict diffusion coefficients when the solute in question is present at vanishingly small concentrations (i.e. infinite dilution). While one might first presume that such formulas are of limited value in moderately strong electrolyte solutions such as marine waters, intuition fails in this case. In fact, infinite dilution coefficients play an integral role in the theories for stronger solutions, and they are reasonably accurate approximations for all but the major ions (see below).

Uncharged Particles. Theories exist for dilute solutions or suspensions of both charged and uncharged particles. Uncharged particles are usually treated as spheres moving in a water continuum. A sphere moving through this continuum will be subject to a frictional force, f,

$$f = f\upsilon_i \tag{4.43}$$

where f is a friction coefficient and υ_i is the particle's velocity under unit force, i.e. its mobility. The force that causes the particle to move is the negative gradient in the chemical potential (per particle). If the driving force and friction are in balance, then

$$-\frac{\partial \psi_i}{\partial x} = N f \upsilon_i \tag{4.44}$$

where N is Avogadro's number. If we ignore the effects of non-ideality, then $\psi \equiv \psi°$ + $RTln(C)$, so that

$$\frac{\partial \psi_i}{\partial x} = \frac{RT}{C}\frac{\partial C}{\partial x} = \frac{NkT}{C}\frac{\partial C}{\partial x} \tag{4.45}$$

where k is Boltzmann's constant. Combining Eqs (4.44) and (4.45),

$$\upsilon_i C = -\frac{kT}{f}\frac{\partial C}{\partial x} \tag{4.46}$$

The quantity $\upsilon_i C$ is the velocity of a particle multiplied by the particle concentration, that is to say, its flux. The flux of the particles at infinite dilution is also described by Fick's First Law, Eq. (4.17),

$$F_D = -D^o \frac{\partial C}{\partial x} \tag{4.47}$$

If both these equations describe the same flux, then it is necessary that

$$D^o = \frac{kT}{f} \tag{4.48}$$

A general friction coefficient for spheres in solution is (Bird et al., 1960)

$$f = 6\pi\mu r_o \left(\frac{2\mu + r_o \varpi}{3\mu + r_o \varpi} \right) \tag{4.49}$$

where $\pi = 3.14159...$, μ is the dynamic viscosity, r_o is the radius of the particle, and ϖ is the coefficient of sliding friction between the particle and the fluid continuum. There are two important limiting cases when Eq. (4.49) is substituted into Eq. (4.48). One limiting case of Eq. (4.49) occurs when $\varpi = 0$; i.e. the fluid does not stick to the particles. The result is (Sutherland, 1905, quoted in Cussler, 1984)

$$D^o = \frac{kT}{4\pi\mu r_o} = \frac{RT}{4\pi\mu N r_o} \tag{4.50}$$

which is a good approximation when water molecules and the diffusing particles are alike in size. Alternatively, if $\varpi \to \infty$, then there is no sliding between the particle and the fluid, and consequently (Einstein, 1906, reprinted 1956),

$$D^o = \frac{\mathcal{K}T}{6\pi\mu r_o} = \frac{\mathcal{R}T}{6\pi\mu N r_o} \tag{4.51}$$

This equation is the justly famous Stokes-Einstein equation, and it best describes the situation in which the particles are very much larger than the water molecules.

The difficulty in employing either Eq. (4.50) or Eq. (4.51) is specification of r_o, the particle radius. The uncharged particle must be roughly spherical (e.g. a small solute). Sometimes there exists an experimental value; more often it must be estimated. If the molar volume, V_m, is known, then

$$r_o \approx \left(\frac{3V_m}{4\pi N}\right)^{1/3} \tag{4.52}$$

where N is Avogadro's number. If molecular weight, M_w, and density, ρ, of the particle are known, then

$$r_o \approx \left(\frac{3M_w}{4\pi\rho N}\right)^{1/3} \tag{4.53}$$

For example, Cornel et al. (1986) use Eqs (4.51) and (4.53) to calculate the diffusion coefficients of humic acids. For polymer strings in dilute solution, Cussler (1984) suggests

$$r_o \approx 0.676\left\langle r_g^2\right\rangle^{1/2} \tag{4.54}$$

where the term in brackets on the right-hand side is the root-mean-square radius of gyration of the polymer.

The restriction to spherical particles for the Stokes-Einstein equation was eased by Perrin (1936, quoted in Cussler, 1984) who found that, for prolate ellipsoids (i.e. the shape of American footballs),

$$D^o = \frac{\mathcal{K}T\,ln\left(\dfrac{a+\left(a^2-b^2\right)^{1/2}}{b}\right)}{6\pi\mu\left(a^2-b^2\right)^{1/2}} \tag{4.55}$$

and, for oblate ellipsoids (discs),

$$D^o = \frac{kT \tan^{-1}\left(\left(\frac{a^2 - b^2}{b^2}\right)^{1/2}\right)}{6\pi\mu\left(a^2 - b^2\right)^{1/2}} \tag{4.56}$$

where a and b are the major and minor axis lengths, respectively, of the ellipsoids. Such equations could be useful for calculating the diffusivities of certain large organic molecules in porewaters.

Theoretical expressions, e.g. Eqs (4.50) through (4.56), are really approximations as they idealize both the fluid and the diffusing molecules/particles. Accuracies appear to be limited to 20-30%. In search of better estimates, engineers have developed various empirical correlations. One of the most widely used is that of Wilke and Chang (1955) as corrected by Hayduk and Laudie (1974),

$$D^o = 4.72 \cdot 10^{-9} \frac{T}{\mu V_b^{0.6}} \qquad (cm^2\ s^{-1}) \tag{4.57}$$

where μ is the dynamic viscosity of water (in units of poise), T is the absolute temperature, and V_b is the molar volume of the nonelectrolyte (at the normal boiling temperature of that solute). Values of V_b for some nonelectrolytes are reproduced in Table 4.3, along with experimentally determined values of the diffusion coefficients. A more complete table and the sources of these data are available in Hayduk and Laudie (1974). The V_b are experimental unless otherwise indicated, in which case they are so-called LeBas volumes. Equation (4.57) is claimed to reproduce the diffusivities in Table 4.3 with zero average error and mean absolute error of 5.8%.

Since the publication of Hayduk and Laudie (1974), a few more determinations have been made of the diffusivities of two geochemically important species, O_2 and CO_2 (Verhallen et al., 1984; Leaist, 1987; Ho et al., 1988; Tan and Thorpe, 1992). Figures 4.5 and 4.6 illustrate these newer data, combined with the older data used by Hayduk and Laudie (1974). Both coefficients exhibit essentially linear temperature/viscosity-dependence in the range relevant to early diagenesis, i.e.

$$D^o_{O_2} = (0.2604 + 0.006383(T/\mu)) \times 10^{-5} \qquad (cm^2\ s^{-1}) \tag{4.58}$$

and

$$D^o_{CO_2} = (0.1954 + 0.005089(T/\mu)) \times 10^{-5} \qquad (cm^2\ s^{-1}) \tag{4.59}$$

where T is again the absolute temperature and μ is the dynamic viscosity from Eq. (4.4) in centipoise. The CO_2 equation is relatively tight ($r^2 = 0.95$), but the data for the O_2 correlation have quite a bit of scatter, which leads to an r^2 of only 0.84.

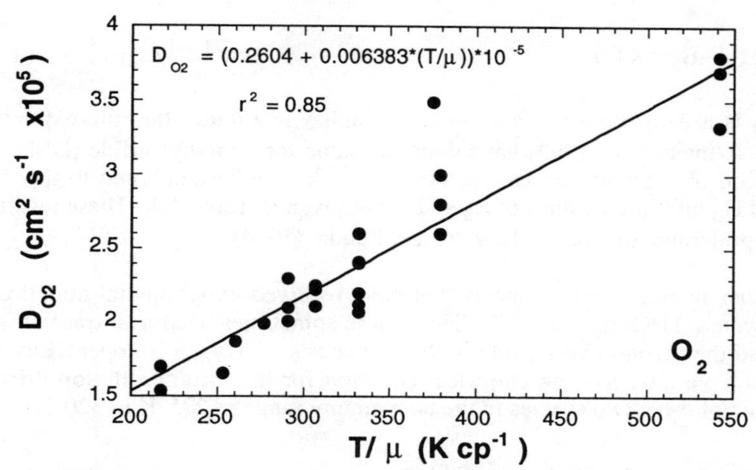

Figure 4.5. Experimental data for O_2 diffusion coefficients as of 1994 as a function of absolute temperature over viscosity, and a linear least-squares fit to these data.

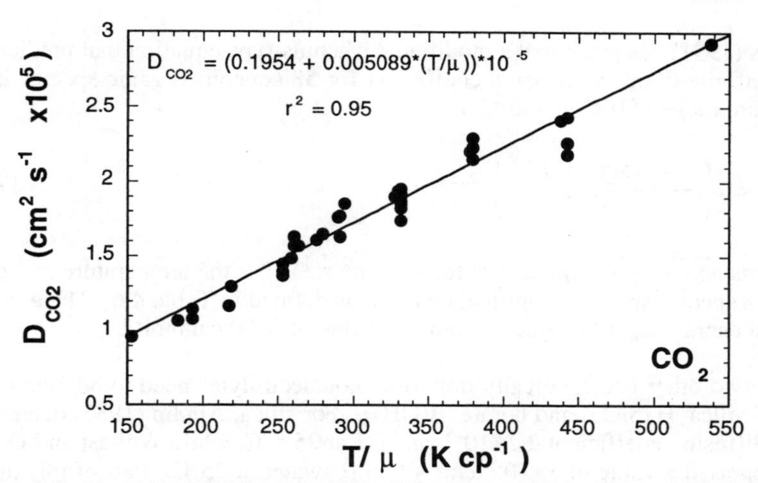

Figure 4.6. Experimental data on CO_2 diffusion coefficients as of 1994 as a function of absolute temperature over viscosity, and a linear least-squares fit to these data.

Recently, Jähne et al. (1987) have re-determined the diffusivities of sparingly soluble gases in water and correlated their diffusion coefficients with an Arrhenius-type equation:

$$D^o = A\, exp\left(-E_a /(RT)\right) \qquad (cm^2\ s^{-1}) \qquad (4.60)$$

where E_a is a data-fitted activation-energy and A is a data-fitted pre-exponential factor. Saltzman et al. (1993) have done the same for dimethylsulfide (DMS), and the equation of Ohsumi and Horibe (1984) for Ar can be converted to this form. Measured D_o and fitted values of E_a and A are given in Table 4.4. These results are probably preferable to those of Hayduk and Laudie (1974).

Two other nonelectrolyte species that have received experimental attention are tritiated water, THO, and $H_2{}^{18}O$. The first is sometimes used as a tracer for irrigation, and the second plays a role in the diagenesis of oxygen isotopes. Easteal et al. (1989) have advanced an empirical equation for the infinite-dilution diffusion coefficients of these two species in the temperature range of 275°K to 320°K,

$$ln(10^9 D^o) = a_1 + a_2\left[\frac{1000}{T}\right] + a_3\left[\frac{1000}{T}\right]^2 \qquad (m^2\ s^{-1}) \qquad (4.61)$$

where a_1, a_2, and a_3 are constants, as given in Table 4.5. Note the different units of D^o in this case.

Oelkers (1991) has presented a modified Arrhenius-type equation that predicts the (tracer) infinite-dilution diffusion coefficient for 58 aqueous organic species in the temperature range of 0°C to 350°C, i.e.

$$D_i^o = A_i\, exp\left(-\frac{E_a + \bar{a}(T-228)^{1.5}}{RT}\right) \qquad (4.62)$$

where A_i is again a pre-exponential factor ($cm^2\ s^{-1}$), T is the temperature in Kelvin, and $\bar{a}$ is a species specific empirical constant as defined in Table 4.6. The apparent activation energy, E_a, is assigned a common value of 3300 cal mol^{-1}.

Lastly, two other geochemically important nonelectrolytes need to be mentioned, dissolved silica, H_4SiO_4, and borate, $B(OH)_3$. For silica, Applin (1987) determined a tracer diffusion coefficient 2.2×10^{-5} $cm^2\ s^{-1}$ at 25.5°C, while Wollast and Garrels (1971) reported a value of 1×10^{-5} $cm^2\ s^{-1}$ in seawater at 25°C. Part of this difference is due to the difference in viscosity of water and seawater (see below), and part is due to the presence of silica dimers at seawater pH (Applin, 1987). Mackin (1986) measured a tracer borate diffusion coefficient of $1.12\pm0.02\times10^{-5}$ $cm^2\ s^{-1}$ at 25°C in solutions of ionic strength 0.5-0.6 M.

Table 4.3. Values of V_b for Eq. (4.57) and Experimental Diffusivities at 25°C.

Species	V_b*	D^o†	Species	V_b	D^o
H_2	28.5	4.40	He	31.9	6.28
NO	23.6	2.34	N_2O	36.0	2.10
N_2	34.7	1.99	NH_3	24.5	2.28
O_2	27.9	2.29	CO	34.5	2.30
CO_2	37.3	1.92	SO_2	43.8	1.59
H_2S	35.2	2.10	Ar	29.2	1.98
Kr	34.9	1.92	Ne	16.8	3.01
CH_4	37.7	1.67	CH_3Cl	50.6	1.49
C_2H_6	53.5	1.38	C_2H_4	49.4	1.55
C_3H_8	74.5	1.16	C_3H_6	69.0	1.44
C_4H_{10}	96.6	0.97	Urea	58.0§	1.38
Methanol	42.5	1.66	Ethanol	62.6	1.24
Glycine	78§	1.06	Alanine	100§	0.91
Serine	108§	0.88	Valine	145§	0.77
Leucine	167§	0.73	Proline	127§	0.88
Dextrose	166§	0.675	Sucrose	325§	0.52
Acids:					
Formic	41.1	1.52	Acetic	63.8	1.19
Proprionic	86.0	1.01	Butyric	108	0.92

* units of $cm^3\ mol^{-1}$; † units of $10^{-5}\ cm^2 s^{-1}$; § LeBas volumes. All values from Hayduk and Laudie (1974). (Reproduced with the kind permission of the American Institute of Chemical Engineers.)

Table 4.4. Parameters for Dissolved Gas Diffusion Coefficients given by Eq. (4.60)§.

Species	E_a*	A†
He	11.70	818
Ne	14.84	1608
Kr	20.20	6393
Ne	14.84	1608
Xe	21.61	9007
Rn	23.26	15877
H_2	16.06	3338
CH_4	18.36	3047
CO_2	19.51	5019
Ar	19.81	7238
DMS	18.1	2000

* units of $kJ\ mol^{-1}$
† units $10^{-5}\ cm^2\ s^{-1}$
§ From Jähne et al. (1987), except Ar from Ohsumi and Horibe (1984) and DMS from Saltzman et al. (1993).

Table 4.5. Parameters for Calculating Diffusion Coefficients of THO and $H_2{}^{18}O$ with Eq. (4.61)†.

Species	a_1	a_2	a_3
THO	3.1460139	0.8253935	-0.4543387
$H_2{}^{18}O$	4.2355605	0.1642818	-0.3525408

† From Easteal et al. (1989). Reproduced with the kind permission of the Royal Society of Chemistry.

Table 4.6. Values of A_i and $\bar{a}$ for Eq. (4.62) as given by Oelkers (1991).

Species	A_i*	$\bar{a}$†	Species	A_i	$\bar{a}$
Methane	575.24	89410.4	Naphthalene:	365.60	127729.4
Ethane	438.27	89410.4	Methyl-	337.38	127729.4
Propane	369.52	89410.4	Ethyl-	304.01	127729.4
Butane	369.52	89410.4	Propyl-	278.01	127729.4
Pentane	295.27	89410.4	Butyl-	257.65	127729.4
Hexane	272.03	89410.4	Pentyl-	241.26	127729.4
Heptane	253.63	89410.4	Hexyl-	227.71	127729.4
Octane	238.59	89410.4	Bipheny	310.46	107970.8
Nonane	225.99	89410.4	Formamide	642.37	134429.7
Decane	215.24	89410.4	Acetamide	516.14	134429.7
Methanol	580.73	123716.6	Formamide	642.37	134429.7
Ethanol	471.39	123716.6	Propanamide	443.78	134429.7
Propanol	406.84	123716.6	Butanamide	395.88	134429.7
Butanol	363.69	123716.6	Pentanamide	361.14	134429.7
Pentanol	332.22	123716.6	Hexanamide	334.43	134429.7
Hexanol	307.94	123716.6	Cyclopentane:	354.87	89001.6
Heptanol	288.45	123716.6	Methyl-	318.98	89001.6
Octanol	272.35	123716.6	Ethyl-	285.58	89001.6
i-Propanol	425.75	150890.4	Protyl-	259.58	89001.6
tert-			Butyl-	239.93	89001.6
Butanol	354.16	129442.1	Cyclohexane	327.48	105757.5
Toluene	394.76	152470.9	Acids:		
Benzene:	430.52	152470.9	Formic	540.06	106618.0
Ethyl-	355.00	152470.9	Acetic	430.51	106618.0
Propyl-	324.31	152470.9	Propanoic	368.10	106618.0
Butyl-	300.37	152470.9	Butanoic	328.73	106618.0
Pentyl-	281.13	152470.9	Pentanoic	299.57	106618.0
Hexyl-	265.27	152470.9	Hexanoic	277.21	106618.0
Acetone	327.99	-12833.8	Ethylene Glycol	411.83	111487.1
Glycine	352.85	82833.9	Glycerol	340.42	104715.1

* units of $cm^2\ s^{-1} x10^5$; † units of cal $mol^{-1}\ K^{1.5}$. (Reproduced with the kind permission of Elsevier Science Ltd.)

Ions and Ion Pairs. The most common way of calculating the diffusion coefficient of a tracer ion at infinite dilution is by use of the so-called Nernst-Einstein expression

$$D_i^o = \frac{RT\lambda_i^\circ}{|z_i|F^2} = \frac{kT\lambda_i^\circ}{|z_i|Ne_o^2} \tag{4.63}$$

where λ°_i is the limiting equivalent conductivity of this i-th ion (Eq. 4.15), z_i is the charge on the ion, F is the Faraday constant, k is Boltzmann's constant, N is Avogadro's number, and e_o is the charge on an electron.

A basis for Eq. (4.63) is supplied by the heuristic derivation found in Bockris and Reddy (1970). They postulate a steady-state experimental situation in which a migrational flux $(F_e)_i$ of a trace ion, Eq. (4.11) with Eq. (4.13), is exactly balanced by a counter diffusive flux, $(F_D)_i$,

$$D_i^o \frac{\partial C_i}{\partial x} = -\frac{C_i\lambda_i^\circ}{|z_i|e_oF}\frac{\partial \Psi}{\partial x} \tag{4.64}$$

In this steady state, the concentration C_i and the potential must be related. The assumption is that this relationship is the Boltzmann distribution,

$$C_i = C_i^o\, exp(-\Psi/(kT)) \tag{4.65}$$

where C°_i is the equilibrium value of the concentration (which need not be specified). Differentiating Eq. (4.65) with respect to x produces

$$\frac{\partial C_i}{\partial x} = -\frac{C_i}{kT}\frac{\partial \Psi}{\partial x} \tag{4.66}$$

If both Eqs (4.64) and (4.66) are to be true, then it is necessary that

$$D_i^o = \frac{kT\lambda_i^\circ}{|z_i|e_oF} \tag{4.67}$$

Finally, $e_o = F/N$ and $R = Nk$, so Eq. (4.67) becomes Eq. (4.63).

Li and Gregory (1974) used Eq. (4.63) to calculate D^o values for various ions. To these values can be added $Al(OH)_4^-$, for which Mackin and Aller (1983) arrived at a D^o value of $1.04 \pm 0.02 \times 10^{-5}$ $cm^2\ s^{-1}$ at 25°C. There is no information, theoretical or experimental, for $B(OH)_4^-$. One could use the ratio of the D° for CO_2 and HCO_3^- as a possible guide (e.g. Boudreau and Canfield, 1988). For those species that have λ° values only at 25°C, a strategy suggested by Li and Gregory (1974) can be followed. They propose that D° for ions that diffuse more slowly than F^- conform fairly well with an equation based on the Stokes-Einstein relation, i.e.

Table 4.7. Linear Regressions[†] of the Infinite-Dilution Diffusion Coefficients D^o for Cations against Temperature[‡] ($cm^2\ s^{-1}$).

+1 Cations	m_o	m_1	r^2
H^+	54.4	1.555	1.0
D^+	31.9	1.402	1.0
Li^+	4.43	0.241	0.99
Na^+	6.06	0.297	0.99
K^+	9.55	0.409	1.0
Rb^+	10.2	0.424	0.99
Cs^+	10.3	0.416	1.0
Ag^+	7.82	0.359	1.0
NH_4^+	9.50	0.413	1.0
+2 Cations	m_o	m_1	r^2
Ba^{2+}	4.06	0.176	1.0
Be^{2+}	2.57	0.140	0.99
Ca^{2+}	3.60	0.179	0.99
Cd^{2+}	3.31	0.152	1.0
Co^{2+}	3.31	0.152	1.0
Cu^{2+}	3.39	0.158	0.99
Fe^{2+}	3.31	0.150	1.0
Hg^{2+}	3.63	0.208	0.98
Mg^{2+}	3.43	0.144	1.0
Mn^{2+}	3.18	0.155	1.0
Ni^{2+}	3.36	0.130	1.0
Sr^{2+}	3.69	0.169	1.0
Pb^{2+}	4.46	0.198	1.0
Ra^{2+}	3.91	0.199	1.0
Zn^{2+}	3.31	0.151	1.0
+3 Cations	m_o	m_1	r^2
Al^{3+}	2.79	0.172	0.98
Ce^{3+}	2.95	0.131	1.0
La^{3+}	2.78	0.136	1.0
Pu^{3+}	2.71	0.120	1.0

† $D° = (m_o + m_1\, t)\cdot 10^{-6}\ cm^2\ s^{-1}$

‡ t in °C

Note: all correlations involved 6 points. Some of the data are synthetic.

Table 4.8. Linear Regressions† of the Infinite-Dilution Diffusion Coefficients D^o for Anions against Temperature‡ ($cm^2\ s^{-1}$).

-1 Anions	m_o	m_1	r^2
OH^-	25.9	1.094	1.0
OD^-	15.3	0.667	1.0
$Al(OH)_4^-$	4.46	0.243	0.99
Br^-	10.0	0.441	1.0
Cl^-	9.60	0.438	1.0
F^-	6.29	0.343	1.0
HCO_3^-	5.06	0.275	0.99
$H_2PO_4^-$	4.02	0.223	0.98
HS^-	10.4	0.273	0.99
HSO_3^-	6.35	0.280	1.0
HSO_4^-	5.99	0.307	1.0
I^-	9.81	0.432	1.0
IO_3^-	4.66	0.252	0.99
NO_2^-	10.3	0.331	0.98
NO_3^-	9.50	0.388	1.0
Acetate$^-$	4.80	0.245	1.0
Lactate$^-$	4.41	0.241	0.99
-2 Anions	m_o	m_1	r^2
CO_3^{2-}	4.33	0.199	1.0
HPO_4^{2-}	3.26	0.177	0.99
SO_3^{2-}	4.53	0.249	0.99
SO_4^{2-}	4.88	0.232	1.0
$S_2O_3^{2-}$	4.82	0.266	0.99
$S_2O_4^{2-}$	3.95	0.187	0.98
$S_2O_6^{2-}$	5.30	0.291	0.99
$S_2O_8^{2-}$	4.88	0.267	0.99
Malate^{2-}	3.36	0.183	0.99
-3 Anions	m_o	m_1	r^2
PO_4^{3-}	2.62	0.143	0.99
Citrate^{3-}	2.67	0.146	0.99

† $D^\circ = (m_o + m_1\ t) \cdot 10^{-6}\ cm^2\ s^{-1}$

‡ t in °C

Note: all correlations involved 6 points. Some of the data are synthetic.

$$\left(\frac{D^\circ\mu}{T}\right)_T = \left(\frac{D^\circ\mu}{T}\right)_{298.15^\circ K} \tag{4.68}$$

where μ is the dynamic viscosity of water, as given by Eq. (4.4) and T is the temperature of interest in °K. For ions that diffuse like F^- or faster, Li and Gregory (1974) found that they are better described by an equation of the form,

$$\left(D^\circ\mu\right)_T = \left(D^\circ\mu\right)_{298.15^\circ K} \tag{4.69}$$

However, such values cannot be expected to be accurate to any better than, say 30%. To simplify the calculation of D^o values at other temperatures, a set of linear correlations is supplied in Tables 4.7 and 4.8, based on recalculation with the equations in Tables 4.1 and 4.2 and Eqs (4.63), (4.68), and (4.69).

The infinite-dilution diffusion coefficients for ion pairs have been considered by Pikal (1971), Johnson (1981), and Applin and Lasaga (1984). The formula presented by Applin and Lasaga (1984) is relatively simple to use, i.e.

$$D^o = \frac{1}{2}\left[\sqrt{D_c^o D_a^o} + \frac{kT}{2\pi\mu a_o b}\left(1+\frac{3}{b}\right)\right] \tag{4.70}$$

where D_c^o is the infinite-dilution diffusion coefficient for the cation in the pair, D_a^o is the infinite-dilution diffusion coefficient for the anion in the pair, a_o is an ion-pair size-parameter (see Table 4.9), and

$$b = \frac{|z_c z_a| e_o^2}{\varepsilon k T a_o} \tag{4.71}$$

where z_c is the charge on the cation in the pair, z_a is the charge on the anion in the pair, and ε is the dielectric constant of water (see CRC, 1993). Applin and Lasaga (1984) use mutual diffusion data (explained below), rather than infinite-dilution data, but the results should not be very different.

Table 4.9. Values of the Size-Parameter in Eqs (4.70) and (4.71). (10^{-8} cm)

Pair	a_o	Pair	a_o
$MgSO_4^\circ$	3.64	$CaCl^+$	5.0
$ZnSO_4^\circ$	4.0	$BaCl^+$	4.0
$NaSO_4^-$	3.3	$NaCl^\circ$	3.0
$MgCl^+$	5.0	KCl°	3.0

4.2.4 Diffusion in Multicomponent Porewaters

Diagenetic fluids are routinely neither simple in composition nor infinitely dilute. Porewaters contain many different ionic and molecular solutes, and seawater-based interstitial waters have ionic strengths of about 0.7. Diffusion in such moderately concentrated multicomponent solutions is only a slightly more involved process than it is at infinite dilution. However, this is a topic usually made unnecessarily complicated by "an excess of theoretical zeal" (Cussler, 1984, p.196). In fact, the bottom line is that diffusion in most porewaters is sufficiently like diffusion in dilute solutions that we can normally ignore the complications arising from the multi-component interactions, except when the main electrolytes (Cl^- and Na^+) are involved. Most of this section is simply a proof of this statement.

Some workers in this field prefer to include the effects of reversible reactions, e.g. formations of ion pairs, in the equations for multicomponent diffusion, e.g. Lasaga (1979) and Katz and Ben-Yaakov (1980). Like Cussler (1984), I do not favor this approach, as I believe that such reactions should be treated separately to avoid obscuring the equations (see Chapter 3 for my treatment of reactions). Nor do I see the need for the introduction of the formalism of irreversible thermodynamics (e.g. deGroot and Mazur, 1962).

Multicomponent diffusion must be defined more precisely than is the case for infinite dilution. When considering more concentrated solutions, aficionados distinguish different types of diffusion coefficients (Tyrrell and Harris, 1984):

i) *Mutual or Inter-diffusion* coefficients that describe diffusion between two solutions of different composition (see Lobo, 1993);

ii) *Self-diffusion* coefficients that pertain to the diffusion of the solvent (water) molecules when the liquid is pure;

iii) *Intra-diffusion* coefficients that account for the diffusion caused by gradients of labeled and unlabeled ions/molecules/particles of a single type in an otherwise uniform environment; and

iv) *Tracer-diffusion* coefficients that resemble intra-diffusion, but do not require the presence of the unlabeled species.

For solutes at low concentrations with a background (supporting) electrolyte (e.g. NaCl), all these coefficients are essentially identical to infinite-dilution coefficients, and we shall see that this is the case for most diagenetic problems.

Formally, diffusion is related to the difference between the velocity of the solution component of interest and some other reference velocity, and this relativity introduces the concept of diffusional reference frames (see Kirkwood et al., 1960; Bird et al., 1960; Himmelblau, 1964; Brady, 1975; Anderson and Graf, 1976). (These diffusion reference frames should not be confused with coordinate reference frames of the diagenetic equation.) While it is sensible to be aware of these different reference frames, the associated diffusion coefficients are approximately the same for most porewaters systems.

The relative velocities of ions/molecules in a multicomponent solution can be assigned in more than one way. For our purposes, four reference frames are of interest in calculating these velocities: the mass-fixed frame, the water(solvent)-fixed frame, the volume-fixed frame, and the (diffusion) cell-fixed frame. In the mass-fixed frame, the porewaters move with a velocity, u, calculated as

$$u \equiv \frac{1}{\rho} \sum C_i u_i \tag{4.72}$$

where C_i is the mass per unit volume concentration of solute i, u_i is the velocity of that solute relative to the sediment-water interface, and ρ is the porewater density. (Molar concentrations can be used if we simply include division by the molecular weights.) The velocity defined by Eq. (4.72) is the reference velocity for this frame. Notice that this velocity is still measured relative to the sediment-water interface; thus, the term "reference frame for diffusion" does not indicate where we anchor the geometric coordinate system.

Diffusive flux of species i in this mass-fixed reference system, $(F_i)_M$, is given by

$$(F_i)_M = C_i \,(u_i - u) \tag{4.73}$$

This illustrates the notion, mentioned above, that diffusion is the result of differential motions. The mass-fixed frame is also known as the barycentric frame, and it has the property that

$$\sum (F_i)_M = 0 \tag{4.74}$$

Diffusion in the other frames is described in similar ways. For instance, the volume-fixed frame uses the velocity

$$u = \sum C_i u_i \left(\frac{\overline{V}_i}{M_i} \right) \tag{4.75}$$

where $\overline{V}_i$ is the partial molal volume and M_i is the molecular weight of species i. There is no net volume flow in this system, so that

$$\sum \overline{V}_i (F_i)_v = 0 \tag{4.76}$$

where $(F_i)_v$ is the diffusive flux of i-th species in this system.

In the water or solvent-fixed reference frame, u is the velocity of the water component only, and the diffusive flux of water is identically zero, i.e.

$$\left(F_{H_2O}\right)_o = 0 \tag{4.77}$$

where the subscript o indicates the water-fixed frame. Finally, I mention the cell-fixed frame, not because it has real utility to diagenetic systems, but because it is often the laboratory-fixed frame, wherein diffusion coefficients are measured. In the cell-frame, customarily $u = 0$.

Kirkwood et al. (1960) have considered the relationships between the diffusive flux in each of these frames of reference. For the water and mass-fixed frames, they establish that

$$\left(F_i\right)_o = \left(F_i\right)_M - \frac{C_i}{C_{H_2O}}\left(J_{H_2O}\right)_M \tag{4.78}$$

where C_{H_2O} is the concentration of water. Even for the most concentrated ion in seawater, i.e. Cl^- at 19.35 g kg^{-1}, the ratio C_i/C_{H_2O} is only 0.02. Because $(J_{H_2O})_M$ must be, at most, of the same order of magnitude as $(F_{Cl})_o$, then $(F_i)_o$ and $(F_i)_M$ must be within 2% of each other. Similarly for the water and volume-fixed frames, they find that

$$\left(F_i\right)_V = \left(F_i\right)_o - C_i \sum_{j=1}^{n} \overline{V}_j \left(J_j\right)_o \tag{4.79}$$

where n is the number of solutes in the porewater. Note that the index of the concentration in the sum is i and not j. The product of the concentration and the partial molar volumes will be much less than one for all species in normal porewaters; consequently, the summation represents a small correction, and it can usually be ignored. Finally, the cell and water-fixed frames are related by

$$\left(F_i\right)_c = \left(F_i\right)_o + \frac{C_i}{C_{H_2O}}\left(F_{H_2O}\right)_c \tag{4.80}$$

which is the same form as Eq. (4.78); thus, the second term on the right-hand side can normally be ignored.

This near equivalence of fluxes suggests that the associated diffusion coefficients, determined in each of the reference frames, would exhibit a comparable property. Tyrrell and Harris (1984) consider just this question, and they find that this is indeed the case. Their development leads to the conclusion that

$$\left(D_i\right)_o \approx \left(D_i\right)_V \left[1 + \frac{C_i \overline{V}_i}{C_{H_2O}\overline{V}_{H_2O}}\right] \tag{4.81}$$

and

$$(D_i)_M \approx (D_i)_v \left[1 - \frac{C_i \overline{V}_i}{\rho} \left(\frac{M_i}{\overline{V}_i} - \frac{M_{H_2O}}{\overline{V}_{H_2O}} \right) \right] \tag{4.82}$$

The correction terms in the square brackets of each of these equations are very nearly unity, so that the diffusion coefficients are numerically equivalent. Cell-fixed diffusion coefficients are similarly related. Therefore, for numerical calculations with the accuracy normally needed in diagenetic studies, the choice of diffusion reference frame is largely immaterial.

No mention has yet been made of the effects of the charge on the ions. In multi-component ionic solutions, such as porewaters, a given ion is surrounded by a cloud of oppositely charged ions. This creates an electrical potential, Ψ, that balances the charge on the central ion. Now, if an ion moves, it perturbs the potential, creating a potential gradient, and a force is exerted on the moving ion (Bockris and Reddy, 1970). In addition, the perturbed potential will affect all the other ions, influencing their motions.

The total force, f_i, acting on a charged ion during diffusion is not simply due to the chemical potential gradient, but is also due to the electrical potential gradient, Ψ, i.e.

$$f_i = -\frac{\partial \psi_i}{\partial x} + z_i \frac{\partial \Psi}{\partial x} \tag{4.83}$$

where ψ_i is the chemical potential of species i, and z_i is the charge on an ion of species i. The resulting flux of the ion, F_i, from this force is

$$F_i = \upsilon_i C_i f_i \tag{4.84}$$

where υ_i is the mobility of the ion. With Eq. (4.83), Eq. (4.84) becomes

$$F_i = -\upsilon_i C_i \left(\frac{\partial \psi_i}{\partial x} - z_i \frac{\partial \Psi}{\partial x} \right) \tag{4.85}$$

In an electrolyte solution, two other constraints apply. First there is static electroneutrality,

$$\sum z_i C_i = 0 \tag{4.86}$$

which says that porewater is not a charged fluid. Furthermore, there is no current flowing through porewater (unless one is applied),

$$\sum z_i F_i = 0 \tag{4.87}$$

Lasaga (1979) has shown that the contribution from the perturbation in the electrical potential, i.e. the second term in parentheses in Eq. (4.85), is small in porewaters, except perhaps for the main electrolyte species Cl^- and Na^+. Lasaga uses the water-fixed frame for his derivation; however, the result must be valid in any reference frame, and this choice is solely a matter of convenience. (In fact, the mass-fixed reference frame would be slightly more appropriate.)

Lasaga (1979) begins his development by substituting Eq. (4.85) into Eq. (4.87) to get

$$\frac{\partial \Psi}{\partial x} = \frac{\sum_{j=1}^{n} z_j \upsilon_j C_j \frac{\partial \psi_j}{\partial x}}{\sum_{k=1}^{n} z_k^2 \upsilon_k C_k} \tag{4.88}$$

where the upper summation limit, n, is the total number of solutes. Neutral solutes will have charges equal to zero and will not contribute. Substituting this result back into Eq. (4.85) gives

$$F_i = -\upsilon_i C_i \left(\frac{\partial \psi_i}{\partial x} - z_i \frac{\sum_{j=1}^{n} z_j \upsilon_j C_j \frac{\partial \psi_j}{\partial x}}{\sum_{k=1}^{n} z_k^2 \upsilon_k C_k} \right) \tag{4.89}$$

Next we expand ψ_i using its thermodynamic definition, Eq. (4.31), and making the same assumptions and transformations as in Eqs (4.37) through (4.44) to get

$$\frac{\partial \psi_i}{\partial x} \approx \frac{RT}{C_i} \frac{\partial C_i}{\partial x} \tag{4.90}$$

and

$$D_i^o \approx \upsilon_i RT \tag{4.91}$$

so that

$$F_i = -D_i^o \frac{\partial C_i}{\partial x} + z_i D_i^o C_i \frac{\sum_{j=1}^{n} z_j D_j^o \frac{\partial C_j}{\partial x}}{\sum_{k=1}^{n} z_k^2 D_k^o C_k} \tag{4.92}$$

Finally, Eq. (4.86) implies that the gradients are not independent; i.e. upon differentiation,

$$\frac{\partial C_{Cl^-}}{\partial x} = \sum_{j=1}^{n-1} z_j \frac{\partial C_j}{\partial x} \tag{4.93}$$

where C_{Cl^-} is the concentration of the chloride ion. Substitution of Eq. (4.93) into Eq. (4.92) produces Lasaga's final result,

$$F_i = \underbrace{-D_i^o \frac{\partial C_i}{\partial x}}_{\substack{\text{tracer}\\ \text{diffusion}\\ \text{contribution}}} + \underbrace{\frac{z_i D_i^o C_i}{\sum_{k=1}^{n} z_k^2 D_k^o C_k} \sum_{j=1}^{n-1} z_j \left(D_j^o - D_{Cl^-}^o \right) \frac{\partial C_j}{\partial x}}_{\text{electrostatic interaction contribution}} \tag{4.94}$$

The contribution from the second term in Eq. (4.94) can usually be neglected, unless C_i is one of the dominant electrolyte ions, i.e. Na^+ and Cl^-, because

$$\frac{z_i C_i D_i^o}{\sum_{k=1}^{n} z_k^2 C_k D_k^o} << 1 \tag{4.95}$$

That is to say, the movement of minor and trace components of the porewaters does not perturb the electrical potential sufficiently to take this disturbance into account.

It should be noted that assigning infinite-dilution coefficients to each and every ion can lead to inconsistencies and a lack of overall charge balance (Toor and Arnold, 1965). However, if only minor species are moving, this should not matter much. When gradients of the major charge-bearing ions, i.e. Na^+ and Cl^-, are present, then the diffusion coefficients of the species cannot be set independently of each other. For example, consider the system H_2O-NaCl as a simplified analog of seawater, and use the mass-fixed reference frame. The fluxes of H_2O, Na^+, and Cl^- are given by (C in mass per unit volume)

$$\left(F_{H_2O}\right)_M = -D_{H_2O} \frac{\partial C_{H_2O}}{\partial x} \tag{4.96}$$

$$\left(F_{Na}\right)_M = -D_{Na} \frac{\partial C_{Na}}{\partial x} + \upsilon_{Na} C_{Na} \frac{\partial \Psi}{\partial x} \tag{4.97}$$

and

$$\left(F_{Cl}\right)_M = -D_{Cl} \frac{\partial C_{Cl}}{\partial x} - \upsilon_{Cl} C_{Cl} \frac{\partial \Psi}{\partial x} \tag{4.98}$$

In addition, static conservation of volume and Eqs (4.74), (4.87), and (4.86) require that

$$\frac{C_{H_2O}}{M_{H_2O}\overline{V}_{H_2O}} + \frac{C_{Na}}{M_{Na}\overline{V}_{Na}} + \frac{C_{Cl}}{M_{Cl}\overline{V}_{Cl}} = 1 \tag{4.99}$$

$$D_{H_2O}\frac{\partial C_{H_2O}}{\partial x} + D_{Na}\frac{\partial C_{Na}}{\partial x} + D_{Cl}\frac{\partial C_{Cl}}{\partial x} - (\upsilon_{Na}C_{Na} - \upsilon_{Cl}C_{Cl})\frac{\partial \Psi}{\partial x} = 0 \tag{4.100}$$

$$D_{Na}\frac{\partial C_{Na}}{\partial x} - D_{Cl}\frac{\partial C_{Cl}}{\partial x} - (\upsilon_{Na}C_{Na} + \upsilon_{Cl}C_{Cl})\frac{\partial \Psi}{\partial x} = 0 \tag{4.101}$$

$$C_{Na} = C_{Cl} \tag{4.102}$$

respectively, where M_i is the molecular weight and $\overline{V}_i$ is the partial molar volume of species i. Differentiation of Eq. (4.102) further implies that

$$\frac{\partial C_{Na}}{\partial x} = \frac{\partial C_{Cl}}{\partial x} \tag{4.103}$$

Equations (4.101) through (4.103) can be combined with Eqs (4.91), (4.97), and (4.98) to demonstrate that the ions have the same apparent diffusivity after the effects of the electrical potential are taken into account (e.g. Cussler, 1984, p. 150), i.e.

$$\left(F_{Na}\right)_M = -D_e\frac{\partial C_{Na}}{\partial x} = -\frac{2D_{Na}D_{Cl}}{D_{Na} + D_{Cl}}\frac{\partial C_{Na}}{\partial x} \tag{4.104}$$

and

$$\left(F_{Cl}\right)_M = -D_e\frac{\partial C_{Cl}}{\partial x} = -\frac{2D_{Na}D_{Cl}}{D_{Na} + D_{Cl}}\frac{\partial C_{Cl}}{\partial x} \tag{4.105}$$

Moreover, by differentiating Eq. (4.99) and combining this with Eqs (4.100) and (4.103), there results the equation

$$D_e = \frac{2D_{Na}D_{Cl}}{D_{Na} + D_{Cl}} = \frac{M_{H_2O}\overline{V}_{H_2O}\left(M_{Na}\overline{V}_{Na} + M_{Cl}\overline{V}_{Cl}\right)}{2M_{Na}\overline{V}_{Na}M_{Cl}\overline{V}_{Cl}}D_{H_2O} \tag{4.106}$$

This last equation says that the apparent diffusivity of the salt ions, D_e, and the (mutual) diffusion coefficient of water are not independent of each other. For a recent complete treatment of multicomponent diffusion, see Felmy and Weare (1991a,b).

If the diffusional reference frame and coulomb-type couplings aren't important for most diagenetic species, what does one need to consider in going from infinite dilution to real porewater solutions of finite strength? Li and Gregory (1974) argue that the change in viscosity is the primary consideration, calculated as

$$\frac{D_i^{sw}}{D_i^o} \approx \frac{\mu_o}{\mu_{sw}} \tag{4.107}$$

where the script sw indicates a (saline) porewater value and the script o indicates an infinite-dilution value. They note that Eq. (4.107) appears to overcorrect the diffusivity for Na^+ and Ca^{2+}. The dynamic viscosity at any temperature and salinity can be calculated from Eq. (4.4). To illustrate the magnitude of this effect, Table 4.10 presents some values of μ_o/μ_{sw} for porewaters composed of normal seawater at S = 35. The adjustment is modest at no more than 7%.

Table 4.10. Relative Viscosity of Porewater at S=35.

T (°C)	μ_o/μ_{sw}
0	0.9506
5	0.9470
10	0.9437
15	0.9397
20	0.9364
25	0.9319

4.2.5 Diffusion in Pores and Tortuosity

It has long been recognized that diffusion in a sediment-water mixture is slower than in an equivalent volume of water because of the convoluted path the ions and molecules must follow to circumvent solid sediment particles (Fig. 3.13). In Chapter 3, I indicated that both theory and dimensional reasoning (e.g. Petersen, 1965; Ullman and Aller, 1982) suggest that the apparent molecular/ionic diffusivity, D′, in sediments is related to the free-solution diffusion coefficient through an expression of the form

$$D' = \frac{D^{sw}}{\theta^2} \tag{4.108}$$

where θ is the *tortuosity* (dimensionless).

Traditionally in diagenesis, the tortuosity is defined as the ratio of the (average) incremental distance, dL, that an ion/molecule must travel to cover a direct distance, dx, in the direction of diffusion, i.e. dL/dx (e.g. Berner, 1980). When diagenetic treatments of this problem are compared with those in other fields (i.e. engineering and hydrology), some confusion may arise because of differences in the definition of tortuosity and the use of a related term, the *tortuosity factor* (Epstein, 1989). Specifically, Carman (1937) defined the "tortuosity factor" as $(dL/dx)^2$ or θ^2. However, this distinction is not always made, with some authors labeling θ^2 as the "tortuosity". Readers of this literature can become understandably confused; therefore, it is important to verify what definition a particular author has given for tortuosity. This book follows Berner's (1980) convention, which is Eq. (4.108).

While the traditional interpretation of θ has been dL/dx, this parameter must account for more than just the added path length in porous media, i.e. the left-hand diagram in Fig. 4.7. As argued by Petersen (1965) and van Brakel and Heertjes (1974), the cross-section of a given diffusional path will vary along its length. Where passages narrow, the ions must come together, decreasing the potential flux, as in the right-hand diagram of Fig. 4.7. *Constrictivity* is the term applied to this phenomenon, and it must be contained within the tortuosity, as defined by Eq. (4.108). A separate coefficient is sometimes employed for constrictivity (e.g. van Brakel and Heertjes, 1974), but I will not do so in this text.

Direct measurement of tortuosity does not appear to be possible. Nevertheless, there are at least two experimental procedures that lead to indirect estimation of this parameter. One method is to measure the diffusion coefficient of a chosen non-reactive species in free solution and in a sediment of known porosity. Equation (4.108) can then be employed to arrive at θ^2. Tortuosity can also be related to a measured quantity called the "formation factor", f, which is obtained from electrical resistivity measurements (McDuff and Ellis, 1979),

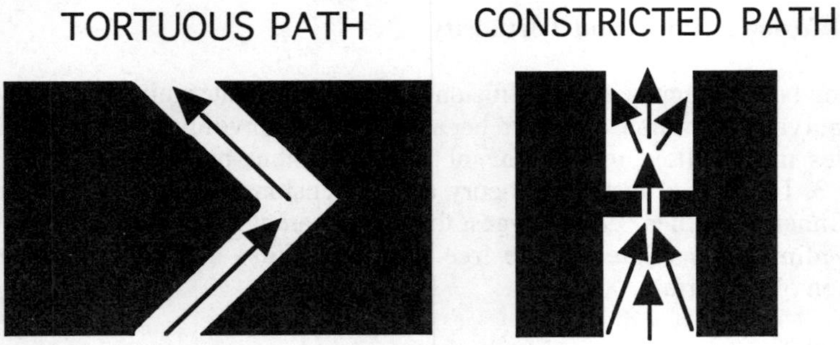

TORTUOUS PATH CONSTRICTED PATH

Figure 4.7. Two types of geometric "deviations" that can cause a decrease in the apparent diffusion coefficient of an ion or molecule in porewaters of aquatic sediments.

$$f = \frac{\text{bulk sediment specific electrical resistivity}}{\text{resistivity of porewater alone}} \qquad (4.109)$$

The relationship to tortuosity is assumed to be (Lerman, 1979; Berner, 1980)

$$\theta^2 = \varphi f \qquad (4.110)$$

Numerical Estimation. Because tortuosity is an essential element in evaluating fluxes in sediments, some methodology is needed for its calculation in the absence of formation factors or D°/D´ measurements. A general numerical property of tortuosity is that

$$\theta^2 \geq 1 \qquad (4.111)$$

Thus, any mathematical technique for its estimation needs to obey this inequality. In addition, it is reasonable to expect that as the porosity φ approaches 1,

$$\lim_{\varphi \to 1}(\theta) = 1 \qquad (4.112)$$

Both theoretical and empirical relationships exist for calculating tortuosity. Theoretical methods are based on some model of the structure of sediments. These relations have the advantage that they generally do not contain any adjustable parameters, but these models are highly idealized. Probably the simplest of such models consists of a collection of randomly oriented pore-capillaries cutting through a solid body. One such capillary is illustrated in Fig. 4.8.

For a given capillary, as in Fig. 4.8, the deviation in the diffusive path, dL, is related to the distance traveled in the x-direction by $dL = dx/cos(\alpha)$. The average path deviation for all the capillaries is

$$\overline{dL} = dx \int_0^{\frac{\pi}{2}} \frac{g(\alpha)}{cos(\alpha)} d\alpha \qquad (4.113)$$

where $g(\alpha)$ is the distribution function for angles of the capillaries, and

$$\int_0^{\frac{\pi}{2}} g(\alpha) d\alpha = 1 \qquad (4.114)$$

The second mean-value theorem for integrals, Eq. (2.52), gives that

$$\overline{dL} = \left(\overline{cos(\alpha)}\right)^{-1} dx \int_0^{\frac{\pi}{2}} g(\alpha) d\alpha = \left(\overline{cos(\alpha)}\right)^{-1} dx \qquad (4.115)$$

where $\overline{cos(\alpha)}$ is the mean value of $cos(\alpha)$ for the chosen distribution of angles $g(\alpha)$.

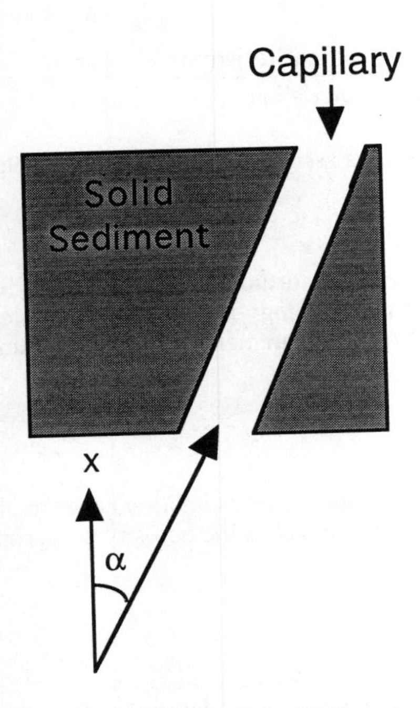

Figure 4.8. A single capillary of the random capillary model of a sediment. The capillary makes an angle α with the x-axis, which is also the macroscopic direction of the concentration gradient.

If the distribution of the capillary angles is truly random and there are a very large number of such capillaries per unit volume, then the mean capillary angle should be 45°; thus

$$\theta = \overline{\frac{dL}{dx}} = \sqrt{2} \approx 1.44 \qquad (4.116)$$

This is an oft-quoted result, but it should be accepted only as a guide, as the random capillary model is not a unique representation of sediment micro-structure. Using a different idealized sediment structure, Dykhuizen and Casey (1989) relate θ to the square root of 3. Tortuosities calculated from formation factors and $D°/D'$ ratios can be both smaller and greater than these values. In fact, experimentally determined values seem to correlate with the porosity, which leads us to theoretical models for θ that depend on φ.

The literature contains many φ-θ relationships. After eliminating those relations that did not satisfy Eq. (4.112), that had singularities at some non-zero φ, or that had θ^2 decreasing with φ, one obtains the compilation given in Table 4.11. It is not necessary to present any of the background theory for these relationships. The important point is whether or not they explain tortuosity data. Figure 4.9 displays θ^2

vs φ data for fine-grained sediments collected from various marine and lacustrine environments (see figure caption).

It is fair to say that none of the theoretical equations in Table 4.11 does a very good job of describing the data in Fig. 4.9. Any theoretical model will constitute at best an imperfect representation of natural sedimentary fabrics, and it is probably unfair to ask these equations to describe the observed φ-θ data. The alternative is to turn to empirical relationships which are appreciably better at this task.

The simplest empirical models resemble the equations of Table 4.11, but contain one adjustable parameter. Let us focus on three that have been used in the past with reasonable success: 1) Archie's Law,

$$\theta^2 = \varphi^{1-m} \tag{4.117}$$

where m is a fitting parameter and not necessarily an integer; 2) the Burger-Frieke equation (Low, 1981),

$$\theta^2 = \varphi + a(1-\varphi) \tag{4.118}$$

where a is a fitting parameter; and 3) a modified Weissberg relation,

$$\theta^2 = 1 - b \cdot ln(\varphi) \tag{4.119}$$

Table 4.11. Theoretically Based Tortuosity vs Porosity Relations.

Relation	Source(s)
$\theta^2 = (3-\varphi)/2$	Maxwell (1881)[1], Neale and Nader (1973)[2]
$\theta^2 = 2-\varphi$	Rayleigh (1892)[1]
$\theta^2 = \varphi^{-1/2}$	Bruggemann (1935)[1]
$\theta^2 = \varphi^{-1/3}$	Millington (1959)[3]
$\theta^2 = 1-1/2\ ln(\varphi)$	Weissberg (1963)[2,3], Ho and Strieder (1981)
$\theta^2 = 1-ln(\varphi)$	Tsai and Strieder (1986)[4]
$\theta^2 = \varphi/(1-(1-\varphi)^{1/3})$	Beekman (1990)
$\theta^2 = ((2-\varphi)/\varphi)^2$	Mackie and Meares (1955)

1 See also Petersen (1958)
2 See also Akanni et al. (1987)
3 See also van Brakel and Heertjes (1974)
4 Quoted in Tomadakis and Sotirchos (1993)

where b is the adjustable constant in this case. Archie's Law has been widely used for sands, with m = 2, and Ullman and Aller (1982) have applied it to their data from muds to find that m=2 if $\varphi < 0.7$, and m = 2.5-3 if $\varphi > 0.7$. Unaware of the previous existence of Eq. (4.118), Iversen and Jørgensen (1993) advanced an equivalent equation: $\theta^2 = 1 + c(1-\varphi)$, where c is an empirical constant; however, this relation can be recognized as the Burger-Frieke equation if one sets a = 1 + c.

Figure 4.10 illustrates the best fits of Eqs (4.117), (4.118), and (4.119) to the same data as in Fig. 4.9, while Table 4.12 summarizes the corresponding parameter values and the statistics of these fits. All three equations provide statistically significant correlations to this data and explain in excess of 50% of the variance. The modified Weissberg equation edges out Archie's Law to supply the best fit amongst the three equations tested.

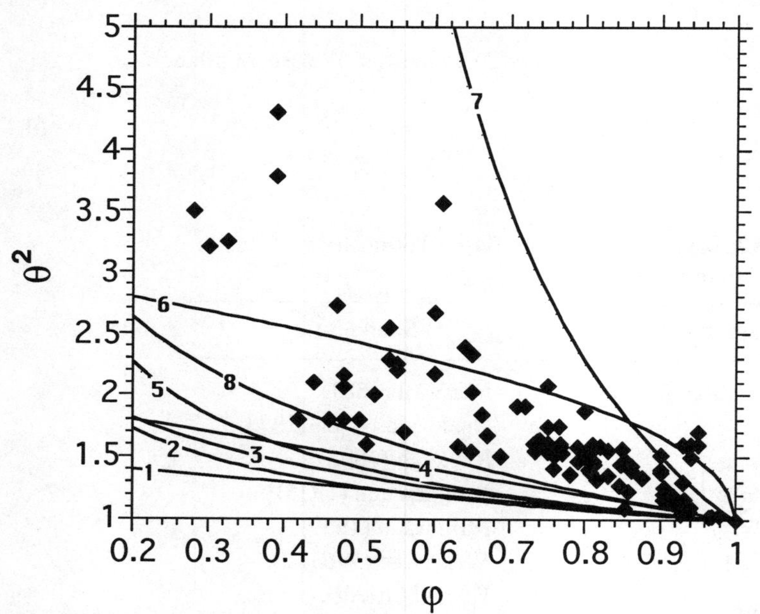

Figure 4.9. Plot of the theoretical tortuosity-porosity relations in Table 4.11 and observed values found in Manheim and Waterman (1974), Li and Gregory (1974), Goldhaber et al. (1977), Jørgensen (1978), Krom and Berner (1980), Ullman and Aller (1982), Sweerts et al. (1991), Iversen and Jørgensen (1993), Archer et al. (1989), and D. Archer (U. Chicago, pers. commun.). The numerical labels correspond to the theoretical relations from: 1-Maxwell, 2-Millington, 3-Weissberg, 4-Rayleigh, 5-Bruggemann, 6-Beekman, 7-Mackie and Meares, and 8-Tsai and Strieder, as given in Table 4.11.

Table 4.12. Summary of Parameter Values and r^2 Values for the Empirical Best Fits to the Tortuosity-Porosity Data in Fig. 4.10.

Equation	Parameter Value	r^2
4.117	m = 2.14±0.03	0.53
4.118	a = 3.79±0.11	0.64
4.119	b = 2.02±0.08	0.65

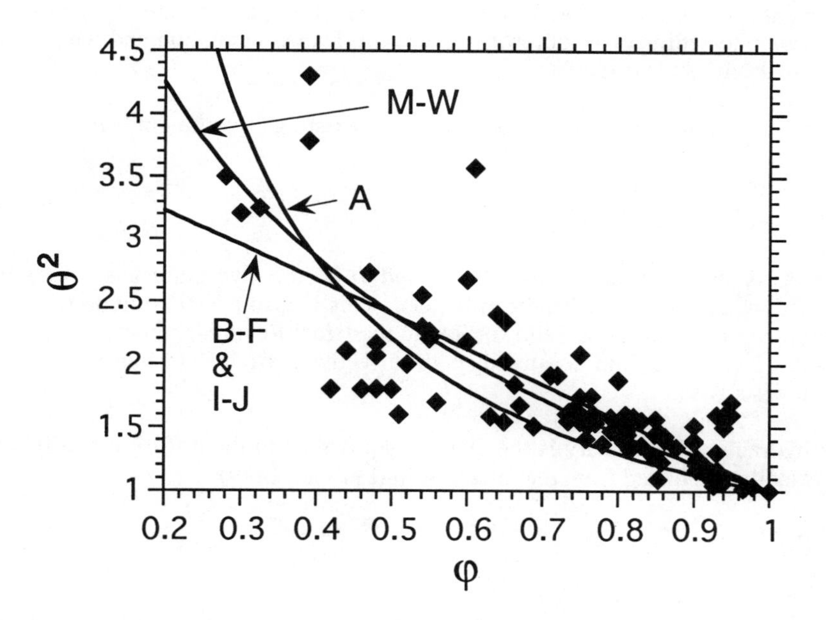

Figure 4.10. Plot of the empirical tortuosity-porosity relations in Table 4.12 and the observed values as found in Fig. 4.9. The key is A for Archie's law (Eq. 4.117), B-F & I-J for Burger-Frieke/Iversen and Jørgensen (Eq. 4.118), and M-W for modified Weissberg (Eq. 4.119).

Given these results, what best can be done when the tortuosity is not measured, and we only have porosity data? If the data of Fig. 4.10 can be taken as universally representative of tortuosities in fine-grained sediments, then θ^2 can be estimated as

$$\theta^2 \approx 1 - 2 \cdot ln(\varphi) = 1 - ln\left(\varphi^2\right) \tag{4.120}$$

This suggestion is made largely because the constant in the M-W equation can be set to a simple integer value, i.e. 2, within the known error on this parameter. Many diagenetic modellers may still be attached to Archie's Law, but it really has no obvious advantage over Eq. (4.120).

4.2.6 Permeability and Dispersion Coefficients

As discussed in Chapter 3, advective flow of porewater can be driven by pressure gradients, caused by gravity waves or flow over sediment-surface topography, and it can be externally impressed as groundwater seepage. The theory of flow through porous media, has been extensively developed (see Bear, 1972; Dullien, 1992). There is simply too much material related to topic for even a cursory review; as a consequence, this section addresses only the two hydrological constitutive equations of relevance to sedimentary porewater transport, Darcy's equation and the Fickian law for hydrodynamic dispersion.

Darcy's Law relates porewater flow, u_D, to the existing pressure gradient,

$$u_D = -\mathcal{K} \frac{\partial h}{\partial x} \tag{4.121}$$

where $\mathcal{K}$ is the hydraulic conductivity and h is the hydraulic head (pressure expressed as an equivalent height of water). This is also a Fickian-type transport equation, and it is said to be valid for low to moderate Reynolds numbers, i.e. $Re = \rho u_D d_p/\mu < 3$ (ρ is the fluid density, d_p is the effective particle diameter, and μ is the dynamic viscosity).

The hydraulic conductivity is the Darcian equivalent to the diffusion coefficient. It is normally calculated from the relationship (Lerman, 1979)

$$\mathcal{K} = \frac{\mathbf{k}\rho g}{\mu} \tag{4.122}$$

where $\mathbf{k}$ is the permeability (units: inverse length), μ is the dynamic viscosity, and g is the acceleration due to gravity. Like the tortuosity, the permeability is thought to be related to the geometry (fabric) of the porous media (Bennett et al., 1989). Measured values of $\mathbf{k}$, and therefore $\mathcal{K}$, span a wide range of values in geological materials (see Lerman, 1979, his Fig. 2.4), but tend to be relatively small for fine-grained (clayey) sediments. Given the magnitudes of $\mathbf{k}$, it is fair to conclude that Darcian flow is important only for sandy sediments with little clay component.

The permeability can also be calculated using various theoretical or semi-theoretical equations (see Bear, 1972; Dullien, 1992; Lerman, 1979). A triplet of the more widely quoted of these are the Carman-Kozeny equation,

$$\mathbf{k} = \frac{d_p^2}{180}\left(\frac{\varphi^3}{(1-\varphi)^2}\right) \tag{4.123}$$

the Blake-Kozeny equation,

$$\mathbf{k} = \frac{d_p^2}{150}\left(\frac{\varphi^3}{(1-\varphi)^2}\right) \tag{4.124}$$

and the Rumpf-Gupte equation,

$$\mathbf{k} = \frac{d_p^2\,\varphi^{5.5}}{5.6} \tag{4.125}$$

where d_p is the (mean) diameter of the sediment particles. None of these equations has been well established for very high porosities ($\varphi>0.8$), as found in muddy surficial sediments. (Alternatively, empirical correlations have been offered (e.g. Bryant et al., 1975; Clukey and Silva, 1982), but beware, these are probably quite site-specific and none use data with $\varphi>0.75$.)

Recently, Hsu and Chang (1991) have noted that, for a dilute array of spheres,

$$\underset{\varphi\to 1}{\text{limit}}(\mathbf{k}) = \frac{d_p\varphi^2}{18(1-\varphi)} \tag{4.126}$$

and that Eq. (4.123) provides a very good value of $\mathbf{k}$ as $\varphi \to 0.4$ (i.e. a sand-like limit). They then propose a blended formula for spheres,

$$\mathbf{k} = \frac{d_p^2\varphi^3}{180(1-\varphi)^2}\left[1-\exp\left(-10\left(\frac{(1-\varphi)}{\varphi}\right)\right)\right] \tag{4.127}$$

which gives Eqs (4.123) and (4.126) in the limits as φ approaches 0.4 and 1.0, respectively. This last equation has not been tested experimentally, and its applicability to natural sediments remains an open question.

Porewater flow in sediments can produce *dispersion*, as discussed in Chapter 3. The resulting flux for a solute i is given by

$$\left(F_{\text{dispersion}}\right)_i = -\varphi D^* \frac{\partial C}{\partial x} \tag{4.128}$$

where D^* is the effective *dispersion coefficient,* which is sometimes made to include molecular diffusion, but is not treated as such here.

Is there a theoretical basis for this Fickian representation of dispersion? There are many such theories (see Bear, 1972; Phillips, 1991), but one is worth repeating here because it represents an alternative basis for the derivation of the other Fickian transport laws, including molecular diffusion. The derivation is in two parts: the first part confirms that dispersion can be characterized by a diffusion coefficient and the second derives Eq. (4.129). We begin by adopting a *Lagrangian* point of view, i.e. one in which we follow a chosen element of porewater as it flows. Let us say that this arbitrary element is at some arbitrary initial position $\mathbf{r}$ at some arbitrary initial time, but it is at some new position, $\mathbf{x}(\mathbf{r},t)$, at some later time t, because of porewater flow (Fig. 4.11). The velocity of the fluid element is then $\mathbf{u}(\mathbf{r},t) \equiv d\mathbf{x}(\mathbf{r},t)/dt$.

Each fluid element follows its own path in the pore system; nevertheless, we can define a mean displacement for a collection of fluid elements, i.e.

$$\mathbf{x}(\mathbf{r},t) - \overline{\mathbf{x}}(\mathbf{r},t) = \int_0^t [\mathbf{u}(\mathbf{r},t') - \overline{\mathbf{u}}]dt' \tag{4.129}$$

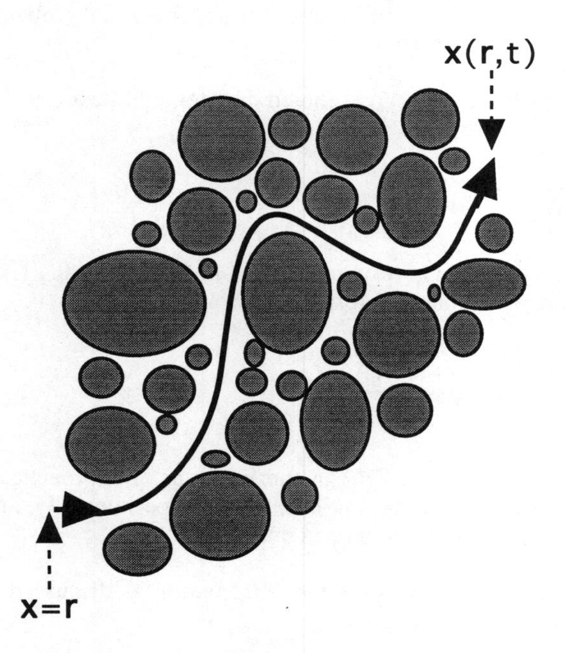

Figure 4.11. Schematic diagram of the flow path followed by a chosen porewater element in time t. (Adapted from Phillips, 1991; reproduced with the kind permission of Cambridge University Press.)

where the overbar indicates mean values for the collection (ensemble average) of elements, and the bold is used to emphasize the vector nature of position. For convenience, let this equation be written as

$$\Delta\mathbf{x} = \int_0^t \Delta\mathbf{u}(\mathbf{r}, t')dt' \tag{4.130}$$

and, upon differentiation,

$$\frac{d\Delta\mathbf{x}}{dt} = \Delta\mathbf{u}(\mathbf{r}, t) \tag{4.131}$$

Now multiply Eqs (4.130) and (4.131),

$$\Delta\mathbf{x}\frac{d\Delta\mathbf{x}}{dt} = \frac{1}{2}\frac{d(\Delta\mathbf{x})^2}{dt} = \int_0^t \Delta\mathbf{u}(\mathbf{r}, t)\Delta\mathbf{u}(\mathbf{r}, t')dt' \tag{4.132}$$

and average this result for the ensemble of fluid elements, i.e.

$$\frac{d\overline{(\Delta\mathbf{x})^2}}{dt} = 2\int_0^t \overline{\Delta\mathbf{u}(\mathbf{r}, t)\Delta\mathbf{u}(\mathbf{r}, t')}\,dt' \tag{4.133}$$

where $\overline{(\Delta\mathbf{x})^2}$ is the mean-squared displacement and $\overline{\Delta\mathbf{u}(\mathbf{r}, t)\Delta\mathbf{u}(\mathbf{r}, t')}$ is the average product of the porewater velocity fluctuations measured at times t and t´. As noted by Phillips (1991), this latter quantity is the statistical covariance.

The term $\overline{\Delta\mathbf{u}(\mathbf{r}, t)\Delta\mathbf{u}(\mathbf{r}, t')}$ is usually considered to be "stationary" (Landahl and Mollo-Christensen, 1986); that is to say, it depends only on the time difference, $\tau = t - t'$, rather than t and t´ individually. Furthermore, it is general practice to write this term as

$$\overline{\Delta\mathbf{u}(\mathbf{r}, t)\Delta\mathbf{u}(\mathbf{r}, t')} \equiv \overline{\Delta\mathbf{u}^2}\ W(\tau) \tag{4.134}$$

where $\overline{\Delta\mathbf{u}^2}$ is called the covariance at t´=t, and $W(\tau)$ is the velocity auto-correlation function. It is also usually the case that $W(0) = 1$ and $W(\tau\rightarrow\infty) = 0$, i.e. that the motions become unrelated as time increases. This means that the integral of W(t) with respect to time should be finite and well-defined; in fact, it's a measure of the time needed for the velocities to de-correlate,

$$T_d = \int_0^\infty W(\tau)\,dt \tag{4.135}$$

which should be independent of t and t´. In fact, T_d is argued to be relatively short and of the order of the particle diameter divided by the mean velocity. Let us now substitute the results given by Eqs (4.134) and (4.135) into Eq. (4.133) to obtain for t $\rightarrow \infty$,

$$\frac{d\overline{(\Delta\mathbf{x})^2}}{dt} = 2\overline{(\Delta\mathbf{u})^2}T_d \tag{4.136}$$

If the right-hand side of Eq. (4.136) is independent of t, then integration gives that

$$\overline{(\Delta\mathbf{x})^2} = 2D^*t \tag{4.137}$$

where

$$D^* \equiv \overline{(\Delta\mathbf{u})^2}T_d \tag{4.138}$$

Equation (4.138) is identical in form to Einstein's relation. Eq. (4.25), and indicates that dispersion is a diffusive process. To further appreciate this link, consider Fig. 4.12. Our porewater element is now moving through a locally linear concentra-

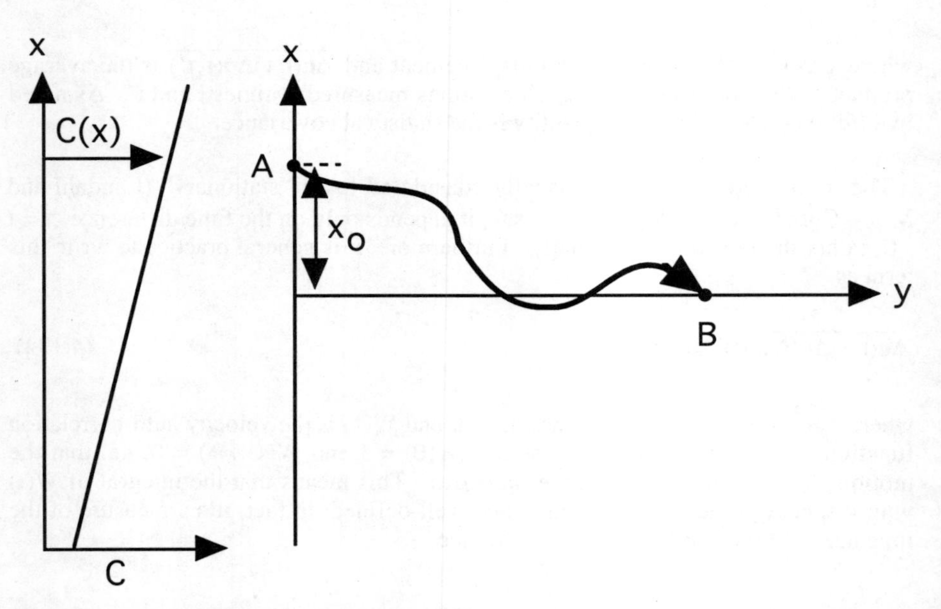

Figure 4.12. Hypothetical path of a porewater element that starts at point A and finishes at point B at time t and that crosses a mean, linear concentration gradient. (Adopted from Phillips, 1991; reproduced with the permission of Cambridge University Press.)

tion gradient. The extra flux at an arbitrary point B caused by the displacement of an element from another close but arbitrary point A, is $\varphi \Delta\mathbf{u} \Delta C$, where ΔC is the difference in concentration between points A and B and

$$\Delta C = -x_o \frac{dC}{dx} \tag{4.139}$$

because of the linear gradient. The total flux from dispersion is the average for all elements, i.e.

$$\left(F_{dispersion}\right)_i = \varphi \overline{\Delta\mathbf{u} \Delta C} = -\varphi \overline{\Delta\mathbf{u}\, x_o} \frac{dC}{dx} \tag{4.140}$$

or, because

$$x_o = \int_0^t \Delta\mathbf{u}\, dt' \tag{4.141}$$

then

$$\left(F_{dispersion}\right)_i = -\varphi \int_0^t \overline{\Delta\mathbf{u}(t) \Delta\mathbf{u}(t')}\, dt' \frac{dC}{dx} \tag{4.142}$$

or, with Eqs (4.132) and (4.133) (Phillips, 1991),

$$\left(F_{dispersion}\right)_i = -\varphi \overline{\Delta\mathbf{u}^2}\, T_d \frac{dC}{dx} \tag{4.143}$$

which, with definition (4.138), reproduces the Fickian dispersion law, Eq. (4.128). Note that, while this derivation is formulated for dispersion, the velocity deviations could also be due to molecular motions, in which case this also constitutes an alternative derivation of Fick's First Law.

The dispersion coefficient, unlike the molecular diffusion coefficient, is not isotropic. The dispersion coefficient in the direction of flow is approximately ten times larger than in a perpendicular direction (Dullien, 1992).

Measurements of dispersion coefficients using various tracers have revealed that it is a relatively smooth, if complicated, function of a dimensionless parameter called the hydrological Peclet number:

$$Pe_h = \frac{\varphi u_D d_p}{(1-\varphi) D^\circ} \tag{4.144}$$

Unless Pe_h is greater than 1, the dispersion is essentially equal to molecular diffusion. Even for sediments as coarse as $d_p = 10$ mm, u_D must be greater than 8 m yr^{-1} before dispersion can make a significant contribution.

Specific formulas for dispersion are discussed in Bear (1972) and Dullien (1992). A commonly quoted relation between Pe_h and the longitudinal dispersion coefficient, D^*, is that offered by de Ligny (1970),

$$\frac{D^*}{D^\circ} \approx 0.7 + \frac{2.5(Pe_h)^2}{c + Pe_h} \tag{4.145}$$

where D° is the molecular/ionic diffusion coefficient, and c is a constant, equal to 8.8 for regular packings of spheres and 7.7 for irregular packings of the same spheres. For $Pe_h > 10$, the longitudinal $D^* \approx u_D d_p$ (Harrison et al., 1983; Cussler, 1984). The lateral dispersion coefficient can be estimated simply by dividing by 10.

Dispersion also accompanies the porewater motions engendered by surface gravity waves (Chapter 3). Rutgers van der Loeff (1981) provides us with a formula for calculating order-of-magnitude values for the resulting dispersion coefficient at any depth, x, in a 1-D (vertical) model of this phenomenon, i.e.

$$D_w^*(x) = \left(\frac{\pi \mathcal{K} H}{\lambda \omega^{1/2} \cosh(kd)} \right)^2 \exp(-2kx) \tag{4.146}$$

where π is the numerical constant pi, H is twice the mean amplitude of the surface waves, λ is their wavelength, $k \equiv 2\pi/\lambda$ (their wavenumber), ω is their angular frequency, and d is the water depth. This result assumes that the amplitude, a, of the vertical porewater motions is sufficiently small that $ka \ll 1$. A more rigorous means of calculating this dispersion is given by Harrison et al. (1983).

4.2.7 Biodiffusion Coefficients and Mixing Depths

The derivation of the biodiffusion constitutive relation, Eq. (3.45) or (3.47), was considered in Chapter 3 and need not detain us here. There is still no fundamental theory that would enable one to predict values of the biodiffusion coefficient, D_B, in an *a priori* manner, and it's likely that there never will be, considering the complexity of the biological processes that are involved. The alternative is to relate D_B to other geochemical parameters that might be known independently; nevertheless, the existence of any such relationship must not be interpreted as causality.

D_B versus w. Figure 4.13 plots calculated D_B values versus their corresponding w-values from the sources listed in Boudreau (1994). These data are limited to cases in which only one surficial mixed layer was modelled or identified. A power law regression over this wide range of D_B and w values produces as a best fit:

$$D_B = 15.7\, w^{0.6} \tag{4.147}$$

with a correlation coefficient, r, of 0.47, and this relation is plotted as the solid line in that figure (D_B in $cm^2\ yr^{-1}$ and w in $cm\ yr^{-1}$). A test of the significance of this regression indicates that there is less than a 1% chance of error that there is, in reality, no correlation; therefore, in a statistical sense, the correlation is significant. At the same time, Eq. (4.147) explains only 25% of the observed variance in the data, i.e. r^2. This means that there is limited physical significance to this regression.

Most of the data in Fig. 4.13 come from ^{210}Pb analyses. If only these data are used in curve fitting, there results the relationship

$$D_B = 15.7\ w^{0.69} \tag{4.148}$$

with r = 0.53. While this regression is stronger, it still accounts for only about 30% of the variance. The remaining variance is attributable to a variety of causes. These include the approximate nature of the diffusion model, temporal and spatial variability of the environment and biological populations, differential mixing, different tracer time-scales for integrating mixing events, sampling artifacts, and the "length" of the chain of processes and events linking D_B to w.

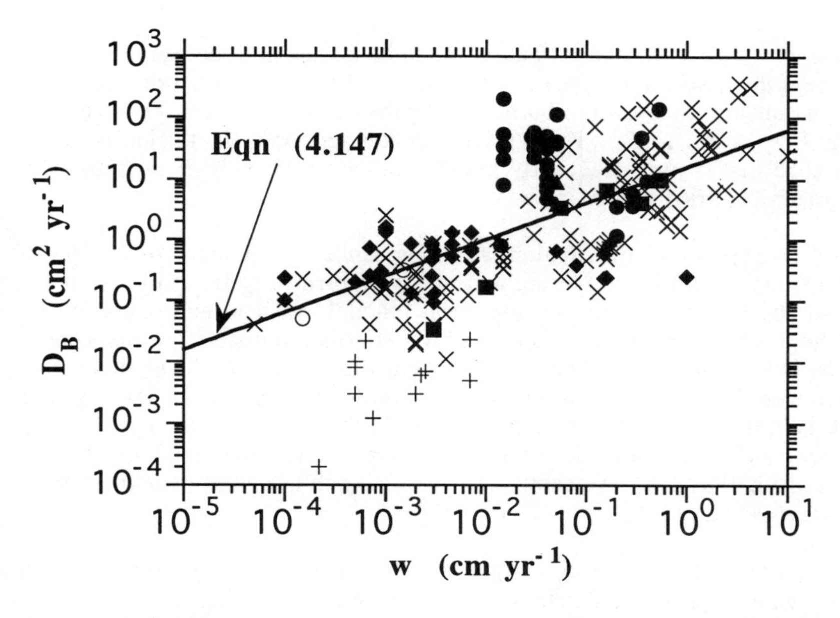

Figure 4.13. Plot of mixing coefficient, D_B, against corresponding burial velocities, w, cited in the sources listed in Boudreau (1994). The solid line is the best power-law fit given by Eq. (4.147). The different symbols identify the tracer, i.e. ◆ for ^{137}Cs and $^{239,240}Pu$, O for ^{32}Si, ● for ^{234}Th and ^{228}Th, X for ^{210}Pb, + for tektites and ash, ▲ for ^{7}Be, and ■ for miscellaneous tracers and techniques. (Reproduced with the kind permission of Elsevier Science Ltd.)

D_B versus Water Depth. Middelburg et al. (in press) have investigated the existence of various predictive empirical relationships for use in marine diagenetic models. They make the justifiable claim that w is not an optimum independent variable for regressions because considerable errors are usually associated with its determination. In addition, the fitting procedure used to calculate D_B values from data may induce correlations with w and lead to propagation of errors. As a consequence, these authors argue that water depth may be a better master variable for D_B.

As a result of this analysis, it was found that, based on ^{210}Pb profiles,

$$D_B = 4.4*10^{(0.762-0.000397*Z)} \tag{4.149}$$

and, from ^{234}Th profiles,

$$D_B = 5.2*10^{(1.754-0.0000751*Z)} \tag{4.150}$$

where D_B is in units of $cm^2\ yr^{-1}$ and the water depth, Z, is in units of meters. Equations (4.149) and (4.150) have associated r^2 of 0.432 (n=132)and 0.241 (n=11), respectively. This r^2 value indicates that Eq. (4.149) explains the variance in its data set better than does Eq. (4.147) with its data; therefore, Eq. (4.149) may be a better predictor.

Mixing Depth versus w. The tracer mixing depth, L, quoted in Boudreau (1994) cannot represent the maximum depths to which infaunal organisms disturb sediments. Specific organisms are known to burrow to depths of over 1 m (e.g. Pemberton et al., 1976; Frey et al., 1978). These mixing depths represent the thickness of the surficial zone that is most frequently mixed; consequently, L is an operationally defined tracer-identified surface mixed depth.

Figure 4.14 displays the distribution of mixing depths as a function of w. Visual inspection suggests little correlation, and this is confirmed statistically by an r = 0.032 for a log-linear curve fit. (A log-linear function was chosen because of the limited range of L compared with that of w.) Further statistics on this sample population indicate a mean and median of 9.8 cm with a standard deviation of ±4.5 cm. Therefore, L is independent on the burial velocity. Very likely, the depth of mixing L is limited primarily by the physical difficulty and increasing energy costs of reworking and excavating deeper than 10-15 cm in a sediment medium (Jumars and Wheatcroft, 1989). Consequently, the mixing depth is similar in environments with different w values.

The variance in Fig. 4.14 is results from the same factors as discussed for D_B. In particular, temporal and spatial heterogeneity of benthic assemblages relative to the timing of sampling is probably a major source of variability. Moreover, there is the problem of objectively defining L from the data. Statistical methods for locating L (e.g. Officer and Lynch, 1983; Lynch and Officer, 1984) are rarely used. Normally L is chosen by eye-balling a perceived break in the tracer data. This somewhat arbitrary procedure can act to both increase and decrease the variability, depending on

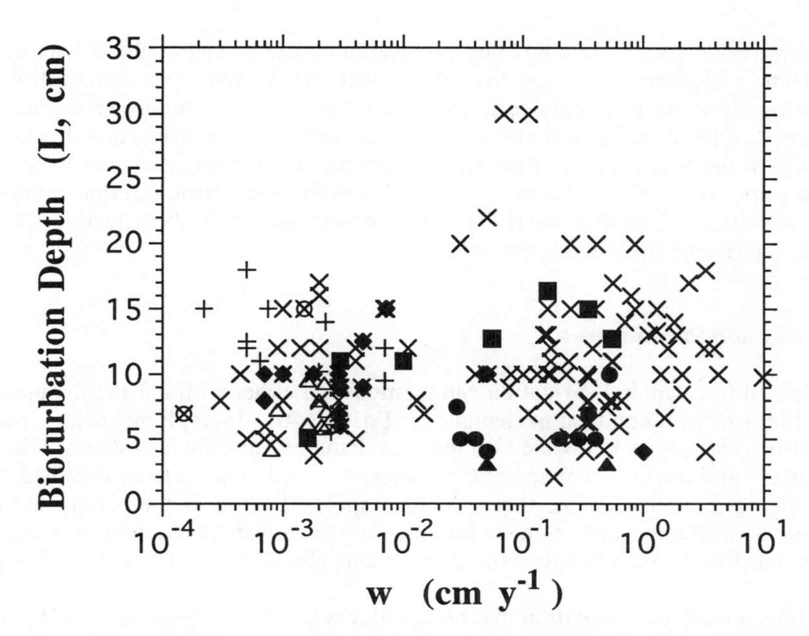

Figure 4.14. Mixing layer depth, L, plotted against corresponding burial velocities for the data in Boudreau (1994). L represents the most frequently mixed zone in terms of the specific tracer used in its identification. The key for the symbols is the same as in Fig. 4.13, with Δ added for ^{14}C. (Reproduced with the kind permission of Elsevier Science Ltd.)

the nature of the data and the prejudices of the investigator. Sampling artifacts also play a significant role in adding variability; e.g. coarse resolution can both under and overestimate the value of L.

4.3 Burial Velocity

The factors that affect the accumulation of sediments in aqueous environments have been discussed extensively elsewhere (e.g. Allen, 1970; Kennett, 1982). Middelburg et al. (in press) have offered a predictive relation for marine environments, based on water depth, Z, i.e.

$$w = 3.3 * 10^{(-0.875-0.000435*Z)} \tag{4.151}$$

where w is in unit of cm yr^{-1} and Z is in meters ($r^2 = 0.615$, n=220).

4.4 Source/Sink-Type Parameters

This category includes kinetic rate laws, irrigation constants and nonlocal exchange coefficients. Unfortunately, at this time, little is known experimentally or observationally about nonlocal coefficients; consequently, they cannot be discussed further here. Only a bit more is known about irrigation rate constants, and this topic is covered in the next section. Rate expressions and their constants seem to suffer from the opposite problem. There seems to be a plethora of forms and rate constant values, so much so that they need their own monograph to do them justice. This book can cover only their basic principles.

4.4.1 Irrigation Parameters

As explained in Chapter 3, irrigation can be modelled either with a 3-D tube model (Aller, 1980) or as a nonlocal exchange, i.e. Eq. (3.140). In applying Aller's tube model, it is necessary to know the average tube diameter, spacing, and depth. These are organism and environment-specific parameters, and they need to be supplied for the site being modelled. Note that it is not important to know the actual rate of burrow-water exchange, unless one is investigating phenomena on a time scale equal to or shorter than the time between irrigation events (Boudreau and Marinelli, 1994).

The 1-D model for irrigation by nonlocal exchange is characterized by an exchange function, $\alpha(x)$, which has units of inverse time. This model has been applied by Emerson et al. (1984) to some continental margin sediments with $\alpha(x)$ as a constant. They obtained a value of 6.3 yr^{-1} through modelling dissolved silica. Martin and Sayles (1987) and Martin and Banta (1992) speculated that irrigation might decrease in intensity with depth because the number of tubes should decrease with sediment depth. They employed the form $\alpha(x) = \alpha_o \, exp(-\alpha_1 x)$, where α_o and α_1 are fitting parameters, to data from Buzzards Bay, Massachusetts, USA. Martin and Sayles (1987) determined α_o values from 12 to 170 yr^{-1} with ^{222}Rn as a tracer, with α_1 varying from 0 to 0.4 cm^{-1}. Martin and Banta (1992) obtained experimental values of α_o from 8 to 300 yr^{-1} for Br^- exchange and 6.9 to 102 yr^{-1} with ^{222}Rn as the tracer, with α_1 varying from -0.04 to 1.0 cm^{-1}.

Irrigation parameters appear in two other forms in the literature, and these can be shown to be equivalent to the exchange coefficient of Eq. (3.137). The first is that reported by Christensen et al. (1987), which has as its equivalence,

$$\alpha = \mathsf{B} \, D_{sed} \tag{4.152}$$

where B is their "diffusive" irrigation constant (cm^{-2}), and D_{sed} is the solute diffusion coefficient, corrected for tortuosity (i.e. D').

Christensen et al. (1987) apply their model to solute data from the Washington Continental Shelf (water depth ≤ 150m), and they find that α is generally greater in the upper 10 cm than in the next 20 cm. Calculated α values from their parameter values range from about 19.5 to 116 yr^{-1} in the top 10 cm and 6 to 17 yr^{-1} in the next 20 cm. These results are not unlike those of Martin and Sayles (1987) and Martin and Banta (1992), quoted above. Christensen et al. (1987) also examined one slope sediment (1800m water depth) and obtained a much lower value of $\alpha \approx 2$ yr^{-1} for the upper 30 cm. This suggests that α decreases with water depth, but we do not have sufficient data to establish a quantitative relationship.

The presentation given by Eq. (4.152) indicates that irrigation rates obtained for one solute can be applied to another by scaling through their respective diffusion coefficients; however, a comparison of Eqs (3.141) and (4.152) would suggest that B is, in part, a function of the radial distance at which the mean concentration of a given solute occurs, $\bar{r}$. Two solutes will have different values of $\bar{r}$ unless their diffusivities, sources and sinks, and boundary conditions are proportionally related. Such a situation is unlikely, and assigning irrigation coefficients by scaling with the ratio of diffusivities is only approximate.

The other type of irrigation data found in the literature might at first seem unrelated to the nonlocal exchange function, but this is purely an illusion; this is the exchange velocity, v_I, introduced by Hammond and Fuller (1979). These authors relate the measured flux out of a sediment due to irrigation, F_I, to an apparent porewater exchange velocity across the sediment-water interface, v_I,

$$F_I = v_I (C_1 - C_o) \tag{4.153}$$

where C_1 is the average solute concentration in the irrigated zone and C_o is the value in the overlying waters.

By comparison, the flux at the sediment-water interface attributable to nonlocal porewater irrigation is the integral of the product $\alpha(x)(C(x) - C_o)$ over the interval of irrigation, $0 \leq x \leq L_I$; that is to say, for constant α,

$$F_I = \alpha \int_0^{L_I} \left(C(x) - C_o\right) dx \tag{4.154}$$

However, with the Mean Value Theorem for Integrals of Chapter 2, this becomes

$$F_I = \alpha L_I \left(C_1 - C_o\right) \tag{4.155}$$

and, consequently, we see that

$$v_I = \alpha L_I \tag{4.156}$$

which shows how the two model parameters are actually related.

Hammond and Fuller (1979) report a value of $v_I = 3$ cm d^{-1}, and their irrigated zone appears to be about 30 cm in their San Francisco Bay sediment cores, based on the extent of ^{222}Rn disequilibrium. This v_I translates into an α value of 36.5 yr^{-1}, which compares favorably with the other results discussed above for similar environments. Archer and Devol (1992) also utilize the v_I formulation for irrigation in their study of benthic oxygen fluxes on the Washington Shelf and Slope. They establish that v_I, and by inference α, is dependent not only on water depth, but also on the bottom water O_2 concentration. Multiple regression analysis will probably be necessary in any attempt to link the value of α or v_I to environmental conditions.

4.4.2 Reaction Kinetics

The kinetics of diagenetic reactions is also a subject worthy of its own monograph (e.g. Brezonik, 1994). The emphasis here is on the *mathematical forms* that appear in the conservation equations, although the kinetics and associated rate parameters for a few geochemically important reactions will be discussed.

Kinetic rate laws can represent mathematical models of reaction schemes, or they can be purely empirical constructions for the purpose of data fitting. Their validity is hardly diminished by being of the latter type, but the former are more intellectually satisfying and, in principle, testable. As kinetic expressions are introduced below, some attempt is made to relate them to kinetic schemes, and in particular, those forms used to describe organic matter oxidation.

General Forms. Kinetic terms constitute sources or sinks of species and appear as one contribution to the *total or material derivative* that defines diagenesis (see Chapters 2 and 3). Such terms can be either linear or nonlinear. Nonlinear kinetics are often at the root of the difficulties encountered while solving transport-reaction models. A *linear* kinetic term is one that contains only linear functions of concentration. A common form is linear production/decay,

$$\left(\frac{DC}{Dt}\right)_{reaction} = \pm kC \tag{4.157}$$

where k is a rate constant (units of inverse time). As written, this equation relates to a homogeneous reaction in the form of a simple uni-directional conversion,

$$C \rightarrow \text{Product(s)} \tag{4.158}$$

Linear kinetics can also describe linear pseudo-homogeneous production or consumption of a solid; for example, if the rate of reaction per unit area of solid sediment is R_{het}, and the reactive surface area per unit volume of solid sediment is

$$\frac{A}{V} = \bar{s}B \tag{4.159}$$

where $\bar{s}$ is the specific surface area, then the rate of reaction is

$$\left(\frac{DB}{Dt}\right)_{reaction} = \pm\frac{A}{V}R_{het} = \pm R_{het}\bar{s}B = \pm k'B \tag{4.160}$$

where $k' = \bar{s}R_{het}$ is the pseudo-homogeneous rate constant, in units of inverse time.

Linearity can be preserved in the face of coupling between species if each kinetic term remains linear, e.g. if species C_i is a member of a simple decay series, such as a radioactive-decay series, i.e.

$$\ldots \rightarrow C_{i-2} \rightarrow C_{i-1} \rightarrow C_i \rightarrow C_{i+1} \rightarrow C_{i+2} \rightarrow \ldots \tag{4.161}$$

thus, for the i-th member of the sequence,

$$\left(\frac{DC_i}{Dt}\right)_{reaction} = k_{i-1}C_{i-1} - k_i\, C_i \tag{4.162}$$

The presence of prescribed functions of space or time in a kinetic term, including constants, in no way affects the linearity. Thus,

$$\left(\frac{DC}{Dt}\right)_{reaction} = \pm k(C_{eq} - C) \tag{4.163}$$

where C_{eq} is a known equilibrium concentration, is still linear.

Nonlinear kinetics arise because of nonlinear functions of concentration itself or through nonlinear coupling. As an example, consider

$$\left(\frac{DC_1}{Dt}\right)_{reaction} = -kC_1^{n_1}C_2^{n_2}C_3^{n_3}\ldots \tag{4.164}$$

where n_1, n_2, n_3 ... are the orders of reaction with respect to species C_1, C_2, C_3 ..., respectively. This describes a multi-species generalization of the homogeneous reaction given by Eq. (4.154), i.e.

$$n_1\, C_1 + n_2\, C_2 + n_3\, C_3 + \ldots. \rightarrow \text{Product(s)} \tag{4.165}$$

Dissolution (disappearance) of a solid species is a natural example of nonlinearity caused by coupling. This process is often represented by a rate law of the form

$$\left(\frac{DB}{Dt}\right)_{reaction} = -k\left(C_{eq} - C\right)B \tag{4.166}$$

where C is the concentration of the solute (i.e. dissolved silica), C_{eq} is an equilibrium value, and B is the concentration of the dissolving solid (e.g. opal, see Schink and Guinasso, 1977b). Equation (4.166) is the pseudo-homogeneous representation of the heterogeneous reaction:

$$B \leftrightarrow C \tag{4.167}$$

If both B and C are true dependent variables, then Eq. (4.166) is nonlinear. The forward heterogeneous rate (dissolution) is a constant, k_f, the backward rate (precipitation) is k_b C, and the net rate is their difference. The net rate of production per unit volume of solid is then

$$\left(\frac{DB}{Dt}\right)_{reaction} = -\frac{A}{V}\left(k_f - k_b C\right) \tag{4.168}$$

Using Eq. (4.156), together with the well-established fact that $k_f/k_b = K_{eq}$ and that $K_{eq} = C_{eq}$ in this situation, Eq. (4.168) becomes

$$\left(\frac{DB}{Dt}\right)_{reaction} = -\bar{s}k_b\left(C_{eq} - C\right)B \tag{4.169}$$

and Eq. (4.166) is recovered if we set $k = \bar{s}k_b$. There are other examples of Eq. (4.166) in which (C_{eq}-C) is raised to some power n, but this is often an experimental expedient.

The potential for nonlinear couplings abounds in sediment studies. Of particular interest in this respect is organic matter (and organic compound) decomposition or oxidation. This process is often described by so-called *Monod kinetics*,

$$\left(\frac{DG}{Dt}\right)_{reaction} = -kG\frac{Ox}{K_{ox} + Ox} \tag{4.170}$$

where G is the conventional symbol for the organic matter concentration, K_{ox} is a *saturation constant*, and Ox is the concentration of the particular oxidant being utilized (e.g. O_2, NO_3^-, $SO_4^=$, etc.). This expression says that the reaction is first order in the organic matter concentration and hyperbolic in the concentration of the oxidant (see Fig. 4.15). The latter means that the reaction is essentially independent of Ox when Ox >> K_{ox}, and it is first order in Ox when Ox << K_{ox}. The original Monod model dealt only with the organic matter as a substrate; however, its extension to two species situations, such as that given above, is quite common.

It is worthwhile considering a possible theoretical basis for Eq. (4.170), as this constitutes another aspect of model building. One approach starts by recognizing that organic matter decay is a pseudo-homogeneous reaction. We can hypothesize at the surface of the organic matter a reaction in which an oxidant molecule/ atom/ion, Ox, attaches to a point of reaction, G·, on this surface to produce a potentially degradable complex, G·Ox (through the mediation of micro-organisms). The formation of the complex is a reversible reaction; thus,

$$G\cdot + Ox \underset{k_{-1}}{\overset{k_1}{\leftrightarrow}} G\cdot Ox \overset{k_2}{\rightarrow} P + G\cdot \tag{4.171}$$

This last equation resembles a *Michaelis-Menten* kinetic scheme, but the surface site appearing as a product is not the same as the catalytic site of the Michaelis-Menten theory. Instead, G· signifies the uncovering/exposure of another (new) site for attack, as the surface gets eaten away by the reaction. The loss of total organic matter (per unit volume) is then given by,

$$\left(\frac{DG}{Dt}\right)_{reaction} = -\frac{A}{V}k_2[G \cdot Ox] \tag{4.172}$$

where [ ] indicates concentration per unit area of reactive solid.

The trouble with Eq. (4.172) is that it makes the rate dependent on the unmeasurable quantity, i.e. [G·Ox]. To rid ourselves of this annoying feature, we construct a balance for unoccupied and occupied sites per unit area per unit time, i.e.

$$\frac{D[G\cdot]}{Dt} = -k_1\,Ox\,[G\cdot] + (k_{-1} + k_2)[G \cdot Ox] \tag{4.173}$$

and

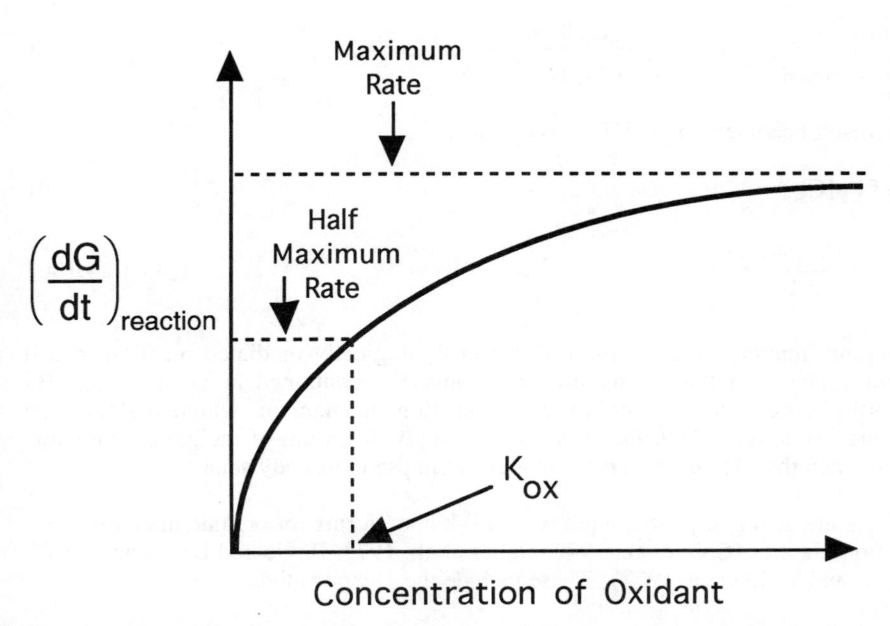

Figure 4.15. Illustration of the dependence of the rate of organic matter decay on the oxidant concentration with the Monod kinetics of Eq. (4.170). The ordinate of this figure is DG/Dt of that equation.

$$\frac{D[G\cdot Ox]}{Dt} = k_1\,Ox\,[G\cdot] - (k_{-1} + k_2)[G\cdot Ox] \tag{4.174}$$

By subtracting these two equations, we can derive that the total number of sites, per unit, area must be a constant, say $[G_o]$, which need not be known, i.e.

$$[G\cdot] + [G\cdot Ox] = [G_o] \tag{4.175}$$

In addition, we can assume that there is a pseudo-steady state in the number of sites with time, so that from either Eq. (4.173) or (4.174),

$$[G\cdot] = \frac{(k_{-1} + k_2)[G\cdot Ox]}{k_1\,Ox} \tag{4.176}$$

This result substituted into Eq. (4.175) produces,

$$[G\cdot Ox] = \frac{[G_o]k_1\,Ox}{(k_{-1} + k_2) + k_1\,Ox} \tag{4.177}$$

Equation (4.177) can be substituted into Eq. (4.172), and Eq. (4.159) invoked to arrive at,

$$\left(\frac{DG}{Dt}\right)_{reaction} = -\bar{s}\,k_2 G \frac{[G_o]k_1\,Ox}{(k_{-1} + k_2) + k_1\,Ox} \tag{4.178}$$

This result becomes Eq. (4.170) if we define,

$$k \equiv \bar{s}\,k_2[G_o] \tag{4.179}$$

$$K_{Ox} \equiv \frac{(k_{-1} + k_2)}{k_1} \tag{4.180}$$

Organic matter decomposition is a microbiologically mediated reaction, and it might appear odd that no mention of biomass is contained in Eq. (4.170). By adopting a slightly more complicated reaction mechanism, which includes the biomass, Boudreau (1992) showed that it simply drops out of the governing equations, when the site numbers per unit area are in pseudo-steady state.

There are also many other equally useful kinetic forms for organic matter decomposition kinetics (Chiu et al., 1972; Dabes et al., 1973; Boyle and Berthouex, 1974; Nihtilä and Virkkunen, 1977). These include the *Teissier* model

$$\left(\frac{DG}{Dt}\right)_{reaction} = -kG\left(1 - exp\left(-\frac{-Ox}{K_{ox}}\right)\right) \tag{4.181}$$

which has much the same behavior as the Monod formulation. The *Contois* model,

$$\left(\frac{DG}{Dt}\right)_{reaction} = -kG\frac{Ox}{K_{ox}G + Ox} \tag{4.182}$$

allows for the possibility that the saturation effect decreases with lowered G concentration, i.e. organisms adapt to starvation and become more resilient. The alternative *Moser* model states that

$$\left(\frac{DG}{Dt}\right)_{reaction} = -\frac{kG}{1 + K_{ox}(Ox)^{-m}} \tag{4.183}$$

where m is a positive number, which can be recognized as a more general form of the Monod equation. Finally, for this listing, there are *Blackman* kinetics,

$$\left(\frac{DG}{Dt}\right)_{reaction} = \begin{cases} -kG & \text{if } Ox > K_{ox} \\ -kG\dfrac{Ox}{K_{ox}} & \text{if } Ox \le K_{ox} \end{cases} \tag{4.184}$$

which approximate the Monod expression using piecewise linear dependence with respect to the oxidant concentration. An alternative reduced Blackman-form is

$$\left(\frac{DG}{Dt}\right)_{reaction} = \begin{cases} -kG & \text{if } Ox > K_{ox} \\ -k'Ox & \text{if } Ox \le K_{ox} \end{cases} \tag{4.185}$$

where $k' = kG$ when $Ox = K_{ox}$, which makes Eq. (4.184) independent of G after $Ox < K_{ox}$. Each of these relations has its advantages and disadvantages, and at least Eq. (4.181) has apparently a respectable theoretical basis, but the actual choice of rate law in a given situation remains the discretion of the user.

Inhibition. This is the suppression of a reaction by the presence of another particular species. While examples of this situation are plentiful, arguably the most important in diagenesis is the inhibition of oxidant utilization. The famed sequence of oxidants, i.e. O_2, then NO_3^-, then MnO_2, then Fe-hydroxides, and then $SO_4^=$, has been attributed to this inhibition. To mathematically simulate inhibition, Humphrey (1972) and Blanch (1981) offer a variety of possible expressions. If decay of species G using oxidant Ox is inhibited by species I, then three possibilities are:

$$\left(\frac{DG}{Dt}\right)_{reaction} = -kG\left(\frac{Ox}{K_{ox} + Ox}\right)\left(\frac{K_I}{K_I + I}\right) \tag{4.186}$$

$$\left(\frac{DG}{Dt}\right)_{reaction} = -\frac{k\,G\,Ox}{K_{ox} + Ox + \dfrac{K_{ox}}{K_I}I} \tag{4.187}$$

$$\left(\frac{DG}{Dt}\right)_{reaction} = -kG\left(\frac{Ox}{K_{ox} + Ox}\right) exp\left(\frac{I}{K_I}\right) \tag{4.188}$$

where K_I is an inhibition constant in each case; K_I represents some critical concentration of the inhibitor above which it increasingly suppresses the reaction.

As in Eq. (4.170), it is possible to design reaction schemes to justify these formulations; in particular, Eq. (4.187) follows from a simple modification of the organic-matter decay scheme presented above†. Let utilization of oxidant Ox(II) be inhibited by the presence of a more readily used oxidant, Ox(I); then this behavior is expressed by the following set of reactions:

$$G\cdot + Ox(II) \underset{k_{-1}}{\overset{k_1}{\leftrightarrow}} G\cdot Ox(II) \tag{4.189}$$

$$G\cdot Ox(II) + Ox(I) \overset{k_2}{\rightarrow} G\cdot + Ox(I) + Ox(II) \tag{4.190}$$

$$G\cdot Ox(II) \overset{k_3}{\rightarrow} P + G\cdot \tag{4.191}$$

Reaction (4.190) characterizes the destruction (inhibition) of the organic-matter-Ox(II) complex by the other oxidant Ox(I). Admittedly, this is far too simplistic and lacking in biochemical reality to be a true representation of the actual reactions involved; nevertheless, it has the merit of simplicity, and if not abused, may actually be instructive.

The idea is to write an equation for the removal of organic matter by reaction with Ox(II) alone given all three of these reactions. This is then given by

$$\left(\frac{DG}{Dt}\right)_{\substack{reaction \\ with\ Ox(II)}} = -\frac{A}{V} k_3[G\cdot Ox(II)] \tag{4.192}$$

where [ ] indicates concentration per unit area. To rid ourselves of the unmeasurable [G·Ox(II)], we again construct balances for unoccupied and occupied sites per unit area per unit time, i.e.

$$\frac{D[G\cdot]}{Dt} = -k_1\, Ox(II)[G\cdot] + (k_{-1} + k_3 + k_2 Ox(I))[G\cdot Ox(II)] \tag{4.193}$$

and

† K. Soetaert (pers. commun.) has made the interesting suggestion that this situation could also (preferably?) be modelled as a set of simple competitive reactions.

$$\frac{D[G \cdot Ox(II)]}{Dt} = k_1\, Ox(II)[G\cdot] - (k_{-1} + k_3 + k_2 Ox(I))[G \cdot Ox(II)] \tag{4.194}$$

By subtracting these two equations, we can derive that the total number of sites per unit area must again be a constant, say G_o,

$$[G\cdot] + [G\cdot Ox(II)] = G_o \tag{4.195}$$

In addition, we can assume that there is a pseudo-steady state in the number of sites per unit area with time, so that from either Eq. (4.193) or (4.194),

$$[G\cdot] = \frac{(k_{-1} + k_3 + k_2 Ox(I))[G \cdot Ox(II)]}{k_1\, Ox(II)} \tag{4.196}$$

This substituted into Eq. (4.195) produces

$$[G \cdot Ox] = \frac{G_o k_1\, Ox(II)}{k_{-1} + k_3 + k_1\, Ox(II) + k_2 Ox(I)} \tag{4.197}$$

Substituting Eq. (4.197) into Eq. (4.192), and invoking Eq. (4.159), we obtain

$$\left(\frac{DG}{Dt}\right)_{\substack{\text{reaction}\\ \text{with Ox(II)}}} = -\bar{s}\, k_3 G \frac{G_o k_1\, Ox(II)}{k_{-1} + k_3 + k_1\, Ox(II) + k_2 Ox(I)} \tag{4.198}$$

This result becomes Eq. (4.192) if we define k and K_{ox} in the same ways as in Eqs (4.179) and (4.180), and K_{ox}/K_I as k_2/k_1.

Even with this theoretical justification, the choice of any of Eqs (4.186) through (4.188) is fairly arbitrary. Both Van Cappellen et al. (1993) and Boudreau (1996) successfully utilize Eq. (4.186) for modelling organic matter decay in sediments. In their treatment, the rate of organic matter oxidation has the following components:

$$\left(\frac{DG}{Dt}\right)_{O_2} = -kG \frac{O_2}{K_{O2} + O_2} \tag{4.199}$$

$$\left(\frac{DG}{Dt}\right)_{NO_3} = -kG \frac{NO_3}{K_{NO3} + NO_3} \frac{K'_{O2}}{K'_{O2} + O_2} \tag{4.200}$$

$$\left(\frac{DG}{Dt}\right)_{MnO_2} = -kG \frac{MnO_2}{(MnO_2)_o} \frac{K'_{O2}}{K'_{O2} + O_2} \frac{K'_{NO3}}{K'_{NO3} + NO_3} \tag{4.201}$$

$$\left(\frac{DG}{Dt}\right)_{Fe(III)} = -kG\frac{Fe(III)}{(Fe(III))_o}\frac{K'_{O2}}{K'_{O2}+O_2}\frac{K'_{NO3}}{K'_{NO3}+NO_3}\frac{1}{1+\frac{K'_{MnO_2}MnO_2}{(MnO_2)_o}} \tag{4.202}$$

$$\left(\frac{DG}{Dt}\right)_{SO_4} = -kG\frac{SO_4}{K_{SO4}+SO_4}\frac{K'_{O2}}{K'_{O2}+O_2}\frac{K'_{NO3}}{K'_{NO3}+NO_3} \cdot \frac{1}{1+\frac{K'_{MnO_2}MnO_2}{(MnO_2)_o}}\frac{1}{1+\frac{K'_{Fe(III)}Fe(III)}{(Fe(III))_o}} \tag{4.203}$$

where k is a generic rate constant, K_{O2}, K_{NO3}, K_{SO4} are Monod-type saturation constants for O_2, NO_3, SO_4, and K'_{O2}, K'_{NO3}, K'_{MnO2}, $K'_{Fe(III)}$ are inhibition constants caused by the effects of O_2 on NO_3, etc. These formulations allow all reactions to be coeval, but the rates of the latter reactions in the sequence are suppressed by the presence of unused oxidants involved in the previous reactions. (In addition, Canfield (1994) has suggested the possibility of nonlinear couplings between the decay reactions of various types of organic matter.)

4.4.3 Rate Constants for Organic Matter Decomposition

To this date, no *a priori* method for predicting the values of rate constants for the vast majority of diagenetic reactions has emerged; consequently, empirical methodologies are forced upon us, whether to our advantage or detriment. Of all the rate constants, that for organic matter decomposition is of greatest interest to sedimentary chemistry. Much effort has already been expended in trying to predict its value(s) for given environmental conditions. Furthermore, it has been realized that there is not a single organic matter type in sediments, but many. The data suggest that there is a reactive fraction, k_1, that decays within the top 10-20 cm of sediments, a more refractory component, k_2, that oxidizes on a scale of about ten to one hundred times longer, and a third that is largely inert. The k_2 material is largely, if not completely, subject to anoxic decomposition. There is also a highly reactive component, say k_o, that decomposes on a seasonal time scale, largely at the sediment-water interface and in the top centimeter of sediments, i.e. k_o = 1-10 yr^{-1}. Predictive correlations have, so far, been applied only to k_1 and k_2.

Toth and Lerman (1977) and Berner (1978) correlated the then-available k_2 values (yr^{-1}) from marine sediments with the compacted burial velocity, w (cm yr^{-1}),

$$k_2 = 0.04\ w^2 \tag{4.204}$$

Tromp et al. (1995) have revisited the relationship between w and the anoxic decay constant (assumed to be k_2) using a larger data set. They have calculated the new, but similar, correlation:

$$k_2 = 0.057\ w^{1.94} \tag{4.205}$$

A similar correlation for the more readily metabolized fraction, k_1 (yr^{-1}), has been longer in coming, because these data are actually more difficult to obtain. Figure 4.16 has been produced by culling the available sources. A power law correlation through these data is given by

$$k_1 \approx 0.4\ w^{0.6} \tag{4.206}$$

One could easily argue that the correlation given by Eq. (4.206) is somewhat biased because of the particular distribution of the underlying data; however, faced with a total lack of independently measured values for even a small fraction of possible environmental conditions, modellers need to use approximate guides, and this equation supplies one.

Tromp et al. (1995) have recently published a correlation between w and the rate constant for oxic organic matter decay; Tromp et al. (1995) propose that:

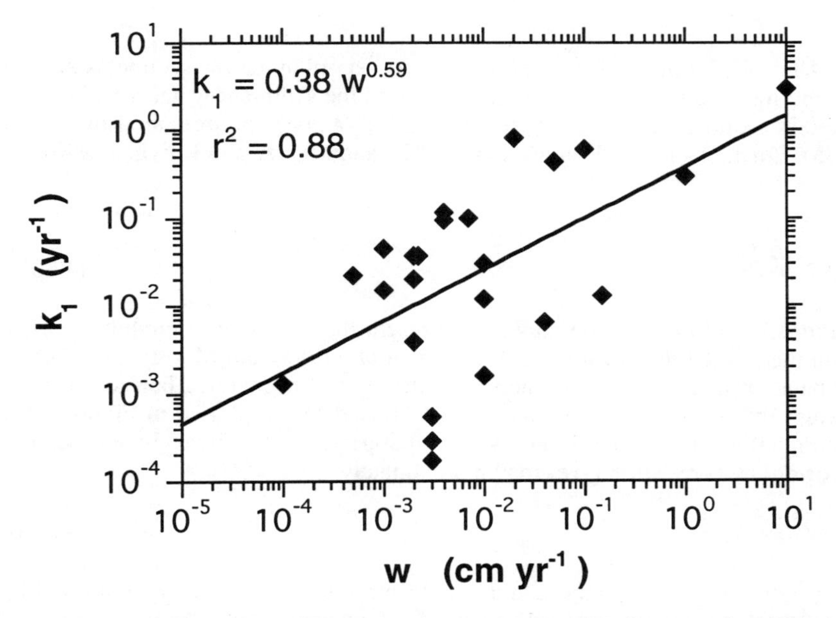

Figure 4.16. A plot of the first-order decay constant, k_1, for organic matter against corresponding values of the burial velocities of the sediment, w. Data sources are Grundmanis and Murray (1982), Martens and Klump (1984), Klump and Martens (1987), Archer et al. (1989), Murray et al. (1978), Emerson et al. (1985), Heggie et al. (1987), Emerson (1985), Berner (1980), Reimers and Suess (1983), and Cwienk (1986, as quoted in Jahnke and Jackson, 1992).

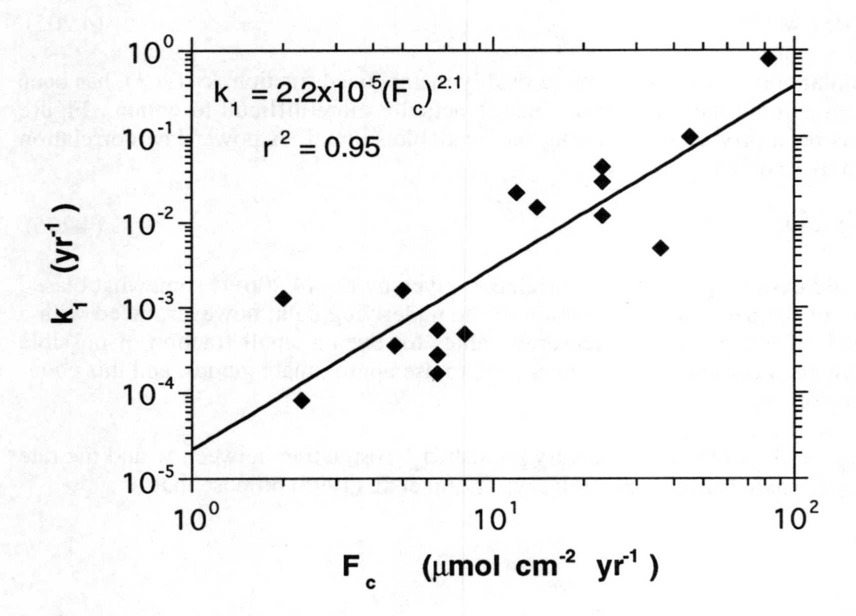

Figure 4.17. A plot of the decay constant, k_1, observed in marine sediments against corresponding values of the organic matter rain to the sediment-water interface, F_c, measured in sediment traps. The correlation is Eq. (4.205). Sources are Emerson et al. (1985), Emerson (1985), Heggie et al. (1987), and Murray and Kuivila (1990).

$$k_1 = 2.97\ w^{0.62} \tag{4.207}$$

Equations (4.206) or (4.207) may not represent the best way to predict k_1, even based on the limited data available. Emerson et al. (1985) and Murray and Kuivila (1990) have argued that k_1 correlates with the flux of organic carbon to the sediment-water interface, F_c (μmol cm^{-2} yr^{-1}). Their data are plotted in Figure 4.17. Even though this type of data is more scarce, it appears to be afflicted by less scatter, and a power law correlation gives that approximately

$$k_1 \approx 2.2\text{x}10^{-5}\ (F_c)^{2.1} \tag{4.208}$$

This correlation with F_c explains a higher percentage of the variability. Values of F_c can be calculated from w values, as explained in Chapter 5.

4.4.4 Rate Constants for Other Geochemically Significant Reactions

Van Cappellen and Wang (1995, 1996) have done an admirable job of tabulating provisional values of the rate constants for many other important diagenetic reactions. These are repeated here for completeness, but the reader should consult these cited papers for details about the source and accuracy of any given value.

The reactions considered by Van Cappellen and Wang (1995) are:

$$Mn^{2+} + 1/2\,O_2 + 2\,HCO_3^- \rightarrow MnO_2 + 2\,CO_2 + H_2O \qquad (4.209)$$

$$Fe^{2+} + 1/4\,O_2 + 2\,HCO_3^- + 1/2\,H_2O \rightarrow Fe(OH)_3 + 2\,CO_2 \qquad (4.210)$$

$$2\,Fe^{2+} + MnO_2 + 2\,HCO_3^- + 2\,H_2O \rightarrow 2\,Fe(OH)_3 + Mn^{2+} + 2\,CO_2 \qquad (4.211)$$

$$NH_4^+ + 2O_2 + 2\,HCO_3^- \rightarrow NO_3^- + 2\,CO_2 + 3H_2O \qquad (4.212)$$

$$H_2S + 2O_2 + 2\,HCO_3^- \rightarrow SO_4^{2-} + 2\,CO_2 + 2H_2O \qquad (4.213)$$

$$H_2S + MnO_2 + 2\,CO_2 \rightarrow Mn^{2+} + 2\,HCO_3^- + S^o \qquad (4.214)$$

$$H_2S + 2\,Fe(OH)_3 + 4\,CO_2 \rightarrow 2\,Fe^{2+} + 4\,HCO_3^- + S^o + 2\,H_2O \qquad (4.215)$$

$$FeS + 2\,O_2 \rightarrow Fe^{2+} + SO_4^{2-} \qquad (4.216)$$

$$CH_4 + 2\,O_2 \rightarrow CO_2 + 2\,H_2O \qquad (4.217)$$

$$CH_4 + CO_2 + SO_4^{2-} \rightarrow 2\,HCO_3^- + H_2S \qquad (4.218)$$

$$Mn^{2+} + 2\,HCO_3^- \rightarrow MnCO_3 + CO_2 + H_2O \qquad (4.219)$$

$$Fe^{2+} + 2\,HCO_3^- \rightarrow FeCO_3 + CO_2 + H_2O \qquad (4.220)$$

$$Fe^{2+} + 2\,HCO_3^- + H_2S \rightarrow FeS + 2\,CO_2 + 2\,H_2O \qquad (4.221)$$

The assumed kinetic rate laws for these reactions can be found in Table 4.13, and the corresponding values of the rate constants for these laws are tabulated in Table 4.14.

4.4.5 Distribution Functions For Reactive Continua

Chapter 3 discusses the use of a reactive continuum as an alternative to a collection of discrete organic matter types. A finite set of discrete types is not necessarily a likely description of natural organic matter distributions. Specifically, we know that

Table 4.13. Rate Laws for Reactions (4.209)-(4.221), as assumed by Van Cappellen and Wang (1995).

Reaction (Eq.. #)	Rate Law
4.209	$k_{MnOx}(Mn^{2+})(O_2)$
4.210	$k_{FeOx}(Fe^{2+})(O_2)$
4.211	$k_{FeMn}(Fe^{2+})(MnO_2)$
4.212	$k_{NH4Ox}(NH_4^+)(O_2)$
4.213	$k_{TSOx}(\Sigma H_2S)(O_2)$
4.214	$k_{TSMn}(\Sigma H_2S)(MnO_2)$
4.215	$k_{TSFe}(\Sigma H_2S)(Fe(OH)_3)$
4.216	$k_{FeSOx}(FeS)(O_2)$
4.217	$k_{CH4Ox}(CH_4)(O_2)$
4.218	$k_{CH4Ox}(CH_4)(SO_4^{=})$
4.219	$k_{MnCO3}\left[\frac{(Mn^{2+})(CO_3^{=})}{K_{MnCO3}}-1\right]$
4.220	$k_{FeCO3}\left[\frac{(Fe^{2+})(CO_3^{=})}{K_{FeCO3}}-1\right]$
4.221	$k_{FeS}\left[\frac{(Fe^{2+})(HS^{-})}{K_{FeS}\,a_H}-1\right]$

where $\Sigma H_2S = H_2S + HS^-$, K_{MnCO3}, K_{FeCO3}, and K_{FeS} are the solubility products for these solids, and a_H is the activity of the hydrogen ion. (Reproduced with the kind permission of Ann Arbor Press, Inc.)

Table 4.14. Rate Constants for Reactions (4.209)-(4.221), as reported by Van Cappellen and Wang (1995) for Various Environments.

Rate Constant†	Environment			
	Deep Sea	Shelf	Coastal	Lacustrine
k_{MnOx}	$1x10^9$	$2x10^9$	$2x10^9$	$2x10^9$
k_{FeOx}	$1x10^9$	$2x10^9$	$2x10^9$	$2x10^9$
k_{FeMn}	$1x10^4$	$2x10^8$	$2x10^8$	$1x10^6$
k_{NH4Ox}	$1x10^7$	$1.5x10^7$	$1.5x10^7$	$2x10^7$
k_{TSOx}	$3x10^8$	$6x10^8$	$6x10^8$	$6x10^8$
k_{TSMn}	$1x10^4$	$1x10^4$	$1x10^4$	$1x10^4$
k_{TSFe}	$1x10^4$	$1x10^4$	$1x10^4$	$1x10^4$
k_{FeSOx}	$2x10^7$	$2.2x10^7$	$2.2x10^7$	$2x10^7$
k_{CH4Ox}	$1x10^{10}$	$1x10^{10}$	$1x10^{10}$	$1x10^{10}$
k_{CH4Ox}	$1x10^{10}$	$1x10^{10}$	$1x10^{10}$	$1x10^{10}$
k_{MnCO3}	0.0	$1x10^{-6}$	$1x10^{-6}$	$1x10^{-6}$
k_{FeCO3}	0.0	$1x10^{-6}$	$1x10^{-6}$	$1x10^{-6}$
k_{FeS}	0.0	$5x10^{-6}$	$5x10^{-6}$	$1x10^{-6}$

† units of Molar yr^{-1}. (Reproduced with the kind permission of Ann Arbor Press, Inc.)

marine organic matter is composed of a wide variety of organic substances, including protein, lignin, cellulose, chitin, plus all of the other products of biosynthesis and the "geopolymerization" process. It would be truly astounding if the decay of all these compounds were to fall neatly into a small finite number of reactive types. More likely, the sensitivity of the measurement methods forcibly divides the true distribution of organic matter types into the distinct categories (Middelburg, 1989). The adoption of a continuous distribution presents, in all probability, a more realistic assumption than its discrete counterpart.

Aris (1968, 1989) and Ho and Aris (1987) have advanced the use of the Gamma distribution as a good all-purpose distribution for reactive continua, i.e.

$$g(k,0) = \frac{g_o k^{\nu-1} exp(-ak)}{\Gamma(\nu)} \tag{4.222}$$

where k is a rate constant, now treated as a continuous variable, $\Gamma(\,)$ is the Gamma Function (Abramowitz and Stegun, 1972, p. 255; Hildebrand, 1976, p. 76; Spanier and Oldham, 1987, p. 411), and a and ν are, in principle, free parameters.

The parameter a in Eq. (4.222) is a measure of the average life-time of the more reactive components of the mixture described by Eq. (3.197). The exponent ν is a nondimensional parameter related solely to the shape of the distribution near k=0. If ν is greater than one, then $g(0,0) = 0$, and a mid-distribution maximum results. If $\nu = 1$, then the Exponential distribution is obtained. Finally, if $0<\nu<1$, then the distribution is weakly singular near $k = 0$, but exponential-like in the limit of large k. Aris (1989) also demonstrates that the Gamma distribution has the interesting and useful property that the apparent order of decay for the mixture remains constant independent of the fraction decomposed.

To illustrate the use of Eq. (4.222) and the continuum model, consider the case of organic matter decomposition that depends only on time (e.g. a laboratory batch experiment or sediment below the zone of bioturbation). The concentration of total organic matter, G(t), is then governed by Eq. (3.198) of Chapter 3, i.e.

$$G(t) = \int_0^\infty g(k,0)\, exp(-kt)\, dk \tag{4.223}$$

If the initial distribution is given by Eq. (4.222), substitution and integration gives (Boudreau and Ruddick, 1991)

$$G(t) = \frac{g_o}{(a+t)^\nu} \tag{4.224}$$

or, if G(0) is the initial total organic matter,

$$G(t) = G(0)\left(\frac{a}{a+t}\right)^\nu \tag{4.225}$$

The rate of disappearance of the total organic matter is the derivative of Eq. (4.225) with respect to time,

$$\frac{DG}{Dt} = -\frac{\nu a^{\nu} G(0)}{(a+t)^{\nu+1}} \tag{4.226}$$

and Eqs (4.224) through (4.226) together lead to

$$\frac{DG}{Dt} = -k_m G^{\frac{\nu+1}{\nu}} \tag{4.227}$$

This equation illustrates explicitly that apparent higher orders of reaction are attainable from small values of ν, even though each member of the continuum decays linearly. The parameter k_m is the apparent rate constant for the decay of the mixture and is equal to $\nu/(g_o)^{1/\nu}$ or

$$k_m = \frac{\nu}{a[G(0)]^{1/\nu}} \tag{4.228}$$

The simple forms offered by Eqs (4.225) and (4.227) are easy to use for curve fitting purposes.

Decomposition experiments in which the (marine) organic matter concentration was monitored accurately during the procedure are not abundant; however, of that limited set, Westrich and Berner (1984) have likely gathered the best data available. The traditional treatment of these data employs exponentials and constants for fitting purposes, i.e. discrete multi-G distributions. Westrich and Berner (1984) required at least two exponential terms plus a constant term to obtain an acceptable fit to the data. This approach involves five arbitrary constants (i.e. a_1, b_1, a_2, b_2, a_3), of which only four are truly independent. Boudreau and Ruddick (1991) illustrate a continuum model for the same data that requires only two independent parameters and attains the same statistical fit to the data. Boudreau (1992) also illustrates the use of the continuum-reaction model in describing depth variable sulfate-reduction rates.

4.5 Thermodynamic Constants

Like kinetics, The thermodynamics of reactions in the aquatic environment is a topic of vast scope that has generated an enormous literature (see Stumm and Morgan, 1996). Again, it is not the aim of this brief section to attempt, in any way, to review this field. My modest goal is to provide the modeller with formulas (constitutive equations) for thermodynamic constants that are most likely to appear in diagenetic equations.

Thermodynamic constants make an appearance in diagenetic transport-reaction models primarily in three principal ways. The first such instance involves constants for equilibrium adsorption (see Berner, 1980). Unfortunately, the state of our understanding of this process is extremely modest and entirely empirical; it is not possible to construct constitutive relations from the data we do have.

The second common occurrence of thermodynamic constants is the solubility product in rate laws governing the precipitation and dissolution of minerals. Of all the minerals occurring in sediments, sufficient information in order to write constitutive equations for the solubility product exists currently for only two minerals, calcite and aragonite. Specifically, for the reaction

$$CaCO_3\text{(calcite or aragonite)} \leftrightarrow Ca^{2+} + CO_3{}^{2-} \tag{4.229}$$

there exists a stoichiometric solubility product of the form:

$$K_{sp}\text{(calcite or aragonite)} = (Ca^{2+})_{eq}(CO_3^{2-})_{eq} \tag{4.230}$$

where (Ca^{2+}) and $(CO_3{}^{2-})$ are equilibrium concentrations of these ions. The term "stoichiometric" is applied to this constant because this approach disregards any complex and ion pair formation. As a result, such constants are functions of temperature, salinity, and pressure, i.e. as determined by Mucci (1983) and Millero (1995),

$$\begin{aligned} log\left(K_{sp}\text{(calcite)}\right) &= -171.9065 - 0.077993\,T + \frac{2839.319}{T} + 71.595\,log(T) \\ &+ \left(-0.7712 + 0.0028426\,T + \frac{178.34}{T}\right)S^{0.5} - 0.07711\,S + 0.0041249\,S^{1.5} \end{aligned} \tag{4.231}$$

and

$$\begin{aligned} \log\left(K_{sp}\text{(aragonite)}\right) &= -171.945 - 0.077993\,T + \frac{2903.293}{T} + 71.595\,log(T) \\ &+ \left(-0.068393 + 0.001727\,T + \frac{88.135}{T}\right)S^{0.5} - 0.10018\,S + 0.0059415\,S^{1.5} \end{aligned} \tag{4.232}$$

where T is the absolute temperature (in Kelvin) and S is the salinity. The effects of pressure on these constants are discussed below.

The final type of thermodynamic constant found in diagenetic equations results from the local equilibrium assumption for fast reversible reactions (see Chapter 3). In such situations, diagenetic models include thermodynamic relations for the species involved in these fast reactions. Such reactions are almost invariably acid-base dissociations in porewaters. The porewater acid-base reactions that have been studied extensively are listed in Table 4.15.

Table 4.15. Common Acid-Base Reactions Occurring in Sediment Porewaters and Corresponding Equilibrium Relations.

Reaction	Equilibrium Relation†
$CO_2(aq) + H_2O \leftrightarrow HCO_3^- + H^+$	$K_{C1}(CO_2) = (HCO_3^-)(H^+)$
$HCO_3^- \leftrightarrow CO_3^{2-} + H^+$	$K_{C2}(HCO_3^-) = (CO_3^{2-})(H^+)$
$B(OH)_3 + H_2O \leftrightarrow B(OH)_4^- + H^+$	$K_B(B(OH)_3) = (B(OH)_4^-)(H^+)$
$H_3PO_4 \leftrightarrow H_2PO_4^- + H^+$	$K_{P1}(H_3PO_4) = (H_2PO_4^-)(H^+)$
$H_2PO_4^- \leftrightarrow HPO_4^{2-} + H^+$	$K_{P2}(H_2PO_4^-) = (HPO_4^{2-})(H^+)$
$HPO_4^{2-} \leftrightarrow PO_4^{3-} + H^+$	$K_{P3}(HPO_4^{2-}) = (PO_4^{3-})(H^+)$
$NH_4^+ \leftrightarrow NH_3 + H^+$	$K_{NH}(NH_4^+) = (NH_3)(H^+)$
$H_2S \leftrightarrow HS^- + H^+$	$K_S(H_2S) = (HS^-)(H^+)$
$H_2O \leftrightarrow OH^- + H^+$	$K_w(H_2O) = (OH^-)(H^+)$

Concentrations are in moles kg^{-1} units.

Millero (1995) has established a set of K versus S and T relations for the equilibrium constants in Table 4.15, i.e.

$$ln(K_{C1}) = 2.18867 - \frac{2275.036}{T} - 1.468591\,ln(T) + (-0.138681 - \frac{9.33291}{T})S^{0.5} + 0.0726483\,S - 0.00574938\,S^{1.5} \tag{4.233}$$

$$ln(K_{C2}) = -0.84226 - \frac{3741.1288}{T} - 1.437139\,ln(T) + (-0.128417 - \frac{24.41239}{T})S^{0.5} + 0.1195308\,S - 0.0091284\,S^{1.5} \tag{4.234}$$

$$\begin{aligned} ln(K_B) = &\left(-8966.90 - 2890.51\,S^{0.5} - 77.942\,S + 1.726\,S^{1.5} - 0.0993\,S^2\right)T^{-1} \\ &+ \left(148.0248 + 137.194\,S^{0.5} + 1.62247\,S\right) + 0.053105\,S^{0.5}T \\ &+ \left(-24.4344 - 25.085\,S^{0.5} - 0.2474\,S\right) ln(T) \end{aligned} \tag{4.235}$$

$$ln\left(K_{P1}\right) = 115.54 - \frac{4576.752}{T} - 18.453\,ln(T) + (0.69171 - \frac{106.736}{T})S^{0.5} + \left(-0.01844 - \frac{0.65643}{T}\right)S \quad (4.236)$$

$$ln\left(K_{P2}\right) = 172.1033 - \frac{8814.715}{T} - 27.927\,ln(T) + (1.3566 - \frac{160.34}{T})S^{0.5} + \left(-0.05778 + \frac{0.37335}{T}\right)S \quad (4.237)$$

$$ln\left(K_{P3}\right) = -18.126 - \frac{3070.75}{T} + (2.81197 + \frac{17.27039}{T})S^{0.5} + \left(-0.09984 - \frac{44.99486}{T}\right)S \quad (4.238)$$

$$ln\left(K_{NH}\right) = -0.25444 + 0.0001635\,T - \frac{6285.33}{T} + (0.46532 - \frac{123.7184}{T})S^{0.5} + \left(-0.01992 + \frac{3.17556}{T}\right)S \quad (4.239)$$

$$ln\left(K_{S}\right) = 225.838 - \frac{13275.3}{T} - 34.6435\,ln(T) + 0.3449\,S^{0.5} - 0.0274\,S \quad (4.240)$$

and, finally,

$$ln\left(K_{w}\right) = 148.9802 - \frac{13847.26}{T} - 23.6521\,ln(T) + \left(-5.977 + \frac{106.736}{T} + 1.0495\,ln(T)\right)S^{0.5} - 0.01615\,S \quad (4.241)$$

Pressure corrections can be made to all these constants using the formula advanced by Millero (1995),

$$ln\left(\frac{K_i^P}{K_i^0}\right) = -\left(\frac{\Delta V_i}{RT}\right)P + \left(\frac{\Delta\kappa_i}{2RT}\right)P^2 \quad (4.242)$$

where K_i^0 and K_i^P are the constants at atmospheric pressure and the applied pressure, respectively, R is the gas constant (83.131 mol bar K^{-1}), and ΔV_i and $\Delta\kappa_i$ are the partial molal volume change and compressibility change, respectively. These latter two parameters are calculated using the formulas

$$\Delta V_i = a_0 + a_1 t + a_2 t^2 \quad (4.243)$$

Table 4.16. Parameter Values for the Calculation of ΔV_i and $\Delta \kappa_i$ using Eqs (4.243) and (4.244), respectively. (Values from Millero, 1995.)

Acid or Solid	$-a_0$	a_1	$10^3 a_2$	$-b_0$	b_1
H_2CO_3	25.50	0.1271	0	3.08	0.0877
HCO_3^-	15.85	-0.0219	0	-1.13	-0.1475
$B(OH)_3$	29.48	-0.1622	2.608	2.84	0
H_2O	25.60	0.2324	-3.6246	5.13	0.0794
H_2S	14.80	0.0020	-0.400	-2.89	0.054
NH_4^+	26.43	0.0889	-0.905	5.03	0.0814
H_3PO_4	14.51	0.1211	-0.321	2.67	0.0427
$H_2PO_4^-$	23.12	0.1758	-2.647	5.15	0.0900
HPO_4^{2-}	26.57	0.2020	-3.042	4.08	0.0714
H_3PO_4	14.51	0.1211	-0.321	2.67	0.0427
Calcite	48.76	-0.5304	0	11.76	-0.3692
HCO_3^-	35	-0.5304	0	11.76	-0.3692

(Reproduced with the kind permission of Elsevier Science Ltd.)

and

$$\Delta \kappa_i = b_0 + b_1 t \qquad (4.244)$$

where a_0, a_1, a_2, b_0, b_1, and b_2 are empirical constants as given in Table 4.16 and t is the temperature in degrees Celcius.

4.6 Other Environmental Parameters

The prediction of other environmental parameters proves to be valuable in various types of diagenetic studies (e.g. Tromp et al., 1995; van Cappellen and Wang, 1996; Stoetaert et al., 1996; Boudreau, 1996). Most of these correlations involve marine sediments, but it is hoped that similar correlations may be found to exist for lacustrine or estuarine environments.

A first relation of interest in global diagenetic modelling is that between the water depth to the sediment-water interface (X_{swi}) and the sedimentation rate. Figure 4.18 illustrates a large data set obtained from sediments around the world and from the resulting correlation,

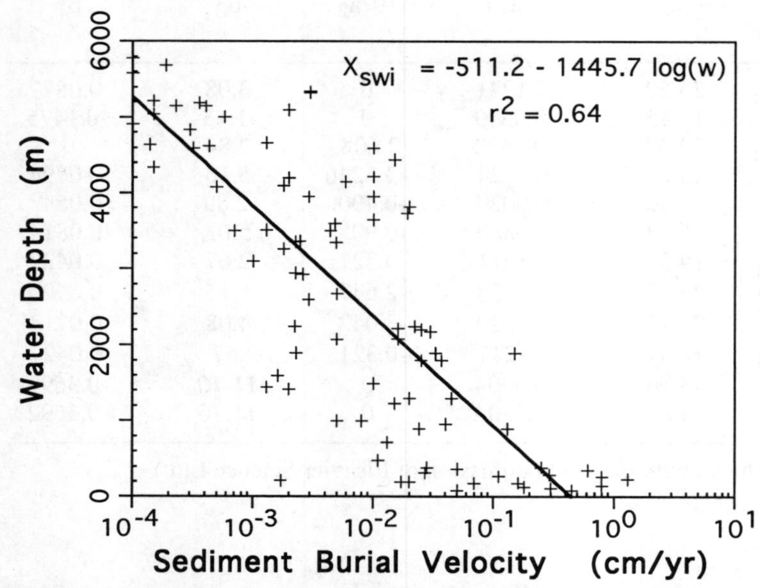

Figure 4.18. Plot of the oceanic water depth (in meters) versus the burial velocity, w, in cm yr^{-1}. (Data supplied, in part, by S. Henrichs, pers. comm.)

$$X_{swi} = -511.2 - 1445.7\,log(w) \tag{4.245}$$

A similar analysis is presented by Tromp et al. (1995), for which they obtain

$$X_{swi} = 2700\,erfc(2.1 + log(w)) \tag{4.246}$$

where w is the burial velocity in cm kyr^{-1} and *erfc*( ) is the Complementary Error function (Chapter 7, Eq. (7.37)).

Once X_{swi} is predicted, it is possible to calculate two other important environmental parameters (see Tromp et al., 1995). The first is the mean bottom temperature at the sediment-water interface,

$$t^{\circ}_{swi} = 1.27 - 1.005x10^{-3}\,X_{swi} + 4.49x10^{-7}\,X^2_{swi} - 1.11x10^{-10}\,X^3_{swi} \tag{4.247}$$

in degrees Celsius, and the bottom water O_2 concentration,

$$(O_2)_{swi} = 251.59 - 0.31\,X_{swi} + 2.92x10^{-4}\,X^2_{swi} - 1.14x10^{-7}\,X^3_{swi} \tag{4.248}$$

5 BOUNDARY CONDITIONS

> For some minutes Alice stood without speaking, looking out in all directions over the country - and a most curious country it was. There were a number of tiny little brooks running straight across it from side to side, and the ground between was divided up into squares by a number of little green hedges, that reached from brook to brook.
>
> (Lewis Carroll, *Through the Looking-Glass*, 1871)

5.1 Introduction

The transport-reaction models developed in Chapter 3 are not complete descriptions of sediment environments. On the one hand, we are faced with the necessity of supplying constitutive relations, as discussed in Chapter 4; on the other hand, the derivation of the diagenetic equations in Chapter 3 also referred at various points to "boundary conditions". Boundary conditions are not an afterthought to a diagenetic problem. From a physico-chemical point of view, boundary conditions place the model in its environmental context. They are essential elements of any properly posed diagenetic problem. Consequently, as much thought must go into their formulation as into the conservation equations themselves.

Mathematically, boundary conditions are needed to resolve a fundamental indeterminacy associated with differential equations. To illustrate this point with an almost trivial example, consider taking the derivative of the two linear algebraic functions

$$f_1(x) = ax + b \tag{5.1}$$

and

$$f_2(x) = ax - b \tag{5.2}$$

where a and b are two (arbitrary) constants. Now, upon differentiation of these two functions, we obtain

$$\frac{df_1(x)}{dx} = a \tag{5.3}$$

and

$$\frac{df_2(x)}{dx} = a \tag{5.4}$$

Thus, the derivatives of the two functions are the same, even though the functions themselves differ by a constant value. If we started with Eqs (5.3) and (5.4) and integrated each, the result for both would have ax as the first term, but how would the b vs -b be retrieved? The answer is, with a boundary or initial condition.

To illustrate, integrate Eq. (5.3) subject to the condition that

$$f_1(x{=}0) = b \tag{5.5}$$

where b is a constant of known value then

$$\int \frac{df_1(x)}{dx} dx = \int a\, dx = ax + c \tag{5.6}$$

where c is an arbitrary integration constant. But Eq. (5.5) requires that

$$f_1(0) = c = b \tag{5.7}$$

The unknown is now specified, and Eq. (5.1) is obtained. To get Eq. (5.2), we would have specified that $f_2(x{=}0) = -b$. In this example, it is somewhat arbitrary whether Eq. (5.5) is called a boundary condition or an initial condition. For non-trivial problems, this distinction is self-evident. Nevertheless, the important point is that the integration of a first-order derivative produces one arbitrary constant that must be resolved with an independent condition.

Let us consider another example, the rather simple conservation equation:

$$D\frac{d^2C}{dx^2} = 0 \tag{5.8}$$

which says that the species is simply subject to steady state diffusion. We can integrate this equation directly to get C(x); i.e. a first integral is

$$\frac{dC}{dx} = A \tag{5.9}$$

where A is an arbitrary constant, and a second integral gives

$$C(x) = Ax + B \tag{5.10}$$

where B is a second arbitrary constant. Therefore, the integration of a steady state diffusion equation (i.e. a second-order ordinary differential equation, ODE) will produce two integration constants. The presence of a lower order derivative in x (i.e. an advection term) or a reaction term will not alter this fact, although the simple type of integration done in this example is normally not possible.

In order to eliminate any confusion, here are two general rules for integration constants:

i) You need the same number of boundary conditions as the order of the highest derivative in each spatial variable; and

ii) You need the same number of initial conditions as the order of the highest derivative in time.

Example 5-1. Consider the steady state advection-decay equation for a solid species in which burial may be a function of depth, i.e. w(x),

$$w(x)\frac{dC}{dx} = -kC \tag{5.11}$$

This first-order ODE needs *only one* boundary condition, i.e. that at $x = 0$. Whether C increases or decreases with depth, x, is solely dependent on the sign of the term on the right-hand side, i.e. negative for a decrease and positive for an increase.

Example 5-2. Consider the steady state advection-diffusion reaction (ADR) equation with both nonlinear power-law production and inhomogeneous production,

$$D\frac{d^2C}{dx^2} - v\frac{dC}{dx} + k(C_s - C)^{2.5} + R_o\, exp(-\alpha x) = 0 \tag{5.12}$$

This equation needs two boundary conditions in total, because it is second order; i.e. a second-order spatial derivative is the highest order of derivative present.

Example 5-3. Now consider the time-dependent, linear diffusion-advection-decay equation,

$$\frac{\partial C}{\partial t} = D\frac{\partial^2 C}{\partial x^2} - v\frac{\partial C}{\partial x} - kC \tag{5.13}$$

This partial differential equation (PDE) needs two boundary conditions and one initial condition.

Example 5-4. How many boundary conditions are needed for the steady state version of Aller's tube model? This model is governed by the equation .

$$D'\frac{\partial^2 C}{\partial x^2}+\frac{D'}{r}\frac{\partial}{\partial r}\left(r\frac{\partial C}{\partial r}\right)+R=0 \quad (5.14)$$

This PDE needs four boundary conditions, two in the x-direction and two in the radial direction, because there are second-order derivatives in both these directions.

5.2 Origin and Types of Diagenetic Boundary Conditions

Boundary conditions are not arbitrary, but reflect knowledge of what happens at the boundaries of the sediment zones being modelled. Like the diagenetic conservation equations developed in Chapter 3, boundary conditions can be derived from well-developed theories (e.g. Standart, 1964; Slattery, 1967, 1980, 1981; Deemer and Slattery, 1978; Gray and Hassanizadeh, 1989; Hassanizadeh and Gray, 1989a,b). The following section of the text presents the various types of diagenetic boundary conditions, illustrating where appropriate their theoretical origins.

There are three broad classes of boundary conditions: a) concentration conditions, b) flux conditions, and c) continuity conditions, also called mixed conditions. The first two types apply to external boundaries, while the third describes what occurs at internal boundaries. An external boundary is one beyond which the model does not extend (e.g. this is often the status of the sediment-water interface). Internal boundaries occur within the modelled region and account for abrupt changes in physical, chemical, or biological properties. Let us now examine in some detail each of these types of boundary conditions.

5.2.1 Concentration or Dirichlet Conditions

As the name implies, this type of boundary condition provides the value of the concentration at some depth in a 1-D model, along some line in a 2-D model, and across a sediment surface in a 3-D model. For example, given a dissolved species in 1-D, we could have that

$$C(x=d) = C_o \quad (5.15)$$

where d is a selected depth§ and C_o is the known concentration at that depth. In 2-dimensions, say x and r,

$$C(r,x=0,t) = C_o(r,t) \quad (5.16)$$

---

§ The position of a boundary condition such as eqn (5.15) can be a function of time, in which case an additional relationship must be added to define this temporal displacement.

where $C_o(r,t)$ is an arbitrary but known concentration function of r and t along the line defined by x = 0, i.e. the sediment-water interface. In three dimensions, one could have

$$C(r=0,x=0,\vartheta) = C_o(\vartheta) \qquad (5.17)$$

where $C_o(\vartheta)$ is an arbitrary function of the radial angle, ϑ, including a constant, over the surface defined by x = 0 and r = 0, i.e. again the sediment-water interface. Equivalent concentration boundary conditions for a solid species are obtained simply by replacing C with B in the three previous equations.

Such concentration conditions usually indicate contact with a well-mixed reservoir of known concentration, such as sediments in contact with overlying seawater when boundary layer effects can be ignored. In the mathematical and modelling literature, such a condition is known as a *Dirichlet condition* or a boundary condition of the *first kind*. When C_o in the above equations is zero, the condition is said to be *homogeneous*. This has implications for the method of solution, as will be discussed in Chapter 7.

5.2.2 Fluxes at External Boundaries

There are many (if not most) situations in which we have information, either theoretical or measured, about the flux at an external boundary of the sediment section we are interested in. Flux boundary conditions also constitute statements of conservation at a boundary (e.g. Hassanizadeh and Gray, 1989a,b). As such, they can be derived by building a proper mass balance at that boundary. If we had need for only 1-D models, then the statement of this balance would be trivial; however, it is preferable to create a form that applies to any geometry or reference system, and then specialize it to a single Cartesian dimension.

Adjacent to an arbitrary external interface of a sediment region, designated by the letter $\mathcal{E}$, place a small highly deformable cap-like volume, $\mathcal{V}$, as shown in Figure 5.1. $\mathcal{E}'$ represents the surface area of $\mathcal{E}$ covered by the cap, and $\mathcal{S}$ is the surface of the cap in the sediment region we are modelling (the stippled area in Fig. 5.1). Note that $\mathcal{S}$ has no thickness and accumulates no mass, and that the position of $\mathcal{E}$ is not necessarily constant with time.

Conservation of mass of a dissolved species within the volume $\mathcal{V}$ balances what accumulates and what is produced/consumed in this volume against the material that fluxes across the surface $\mathcal{S}$ by diffusion or advection and that is supplied/removed at the surface $\mathcal{E}'$ by reaction or by a prescribed flux. This balance can be stated as

$$\underbrace{\iiint_{\mathcal{V}} \left(\frac{\partial \varphi C}{\partial t} - \sum R \right) d\mathcal{V}}_{\text{Total Accumulation \& Reaction in Volume } \mathcal{V}} = \underbrace{\iint_{\mathcal{S}} \varphi(uC - D'\nabla C) \cdot n \, d\mathcal{S}}_{\text{Total Net Flux Outward Across Surface } \mathcal{S}} - \underbrace{\iint_{\mathcal{E}'} \Im \, d\mathcal{S}}_{\text{Total Flux Into } \mathcal{V} \text{ From Surface } \mathcal{E}'} \qquad (5.18)$$

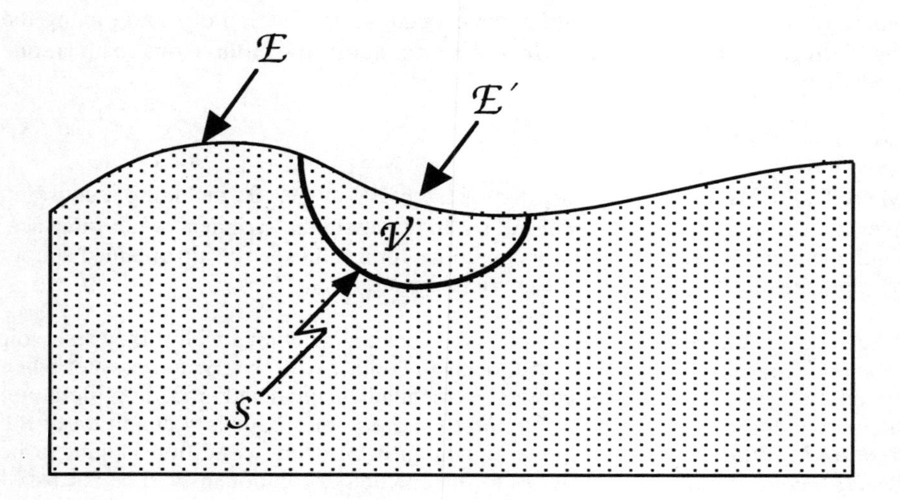

Figure 5.1. Illustration of the small conservation volume $\mathcal{V}$ placed next to an external interface $\mathcal{E}$ and used to derive the flux boundary condition, Eq. (5.20). The position of $\mathcal{E}$ can be a function of time.

where, as in Chapter 3, $\varphi \equiv$ porosity, $\Sigma R \equiv$ net rate of reaction per unit volume in $\mathcal{V}$, $u \equiv$ porewater velocity (vector), $D' \equiv$ effective diffusion coefficient, $n \equiv$ outward unit vector along the surface $\mathcal{S}$ (it points into the sediment zone being modelled), and $\mathfrak{I} \equiv$ the net amount of material supplied/removed along $\mathcal{E}$ from any reactions or "prescribed fluxes" at this surface. $\mathfrak{I}$ can be, in general, a function of time and concentration. Displacement onto the surface $\mathcal{E}$ as a result of nonlocal mixing is included in the $\mathfrak{I}$ term, if present.

Now, deform the cap in such a way that the volume $\mathcal{V}$ goes to zero, but the surface $\mathcal{S}$ is stretched to completely cover the external surface $\mathcal{E}$, i.e. $\mathcal{V} \rightarrow 0$ and $\mathcal{S} \rightarrow \mathcal{E}' \rightarrow \mathcal{E}$. Consequently, the term on the left-hand side of Eq. (5.18) disappears, and the remaining terms give that

$$\iint_{\mathcal{E}} \left[\varphi\left(uC - D'\nabla C\right) \cdot n - \mathfrak{I}\right] d\mathcal{E} = 0 \tag{5.19}$$

This integral will be zero for all cases only if the integrand, the interior part of the integral, is itself always zero, i.e.

$$\varphi\left(uC - D'\nabla C\right) \cdot n = \mathfrak{I} \tag{5.20}$$

along $\mathcal{E}$. This general boundary condition states that the flux that is prescribed or due to reaction on the surface $\mathcal{E}$ (right-hand side) must be balanced by an equal advective and/or diffusive flux (left-hand side). In one Cartesian dimension (i.e. depth only), mass balance reduces to:

$$\left[\varphi\left(uC - D'\frac{\partial C}{\partial x}\right)\right]_d \cdot n = \mathfrak{I} \tag{5.21}$$

where x = d is the arbitrary depth in the sediment of the $\mathcal{E}$ plane. In one dimension, n=+1 if n points downwards into the sediment in the positive x-direction, and n=-1 if n points upwards in the negative x-direction. Equation (5.21) applies only at the depth x = d and has no meaning anywhere else; this equation states that an input (or removal) flux is balanced by the combined effects of diffusion and advection.

For a solid, B, (1-φ), w, and D_B can be substituted for C, φ, u, and D', respectively, in order to get a boundary condition on a solid species, i.e.

$$\left[(1-\varphi)\left(wB - D_B\frac{\partial B}{\partial x}\right)\right]_d \cdot n = \mathfrak{I}_B \tag{5.22}$$

with intraphase mixing. A switch to the interphase mixing form requires a slight modification of this last equation.

Many special cases evolve from Eq. (5.21) or (5.22). First, at the sediment-water interface (x=0, n=+1), if advection is negligible and $\mathfrak{I}$ is a prescribed flux, F_o, then Eq. (5.21) becomes

$$\left[-\varphi D'\frac{\partial C}{\partial x}\right]_0 = F_o \tag{5.23}$$

However, if that same source of material into the sediment volume occurs at the base of the model, say x=L, then n=-1, and the boundary condition is

$$\left[\varphi D'\frac{\partial C}{\partial x}\right]_L = F_L \tag{5.24}$$

Notice the difference in sign between Eqs (5.24) and (5.23), even though both represent sources to the volume. All this fussing with the sign of n reflects the fact that the diffusive flux on the left-hand side of these equations is a vector quantity; i.e. it has a direction. By properly defining n, we assure that the flux is in the right direction.

If F_o=0, then Eq. (5.24) reads

$$\left[\frac{\partial C}{\partial x}\right]_L = 0 \tag{5.25}$$

which says that there is no gradient at x=L and, therefore, *no diffusive flux* across this boundary. Equation (5.25) is an example of a homogeneous boundary condition. Further, if L$\rightarrow\infty$, then the concentration remains bounded at infinity (i.e. the concentration is not infinite). Equations (5.23) through (5.25) are generally known as *Neumann conditions* or boundary conditions of the *second kind.* Neumann conditions involve only the derivative of the concentration, and not the concentration itself.

The $\Im$ term can also represent a (heterogeneous) interfacial reaction; e.g. the plane x=d can be the locus of a chemical precipitation (removal) which is governed by linear kinetics, i.e.

$$\Im = -k_{het}(C_s - C(d)) \tag{5.26}$$

where k_{het} is a heterogeneous rate constant and C_s is a saturation constant ($C_s >$ C(d)). Notice that $\Im$ is negative in this case to account for the removal of C. Given that this reaction occurs at the sediment-water interface (x=0, n=+1), and that we can ignore advection, then the correct boundary condition for this situation reads

$$\left[\varphi D' \frac{\partial C}{\partial x}\right]_0 = k_{het}(C_s - C(0)) \tag{5.27}$$

On the other hand, a similar reaction at the base of the model (x=L, n=-1) would lead to

$$\left[\varphi D' \frac{\partial C}{\partial x}\right]_L = -k_{het}(C_s - C(L)) \tag{5.28}$$

Both Eqs (5.27) and (5.28) are known as *Robin's condition* or *third kind boundary conditions*, and occasionally, they are also known as *surface evaporation conditions.* Such conditions involve the unknown concentration and its derivative at a given depth (or at a line in 2-D or on a surface in 3-D).

5.2.3 Conditions at Internal Boundaries

Sedimentary properties and/or processes need not be continuous. For example, discrete changes in porosity are not uncommon at the macroscopic scale, so that φ has a value φ_1 on one side of an interface and φ_2 on the other. Furthermore, biological mixing seems to occur preferentially in a surface layer of about 10 cm in thickness (Chapter 4), so that $D_B(x) = D_B$ for $x < L$ and $D_B(x) = 0$ for $x > L$. The most convenient formulation of diagenetic equations in this case is to write one version valid for one side of an interface and another set valid for the other side. The equations must then somehow be linked in order to obtain continuity of the solutions across the interface. If neither differential conservation equation contains diffusion terms (only first-order derivative with respect to depth), then it is sufficient to state the continuity of the concentration across the interface at the depth x = d,

$$[C(d)]^- = [C(d)]^+ \tag{5.29}$$

where the superscripts - and + indicate the downward and upward sides of the interface, respectively.

If the equations are both second order, i.e. both contain diffusive terms, then in addition to the concentration continuity condition Eq. (5.29), something must be said about the continuity of fluxes. This statement can be derived in a manner similar to the boundary condition for an external boundary. Specifically, consider an arbitrary internal (non-material)† interface, I, between two sediment regions (Fig. 5.2). Again place a small highly deformable balancing volume astride this interface, as shown in Fig. 5.2. The total volume of the balancing volume is $\mathcal{V}$, with a portion $\mathcal{V}_1$ on one side of the interface and $\mathcal{V}_2$ on the other. I' represents the portion of the surface area of I enclosed in the cap, while $\mathcal{S}_1$ is the surface of the balancing volume in sediment region 1 and $\mathcal{S}_2$ is the surface in region 2. The total surface is $\mathcal{S} = \mathcal{S}_1 + \mathcal{S}_2$.

Conservation of a dissolved species within the total volume $\mathcal{V}$ is given by:

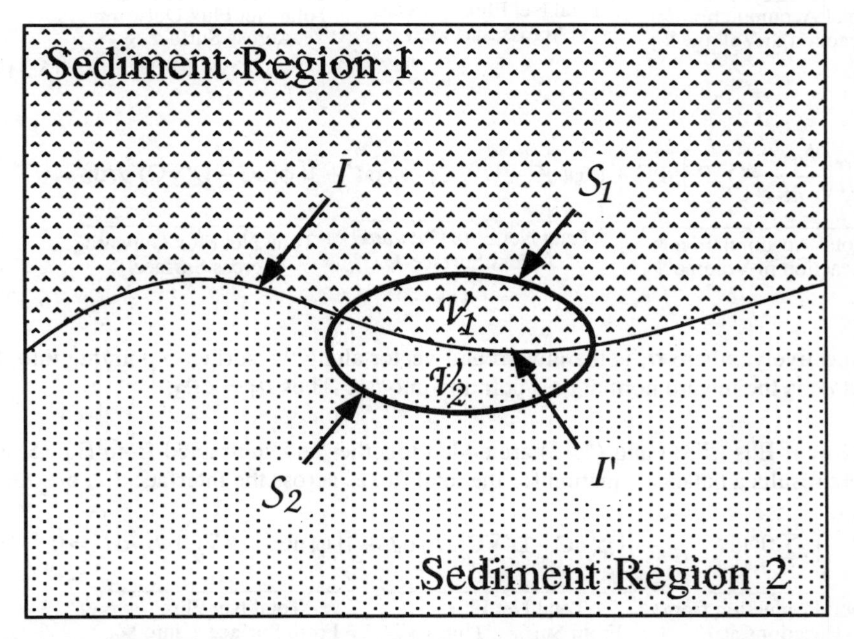

Figure 5.2. Illustration of a small conservation volume, $\mathcal{V}$, surrounding an internal interface, I, that separates two sediment regions with different properties, e.g. different diffusivities or porosities. Note that $\mathcal{V} \equiv \mathcal{V}_1 + \mathcal{V}_2$ and $\mathcal{S} \equiv \mathcal{S}_1 + \mathcal{S}_2$.

---

† This means that the interface is not formed of a distinct set of particles and that it has no mass of its own.

$$\underbrace{\iiint_{\mathcal{V}}\left(\frac{\partial \varphi C}{\partial t}-\sum R\right) d\mathcal{V}}_{\substack{\text{Total Accumulation \&} \\ \text{Reaction in Volume } \mathcal{V}}} + \underbrace{\iint_{I'}\left(\sum \Re\right) dI'}_{\substack{\text{Total Production in } \mathcal{V} \\ \text{From Reactions Along} \\ \text{Interface } I' \text{ Contained} \\ \text{in } \mathcal{V}}} + \underbrace{\iint_{\mathcal{S}} \varphi(uC - D'\nabla C)\cdot n \, d\mathcal{S}}_{\substack{\text{Total Net Flux Outward} \\ \text{Across Surface } \mathcal{S}}} = 0 \qquad (5.30)$$

In Eq. (5.30), the second term accounts for reactions along the interface I. This time the normal n is the inward normal, which explains the change in sign.

At the same time, conservation equations can be written for each sub-volume $\mathcal{V}_1$ and $\mathcal{V}_2$, i.e.

$$\underbrace{\iiint_{\mathcal{V}_1}\left(\frac{\partial \varphi C}{\partial t}-\sum R\right) d\mathcal{V}}_{\substack{\text{Total Accumulation \&} \\ \text{Reaction in Volume } \mathcal{V}_1}} + \underbrace{\iint_{I'} \varphi(uC - D'\nabla C)\cdot n_1 \, dI'}_{\substack{\text{Total Net Flux Outward} \\ \text{Across Surface } I'}} + \underbrace{\iint_{\mathcal{S}_1} \varphi(uC - D'\nabla C)\cdot n \, d\mathcal{S}}_{\substack{\text{Total Net Flux Outward} \\ \text{Across Surface } \mathcal{S}_1}} = 0 \qquad (5.31)$$

and

$$\underbrace{\iiint_{\mathcal{V}_2}\left(\frac{\partial \varphi C}{\partial t}-\sum R\right) d\mathcal{V}}_{\substack{\text{Total Accumulation \&} \\ \text{Reaction in Volume } \mathcal{V}_2}} + \underbrace{\iint_{I'} \varphi(uC - D'\nabla C)\cdot n_2 \, dI'}_{\substack{\text{Total Net Flux Outward} \\ \text{Across Surface } I'}} + \underbrace{\iint_{\mathcal{S}_2} \varphi(uC - D'\nabla C)\cdot n \, d\mathcal{S}}_{\substack{\text{Total Net Flux Outward} \\ \text{Across Surface } \mathcal{S}_2}} = 0 \qquad (5.32)$$

respectively, where n_1 is the inward unit vector along the surface I pointing into $\mathcal{V}_1$, and n_2 is the inward normal on I pointing into $\mathcal{V}_2$. Thus $n_1 = - n_2$.

Now if Eqs (5.31) and (5.32) are added together and subtracted from Eq. (5.30), there results a balance equation for the total fluxes across the interface I', i.e.

$$\underbrace{\iint_{I'}\left(\sum \Re\right) dI'}_{\substack{\text{Total Production from} \\ \text{Reactions at } I'}} - \underbrace{\iint_{I'} \varphi(uC - D'\nabla C)\cdot n_1 \, dI'}_{\substack{\text{Total Net Flux} \\ \text{From Surface } I' \text{ into } \mathcal{V}_1}} - \underbrace{\iint_{I'} \varphi(uC - D'\nabla C)\cdot n_2 \, dI'}_{\substack{\text{Total Net Flux} \\ \text{From Surface } I' \text{ into } \mathcal{V}_2}} = 0 \qquad (5.33)$$

All these integrals have the same limits, and so they can be placed within a single integral. This integral will be equal to zero for all cases if, and only if, the integrand itself is identically zero, i.e.

$$\left[\varphi(uC - D'\nabla C)\cdot n_2\right]^{-} - \left[\varphi(uC - D'\nabla C)\cdot n_2\right]^{+} - \sum \Re = 0 \qquad (5.34)$$

where the superscripts + and - indicate the $\mathcal{V}_2$ and $\mathcal{V}_1$ sides of the interface, respectively. In one Cartesian dimension (i.e. depth only), this balance reduces to:

$$\left[\varphi\left(uC - D'\frac{\partial C}{\partial x}\right)\right]^{-} - \left[\varphi\left(uC - D'\frac{\partial C}{\partial x}\right)\right]^{+} = \sum \Re \tag{5.35}$$

i.e. n_2=+1 in this case. Equation (5.35) says that the total flux on the downward-side of the interface minus that on the upward side must be equal to any production/consumption that occurred at the interface. An example of the use of this equation with $\Sigma\Re \neq 0$ is the reaction between H_2S and O_2 as modelled in Boudreau (1991) and Boudreau and Canfield (1993).

Ii is often the case that there is no interfacial reaction, $\Sigma\Re = 0$, and the porosity is continuous across the interface, $(\varphi)^- = (\varphi)^+$. The interface then represents a boundary separating either two regions where diffusion is different or two distinct chemical regimes. Continuity of porosity implies that $(u)^- = (u)^+$, and Eq. (5.35) reduces to:

$$\left[D'\frac{\partial C}{\partial x}\right]^{-} = \left[D'\frac{\partial C}{\partial x}\right]^{+} \tag{5.36}$$

which says that the diffusive fluxes must be continuous across this interface.

Let us consider another special case of Eq. (5.35), specifically, at the sediment-water interface and with $\Sigma\Re = 0$. As explained in Section 5.3.1 below, there is a zone of water of thickness δ that overlies sediment and in which the concentration is a linear function of height above the interface, i.e.

$$[C(x)]^{-} = C(0) - \frac{(C_\infty - C(0))x}{\delta} \tag{5.37}$$

where C(0) is the concentration at x=0, C_∞ is the concentration in the well-mixed water at x=-δ, and x takes on negative values through this region. From Eq. (5.37),

$$\left[\frac{\partial C}{\partial x}\right]_0^{-} = -\frac{(C_\infty - C(0))}{\delta} \tag{5.38}$$

and Eq. (5.36) becomes,

$$\left[\varphi D'\frac{\partial C}{\partial x}\right]_0^{+} = \beta(C(0) - C_\infty) \tag{5.39}$$

where $\beta \equiv [D]^-/\delta$, which is known as a *mass-transfer coefficient.* This last equation is also a Robin's boundary condition.

Equation (5.34) and its progeny apply equally to solid species (i.e. set $C = B$, $u = w$, $D' = D_B$, etc.) An interesting special case occurs when $(D_B)^- = 0$ in Eq. (5.36). For example, in a two-layer model of diffusional bioturbation, we usually have a mixed layer, $0 < x < L$, where $(D_B)^+ = D_B$, and a non-mixed layer below where $D_B = 0$. If these conditions are substituted into Eq. (5.36), then the boundary condition at the base of the bioturbated zone, $x = L$, is (Guinasso and Schink, 1975)

$$\left[\frac{\partial B}{\partial x}\right]_L = 0 \tag{5.40}$$

which is a Neumann boundary condition.

Remember that in all these cases with an internal interface, one must apply both Eq. (5.29) plus a condition on the flux to resolve the solution's integration constants (see Chapter 6 for some examples).

5.2.4 Boundary Conditions for Unknown Points

In some cases we know that something goes on at a boundary, but we don't know the location of that boundary. For example, during organic matter decomposition, a particular oxidant, such as O_2, may become completely exhausted at some depth. However, the depth where this occurs is part of the solution of the problem, and it is not known *a priori*. Not knowing where the boundary is located means that there is an additional unknown in the problem; as a result, we need to add another equation to solve for this unknown location.

Normally, what is done in this case is to make a statement about both the magnitude of the concentration and the flux at the unknown boundary, say x_o. For example, for O_2 disappearance caused by a zeroth-order reaction in 1-D, i) the concentration at x_o is zero (Bouldin, 1968):

$$C(x_o) = 0 \tag{5.41}$$

ii) but there is additionally no flux of this species across that boundary:

$$vC(x_o) - D\left.\frac{\partial C}{\partial x}\right|_{x_o} = 0 \tag{5.42}$$

Applications of these boundary conditions in diagenetic models can be found in Boudreau and Westrich (1984) and Boudreau and Taylor (1989). Do (1984) provides a detailed mathematical analysis of these boundary conditions.

5.2.5 At a Well-Stirred Reservoir of Limited Size

This situation occurs when the depth of the water column overlying a sediment is similar or smaller than the thickness of the sediment section under consideration, or when a flux-sampling chamber is present at the sediment-water interface. If we assume that the water column or chamber is well-mixed, i.e. ignoring the effects of a diffusive boundary layer (see further below), then the appropriate condition is

$$\frac{V}{A}\frac{\partial C_*}{\partial t} = D\left.\frac{\partial C}{\partial x}\right|_0 \tag{5.43}$$

where C_* is the concentration in the reservoir, V is the volume of the reservoir, and A is the contact area between the sediment and reservoir.

5.2.6 Coupled Boundary Conditions

The presentation given above involved only one species. The reality of diagenesis is that species interact; some species react with each other, while others are produced by reactions between still other species. Some of these interactions occur at boundaries, and the boundary conditions derived above allow for this possibility. As an example, consider two solutes, A and B, with concentrations C_A and C_B, respectively. Species A undergoes first-order consumption at an external boundary at a depth x = d (the bottom), where it produces species B, i.e.

$$A \rightarrow B \tag{5.44}$$

Thus, the boundary condition for species A is:

$$\left[\varphi D'\frac{\partial C_A}{\partial x}\right]_d = -kC_A \tag{5.45}$$

and for species B,

$$\left[\varphi D'\frac{\partial C_B}{\partial x}\right]_d = kC_A \tag{5.46}$$

or, equivalently from Eq. (5.45),

$$\left[-\varphi D'\frac{\partial C_A}{\partial x}\right]_d = \left[\varphi D'\frac{\partial C_B}{\partial x}\right]_d \tag{5.47}$$

Equation (5.46) contains the concentration for species A even though it is a boundary condition on species B; similarly, Eq. (5.47) relates the gradient in two different species. Both these boundary conditions are said to be coupled. There is an infinite variety of such conditions, but they can be deduced from knowledge of the kinetics at the boundary and the essential concept of conservation of mass.

5.2.7 Nonlocal Boundary Conditions

When deriving the equations for conveyor-belt mixing, i.e. Eqs (3.124) and (3.125), a boundary condition, i.e. Eqs (3.126) and (3.127) combined, accounted for the return of the solids to the sediment-water interface. To derive these conditions, start with Eq. (5.22). Next, in the derivation of Eqs (3.124) and (3.125), we stated that the removal of a solid species from a given depth was $K_R(x)B(x)$, where $K_R(x)$ is the removal rate constant. The total amount of solid species removed from the interior of the sediment and dumped to the surface is the integral of $K_R(x)\,B(x)$ over all depths between 0 and L (the base of the mixed layer). The depth integral of $(1\text{-}\varphi)\,K_R(x)B(x)$ is then the component of the surface flux from recycling by the biological conveyor. In addition, there may be a net input from new sedimentating material, $F_o(t)$; thus,

$$\Im_B = F_o(t) + \int_0^L (1-\varphi)\,K_R(x)\,B(x)\,dx \tag{5.48}$$

Furthermore, a diffusive mixing flux is assumed to be absent in this case ($D_B = 0$); consequently, substitution into Eq. (5.22) leads to

$$[(1-\varphi)wB]_0 = F_o(t) + \int_0^L (1-\varphi)\,K_R(x)\,B(x)\,dx \tag{5.49}$$

which is Eq. (3.127). This is an example of a nonlocal boundary condition, as it involves values of the concentration, B (or equivalently the solute concentration C), at more than one depth. The integral has a single value, but it is calculated over an interval of x values.

5.3 Boundary Condition Parameters

5.3.1 Mass-Transfer Coefficients for Solutes

Equation (5.39) contains a parameter, β, called the *mass-transfer coefficie*nt, that has units of a velocity. Multiplying β by the difference in concentration between the sediment-water interface and the overlying water produces a flux that balances the molecular diffusive flux on the right-hand side Eq. (5.39). The obvious questions are, what is the meaning of this velocity, β, and can it be calculated *a priori*?

While the mass-transfer coefficient can have a variety of meanings in different environmental situations, it is primarily used in diagenetic studies to account for the exchange of solutes between porewaters and overlying waters. To give further meaning to this statement, we need to introduce the theory of *boundary-layer mass-*

transfer (consult Bird et al., 1960; Sherwood et al., 1974; Sideman and Pinczewski, 1975; Davies, 1977; Thibodeaux, 1979; Boudreau and Guinasso, 1982; Cussler, 1984; Dade, 1993).

Transport of solutes in a fluid is intimately connected with the motion of that fluid. In a stagnant liquid, molecular diffusion is the only means of solute mass transport. In slowly moving water, advection becomes an additional mode that supplements diffusion. As the flow velocity of the fluid increases, there comes a point at which the motions of the liquid become unstable, and random velocity fluctuations are amplified. This is turbulent flow, and the random motions themselves engender another means of solute transport, called turbulent or eddy diffusion. In the vicinity of the sediment-water interface, all three phenomena interact in such a way that the exchange of solutes between sediment and overlying water can be characterized by a mass-transfer coefficient.

Engineers and physicists have advanced a number of models to explain the mass-transport phenomenon between an infinite flat (impermeable) plate and an overlying turbulently flowing fluid (water in this case). We will examine two of these: classical boundary-layer theory and periodic boundary-layer (bursting) theory.

Classical Boundary-Layer Theory. A turbulent flow over an infinite flat surface can be described by a time-averaged (mean) flow velocity, from which there are, at any moment, random fluctuations. Momentum is transported not only by the mean flow and molecular (kinematic) viscosity, ν, but also by these fluctuations, called *eddies*. To quantify this eddy component of the momentum flux, we can do a Reynolds decomposition and averaging of the equations of motion (e.g. Schlichting, 1968; Sideman and Pinczewski, 1975; Boudreau and Guinasso, 1982). As a result, the component of momentum transport perpendicular to the interface from the eddies, τ', is given by

$$\tau' = -\overline{\left(u'_y u'_x\right)} \tag{5.50}$$

where u'_y and u'_x are the fluctuating velocity components in the direction parallel and perpendicular to the surface, respectively, and the overbar indicates time averaging. (The time average of the product of velocity fluctuations is non-zero, as explained in Chapter 2.)

This transport by random motions can be modelled as a type of diffusion, defined by an eddy viscosity, ε_m (Boussinesq, 1877); thus,

$$\overline{\left(u'_y u'_x\right)} = -\varepsilon_m \frac{\partial \bar{u}_y}{\partial x} \tag{5.51}$$

where $\bar{u}_y$ is the mean velocity in the direction parallel to the plate, i.e. y.

Equation (5.51) is the simplest possible way of resolving the unknown eddy velocity correlation, and it is known as a closure scheme. There are nonetheless more sophisticated methods based on kinetic-energy balances, and some of these have been applied to natural boundary layers (Rahm and Svensson, 1989; Svensson and Rahm, 1991; Dade, 1993). Yet, Eq. (5.51) and the simple treatment that follows appear to be sufficiently accurate for most problems of sediment-water exchange.

In the turbulent flow far above a stationary surface, the eddy viscosity is many orders of magnitude larger than the molecular viscosity, $\varepsilon_m >> \nu$. However, near the wall, friction retards the motion of the fluid. Intermolecular forces cause the fluid to adhere to the surface and, as a consequence, the velocity of the fluid must be zero at the interface. This is referred to as the *no-slip* condition. The forces that act to retard the flow also dampen the velocity fluctuations, such that u'_y and u'_x both disappear at the interface. As a result, τ' must vanish at the wall. There is little to suggest that this is not done in a smooth way, so that ε_m is a continuous function of height x. Furthermore, there must be a height at which $\varepsilon_m = \nu$. This is the top of the so-called *viscous sublayer*, where molecular viscosity becomes more important than eddy viscosity in transporting momentum (Fig. 5.3). Empirical evidence shows that the top of the viscous sublayer occurs approximately at the point where $xu_*/\nu \approx 10$, in which u_* is the shear velocity of the flow, i.e. the square-root of the total shear stress on the surface divided by the fluid's density.

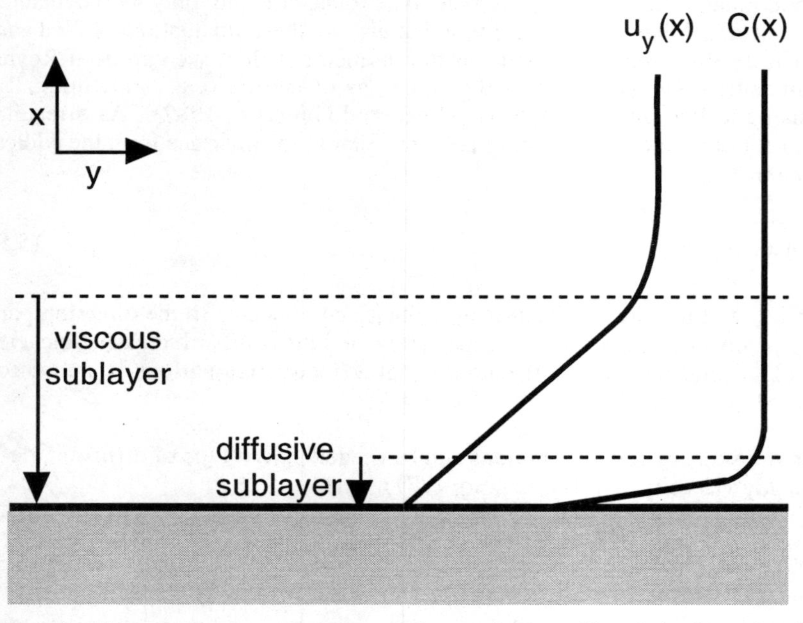

Figure 5.3. Schematic representation of the vertical profile of the lateral velocity component, $u_y(x)$, and the concentration profile, C(x), as a function of height above a solid surface. It is assumed that the solute is not undergoing chemical reactions in the boundary layer, but that it is consumed at the surface. The tops of the viscous and diffusive sublayers are indicated by the labeled dashed lines.

Any solute in a turbulent flow is also transported by the mean advective flow, by molecular diffusion, and by eddy diffusion. The latter results from averaging the product of the velocity and concentration fluctuations, i.e.

$$\overline{\left(u_x' C'\right)} = -\varepsilon_c \frac{\partial \overline{C}}{\partial x} \tag{5.52}$$

where ε_c is the solute eddy diffusivity. There has been much debate about the relationship between ε_m and ε_c, but for most environmental problems, that is largely a detail, and we can consider them equal.

The solute eddy diffusivity, ε_c, is a function of height above the interface, for the same reasons that ε_m changes. However, molecular diffusion coefficients of dissolved species, D, are much smaller than the kinematic viscosity of water, ν, i.e. $D \ll \nu$. This means that the region where molecular viscosity dominates momentum transport extends further from the surface than does the region where molecular diffusion dominates the chemical flux. Although momentum transfer is accomplished primarily by viscous interactions below $xu_*/\nu \approx 10$, the small turbulent motions that persist in this layer continue to be the principal mechanism of mass transfer. At some point that is much closer to the sediment surface, the decreasing intensity of the turbulent motions allows molecular diffusion to finally control mass transfer. The region below this point is the *diffusive sublayer* (Fig. 5.3).

Most boundary layer researchers agree that the ratio of the thickness of the viscous sublayer, δ_V, to the diffusive sublayer, δ_D, is (approximately) proportional to the third root of the ratio of the kinematic viscosity to the molecular diffusivity, usually called the Schmidt number, Sc,

$$\frac{\delta_V}{\delta_D} \propto Sc^{1/3} \tag{5.53}$$

where $Sc = \nu/D$ (Sideman and Pinczewski, 1975). As D is of the order of 10^{-5} cm^2 s^{-1} and ν is about 10^{-2} cm^2 s^{-1}, $Sc \approx 1000$ for surficial waters; thus, the diffusive sublayer is about 10 times thinner than the viscous sublayer.

Within most of the diffusive sublayer, the concentration gradient is linear (Fig. 5.4). This can be proven by considering the conservation equation for a non-reacting species in the lower part of this sublayer. With no net advection, no reaction, and $D \gg \varepsilon_c$, the concentration is governed by

$$D \frac{d^2 C}{dx^2} = 0 \tag{5.54}$$

which can be integrated twice to obtain C as a linear function of x. Confirmation of the time-averaged linearity of concentration profiles in most of the diffusive sublayer was provided by Lin et al. (1953) using optical measurements of changes in refractive index and directly using micro-electrodes (e.g. Jørgensen and Revsbech, 1985; Gundersen and Jørgensen, 1990). Indirect evidence for benthic boundary layers was also provided by the alabaster plate dissolution experiments conducted by Santschi et al. (1983, 1991), as reinterpreted by Opdyke et al. (1987).

As illustrated in Fig. 5.4, the diffusive flux across the diffusive sublayer region of the boundary layer can be approximated as

$$F_D = \frac{D}{\delta_e}(C_o - C_\infty) \tag{5.55}$$

where δ_e is the thickness of an effective diffusive sublayer that has a completely linear gradient and has the same flux as the real diffusive sublayer of thickness, δ_D, (Fig. 5.4). In this equation C_o is the concentration at the interface and C_∞ is that in the overlying turbulent flow region, where there is no perceptible gradient in C.

In practice, we simplify this equation by defining the mass-transfer coefficient as

$$\beta \equiv \frac{D}{\delta_e} \tag{5.56}$$

to obtain,

$$F_D = \beta(C_o - C_\infty) \tag{5.57}$$

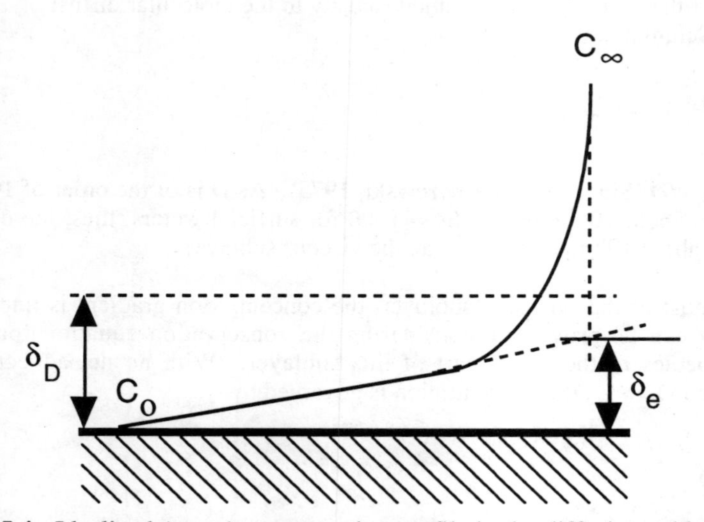

Figure 5.4. Idealized (mean) concentration profile in the diffusive-sublayer region, showing the thickness of the true diffusive sublayer, δ_D, and an effective sublayer of thickness, δ_e, that has a completely linear gradient that produces the same flux.

The mass transfer coefficient has also been called the *piston velocity*, although that term seems more appropriate to gas transfer studies.

Engineers have expended much effort to develop predictive empirical formulas for β based on flow conditions. One of the best of these correlations is that provided by Shaw and Hanratty (1977), i.e.

$$\beta = 0.0889\, u_* \, Sc^{-0.704} \tag{5.58}$$

where u_* is again the shear velocity and Sc is the Schmidt number. For reference, this equation predicts diffusive sublayers of the order of 1 mm for shear velocities of 0.1 cm s^{-1}.

Periodic Boundary Layer Models. The mechanism for eddy dampening near the wall suggested in classical boundary layer theory is not accurate. Flow visualization experiments conducted by Corino and Brodkey (1969), and many others since, show that the lower boundary layer is composed of a mosaic of rolling motions, periodically disrupted by violent movement of water, called *bursts*. The frequency of bursts decreases as the surface is neared. Interestingly enough, when the motions are time-averaged, velocity profiles result that are identical to those of the classic theory.

Mass transfer theories have been developed for this more realistic picture of the interfacial region (e.g. Pinczewski and Sideman, 1974; Wood and Petty, 1983). These theories share common elements with earlier theories of periodic boundary layer structure, called *surface renewal* theories (Danckwerts, 1951), but they utilize contemporary information. The model advanced by Pinczewski and Sideman (1974) is representative of this type, and its development is repeated here.

At high Schmidt numbers, only high energy eddies will penetrate the region of fluid classically called the diffusive sublayer, i.e. about $\delta_D u_*/\nu \approx 1$. When an eddy reaches the surface, it flows along the interface until it is ejected by a "bursting" event, which replaces it with a new eddy. Mass transfer from the eddy to the interface is governed by the equation,

$$u_y \frac{\partial C}{\partial y} = D \frac{\partial^2 C}{\partial x^2} \tag{5.59}$$

with boundary conditions

$$C = C_o \qquad x = 0 \tag{5.60}$$
$$C = C_\infty \qquad x \to \infty \tag{5.61}$$
$$C = C_\infty \qquad y = 0 \tag{5.62}$$

where the point y = 0 is the place the downward-moving eddy first encounters the surface. The lateral velocity, u_y, is given by the equation

$$u_y = \frac{xu_*^2}{\nu} \tag{5.63}$$

in the viscous sublayer. The solution to Eq. (5.59) subject to these boundary conditions is a problem of long standing in chemical engineering (for example, problem 17.J, p. 551, Bird et al., 1960). Without getting into the details of its derivation, this solution requires that the instantaneous mass-transfer coefficient at any point along the portion of the wall covered by an eddy have the value

$$\beta(y) = 0.538\left(\frac{D^2 u_*^2}{y\nu}\right)^{1/3} \tag{5.64}$$

which indicates that the transfer is immediate at the leading edge of the eddy (i.e. zero resistance) and decreases as the concentration boundary layer develops in the eddy with distance along the surface.

The distance an eddy moves among the surface before ejection is y_e. The average mass-transfer coefficient for that eddy on the surface is then

$$\beta = \frac{1}{y_e}\int_0^{y_e} \beta(y)\,dy \tag{5.65}$$

or, with Eq. (5.64),

$$\beta = 0.807\left(\frac{D^2 u_*^2}{y_e \nu}\right)^{1/3} \tag{5.66}$$

The mean horizontal velocity of an eddy in the diffusive boundary layer region is about 0.5 u_*. Pinczewski and Sideman (1974) argue that the time, T_e, an eddy spends at the interface is given by

$$T_e = \frac{K\nu}{bu_*^2} \tag{5.67}$$

where K is an empirical proportionality constant of magnitude 243 s^{-1}, based on the data of Corino and Brodkey (1969) and others. The parameter b in this equation is the fraction of generated eddies that actually penetrate to the wall. Based on observational data, Pinczewski and Sideman (1974) set b = 0.07; i.e. 7% of the eddies make it to the wall.

It is now possible to calculate y_e as

$$y_e = \bar{u}_y T_e \tag{5.68}$$

where the velocity in this equation is the mean velocity of an eddy. With Eq. (5.67),

$$y_e = 1740 \frac{\nu}{u_*} \tag{5.69}$$

Substitution of Eq. (5.69) into Eq. (5.66) produces the final result for the mass-transfer coefficient:

$$\beta = 0.0671\, u_*\, Sc^{-2/3} \tag{5.70}$$

This relation has a coefficient about 16% smaller than that favored by Santschi et al. (1991). A somewhat different treatment of the same problem by Wood and Petty (1983) leads to the relationship

$$\beta = 0.0967\, u_*\, Sc^{-7/10} \tag{5.71}$$

While both Eqs (5.70) and (5.71) are slightly different in form than Eq. (5.58), the predictions from these three equations are quite similar, as illustrated in Fig. 5.5.

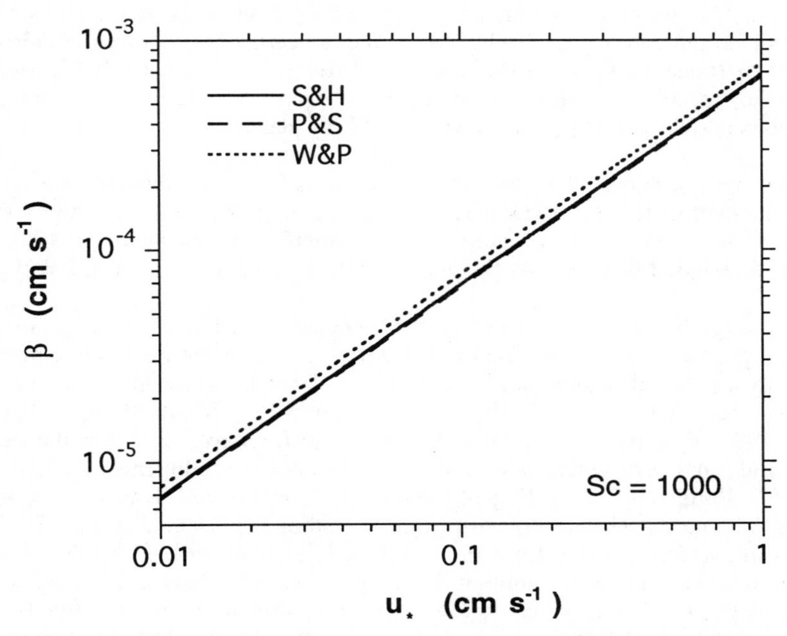

Figure 5.5. Mass-transfer coefficient, β, as a function of the shear velocity, u_*, as predicted by the formulas advanced by Shaw and Hanratty (S&H, Eq. (5.58)), Pinczewski and Sideman (P&S, Eq. (5.70)), and Wood and Petty (W&P, Eq. (5.71)), for a Schmidt number, Sc, equal to 1000.

This near equality of the predictions from these two types of models, classical and periodic, should not come as a surprise. Both have empirical constants that are adjusted to fit the measured mass-transfer data. As they both must explain the same data, they should be numerically equivalent. Secondly, these two theories are not incompatible. To quote Sideman and Pinczewski (1975, p. 51):

> "Although it may first appear that there is little in common between the approach based on the classical concept of an eddy transport coefficient and that based on the concept of surface renewal [bursts], the two concepts are by no means mutually exclusive. Rather, we may consider the former to represent a macro-scale study of the phenomena, while the latter is a micro-scale analysis limited to the viscous transport region. In a sense, these two approaches are complementary." (My addition in square brackets.)

Even more to the point, Davies (1977, p.7) states unequivocally that:

> " ... the 'bursts' of fluid are physically the 'lumps' of fluid envisaged in the eddy-mixing length theory of mean eddy diffusivities."

Therefore, the bursting phenomenon supplies a mechanistic understanding of mass transfer, but it does not negate past work. Furthermore, it is quite proper to employ classical theory as long as the time scale of interest is much longer than the time span of a few bursting events, i.e. a few hundred seconds at most, based on Blackwelder and Haritonidis (1983). As remarked earlier by Townsend (1956), the concept of transport by means of an eddy diffusion is useful in that it yields a successful (predictive) description of mass transfer, even if it bears little resemblance to the actual physical processes taking place.

Before we can equate the flat surface mass-transfer coefficients to those at the sediment-water interface, two issues must be faced: sediments are permeable (if only weakly so for muds), and the sediment-water interface is not necessarily smooth. How do these conditions modify the picture of the boundary layer painted above?

Slip at the Sediment-Water Interface. First, because a sediment bed is permeable, the no-slip condition may not hold at the water-sediment interface. Penetration of flow into a permeable substrate is a problem of some engineering interest (e.g. Beavers and Joseph, 1967; Saffman, 1971; Liu, 1979; Nield, 1983; Haber and Mauri, 1983; Hsu and Cheng, 1991; Sahraoui and Kaviany, 1992; Chellam et al., 1992), and it has been considered in the context of bottom sediments by McClain et al. (1977), Boudreau (1981), Nagaoka and Ohgaki (1990), and Svensson and Rahm (1991). The picture that emerges from these studies is presented in Fig. 5.6. The flow in the sediment away from the interface is Darcian, driven by the pressure gradient related to the water column flow, e.g. geostrophic balance, gravity waves, surface slope. The boundary layer flow above the sediment is linked to this Darcian flow through a transitional layer called a Brinkman layer (after the person who developed the hydrodynamic equations for this layer). In the Brinkman layer, shear from the overlying flow diffuses into the bed by molecular viscosity.

McClain et al. (1977) estimate that the depth of shear penetration (the thickness of the Brinkman layer), ℓ_v, is of the order of

$$\ell_v \sim \mathbf{k}^{1/2} \tag{5.72}$$

where $\mathbf{k}$ is the permeability (units of length squared). Equations (4.124) through (4.126) suggest values for $\mathbf{k}$ of about 10^{-9} cm^2 for a mud of grain size d_p of 10^{-4} cm and $\varphi = 0.9$. This means a penetration of only 3×10^{-5} cm, or about a grain size. At $d_p = 10^{-2}$ cm and the same porosity, $\mathbf{k}$ has increased to around 10^{-5} cm^2. This time, ℓ_v is larger, being of the order of 3×10^{-3} cm, or again something like the grain size. By comparison, diffusive sublayers measure about 0.1 cm in thickness. It would appear that penetration is important only in sands and coarser sediments (e.g. Nagaoka and Ohgaki, 1990); consequently, the classical boundary layer is an accurate description for any sediment containing an appreciable amount of mud. For sands, there are other transport processes (see Chapter 3) that make the question of boundary-layer impedance moot.

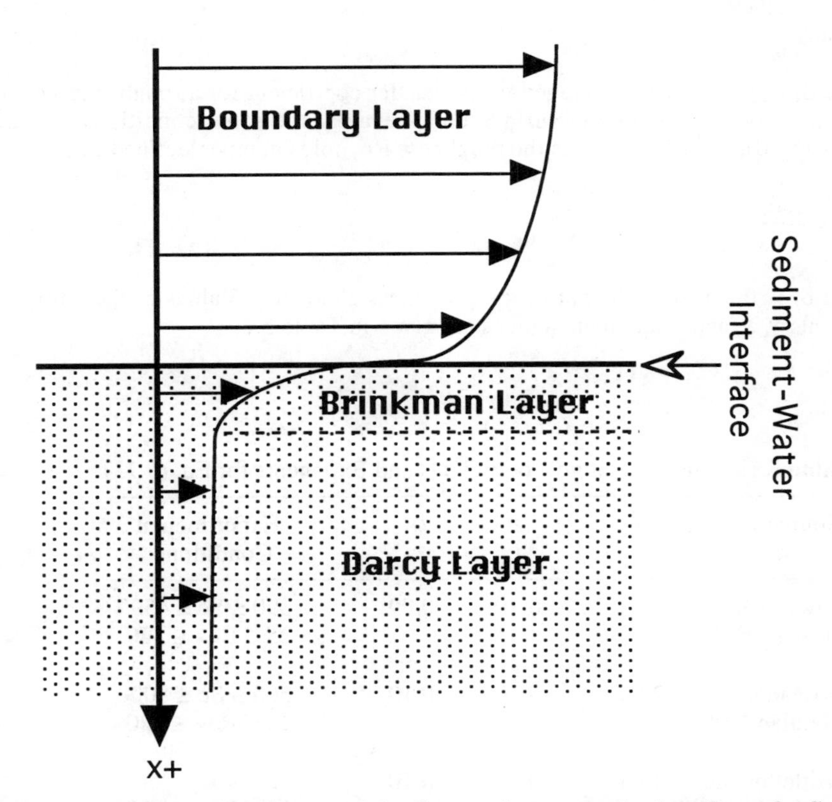

Figure 5.6. Schematic diagram of the flow regimes near the interface of a flowing fluid and a permeable bed. The diagram assumes the existence of a pressure-driven flow.

Interfacial Roughness. Contrary to some literature statements, the presence of roughness does not destroy a boundary layer, either for momentum or mass. Jørgensen and Des Marais (1990) have shown definitively that surface roughness elements have diffusive boundary layers for dissolved O_2, just like the sheltered areas in between. The true effect of roughness is to rob us of the use of Eqs (5.58), (5.70), and (5.71) as accurate predictors of average mass transfer at the surface; in other words, the geometry of the diffusive sublayer becomes so complicated because of the roughness that the smooth surface relations aren't accurate.

Engineers have attempted to find ways to predict the effects of roughness on mass transfer. The unique geometry of random rough surfaces makes this difficult, and the resulting formulas are really situation specific. Keeping this in mind, Boudreau and Scott (1978) and Dade (1993) have searched the literature and discovered three correlations that extend the use of the flat surface equations to the rough surface case. They are of the form

$$\frac{\beta_{rough}}{\beta_{smooth}} = \gamma \, Re_*^a \, Sc^b \tag{5.73}$$

where β_{rough} is the area-averaged mass-transfer coefficient for a rough surface, and β_{smooth} is one for a smooth surface under exactly the same conditions, i.e. Eq. (5.58), (5.70), or (5.71). Re_* is the roughness Reynolds number, defined as

$$Re_* \equiv \frac{u_* \delta_r}{\nu} \tag{5.74}$$

where δ_r is the average height of the roughness elements. Values of the empirical constants γ, a, and b and their sources are given in Table 5.1.

Table 5.1. Some Literature Coefficients and Exponents for Eq. (5.73)

Source	γ	a	b	Experimental Conditions
Dawson and Trass (1972)	1.94	-0.10	0.09	$390 \le Sc \le 4590$ $25 \le Re_* \le 120$
Grimanis and Abedian (1985)	2.00	-0.20	0.10	$500 \le Sc \le 8000$ $25 \le Re_* \le 120$
Postlethwaite and Lotz (1988)	1.64	-0.14	0.10	$Sc = 447$ $30 \le Re_* \le 120$

These formulas lead to increases in the predicted mass-transfer coefficients and, therefore, in solute fluxes with increasing roughness. Dade (1993) points out that this enhancement could be as high as a factor of 2 to 4 for biologically roughened mud surfaces. Still, even in the face of roughness effects, the mass-transfer coefficient approach retains its utility.

5.3.2 Sediment-Water Flux of Organic Carbon

Another important parameter in many boundary conditions is the prescribed flux, i.e. $\Im_B$ in Eqs (5.23) and (5.26). There is as yet no theory that can supply an *a priori* value for such a flux; however, it is possible to obtain empirical correlations for the organic matter flux, $\Im_C$, under oceanographic conditions. These empirical correlations link $\Im_C$ to other more easily measured environmental parameters.

As a first example, Henrichs and Reeburgh (1987) advance the relation

$$log(\Im_C) = 0.69\, log(w_\infty) + 2.27 \tag{5.75}$$

where $\Im_C$ is in units of gC m^{-2} yr^{-1}, and w_∞ is the burial velocity of the sediment after the initial compaction near the surface, in units of cm yr^{-1}. The r^2 for this correlation is 0.95. Alternatively, Tromp et al. (1995) offer the correlation

$$log(\Im_c) = 2.49 + 0.85\, log(w) \tag{5.76}$$

with $\Im_C$ in the same units, but w in cm kyr^{-1}.

Middelburg et al. (in press) supply a predictive empirical relation for $\Im_C$ as a function of oceanic depth, Z (in meters), i.e.

$$\Im_c = 1.8 \times 10^{(-0.509 - 0.000389 * Z)} \qquad (r^2 = 0.69,\ n = 80) \tag{5.77}$$

where $\Im_C$ has units of mmol Carbon cm^{-2} yr^{-1}.

5.3.3 Flux of Oxidants at the Sediment-Water Interface

As with the organic matter flux, no theory exists to predict such fluxes *a priori*, but empirical correlations can be found. One such set of correlations was presented by Canfield (1989,1991,1993), as recalculated by Boudreau and Canfield (1993). It relates the flux of O_2 and SO_4 at the sediment-water interface to the sedimentation, S (units of g cm^{-2} yr^{-1}),

$$\Im_{O_2} = 0.752\, \mathsf{S}^{0.66} \tag{5.78}$$

and

$$\mathfrak{I}_{SO_4} = 0.983\, S^{0.57} \tag{5.79}$$

where both fluxes are in units of mmol cm^{-2} yr^{-1}.

Middelburg et al. (in press) supply three related correlations for the integrated rates of carbon oxidation by oxygen, nitrate, and sulfate reduction as functions of oceano-graphic depth, Z, i.e.

$$\int R_{O_2} = 4.0 \times 10^{(-0.827-0.000305*Z)} \qquad (r^2{=}0.58,\ n{=}35) \tag{5.80}$$

$$\int R_{NO_3^-} = 1.9 \times 10^{(-0.956-0.000520*Z)} \qquad (r^2{=}0.64,\ n{=}38) \tag{5.81}$$

and

$$\int R_{SO_4^=} = 1.3 \times 10^{(-0.721-0.000547*Z)} \qquad (r^2{=}0.76,\ n{=}49) \tag{5.82}$$

where Z is in meters. These formulas can be converted to interfacial fluxes of oxygen, nitrate, and sulfate by adopting a stoichiometry for the various reduction reactions. The canonical (Redfield) values are 138/106, 4/5, and 1/2 for oxygen, nitrate, and sulfate, respectively.

6 ANALYTICAL SOLUTION OF STEADY-STATE 1-DIMENSIONAL DIAGENETIC EQUATIONS

... "Fourteenth of March, I think it was," he said.

"Fifteenth," said the March Hare.

"Sixteenth," said the Dormouse.

"Write that down," the King said to the jury; and the jury eagerly wrote down all three dates on their slates, and then added them up, and reduced the answer to shillings and pence.

(Lewis Carroll, *Alice's Adventures in Wonderland*, 1865)

6.1 Steady State and Uni-Dimensionality

The development in Chapter 3 demonstrates that most diagenetic conservation equations are naturally formulated as multi-dimensional, time-dependent partial-differential equations. Yet, much of diagenesis need not be described in such complicated terms. Two conditions/assumptions lead to substantial simplifications, i.e. *steady state* and *uni-dimensionality.*

As further explained in Chapter 3, *steady state* (with respect to the sediment-water interface) occurs when both concentration and the parameter functions (i.e. diffusion coefficients, burial velocities, etc.) are independent of time. The conditions that permit a steady state to exist in sediments have been investigated extensively by Lerman (1975, 1979), Berner (1980), Boudreau (1986a), and Rabouille and Gaillard (1990). These authors observe that perturbations (random and otherwise) are the rule in natural sedimentary systems, rather than the exception. Nevertheless, a steady state may exist in at least two situations.

First, if the fluctuations have a mean period very much shorter than the other time scales for diffusion, advection, and reaction, then it is possible to time-average over the variability. An example is the derivation of biodiffusion from the biologically induced motions of sediment grains (Chapter 3). Given no other time-dependent forcings, the time-derivative term may then disappear, and a steady state can exist. In this comparison, the time scale for diffusion is defined by the Einstein-Smoluchowski relations, Eq. (4.25), which tell us that this scale is dependent on the square of the size of the system being modelled. The advective time scale is important only where advection is the dominant transport process, and it is given by dividing the size of the system by the advective velocity. Finally, the reactive time scale is given by the inverse of the apparent first-order reaction rate constant.

Secondly, a *pseudo-steady state* can be assumed when the time between fluctuations is much longer than the other process time scales. In this situation, the perturbation decays long before the next disruption to the system. Similarly, if the forcing is continuous, but has a very long relative time scale, then the sediment can be described by a series of steady states (Berner, 1980).

Uni-dimensionality means that we need consider only the changes with depth in a sediment. Berner (1980, p.9) popularized this assumption by noting that "... changes exhibited by sediments on the scale of a few hundred meters (as represented by early diagenesis) are predominantly vertical in nature (e.g., bedding, chemical gradients), and depth is uniquely different from the lateral dimensions ... ." One needs to be careful in applying this assumption, but if it is basically true, then the spatial terms in the lateral directions can be ignored.

Given both steady state and uni-dimensionality, then a species conservation equation reduces to an ordinary differential equation (ODE), e.g.

$$\frac{d}{dx}\left[\varphi(x)D(C,x)\frac{dC}{dx} - \varphi(x)v(x)C\right] - R(\varphi, C, x) = 0 \qquad (6.1)$$

where $\varphi(x)$, $D(C,x)$, $v(x)$, and $R(\varphi,C,x)$ are generic volume fraction, diffusion coefficient, advective velocity, and net reaction rate, respectively. The present chapter discusses the solution of Eq. (6.1) in its various forms.

6.2 Dimensional Analysis: Scaling and Simplification

Before trying to solve a differential transport-reaction equation like Eq. (6.1), it is highly advisable to analyze this model equation(s) in *scaled form*, because you may be solving a more complicated model than you really need to. Some terms may contribute only negligibly to the solution, and a simpler approximate solution both is possible and provides all the needed accuracy. Secondly, solving a problem is more than just trying to generate a concentration profile(s). You should also understand the interplay of processes that lead to the generated profile, and much about this can be discovered from the differential equation(s) itself. Finally, you will discover that the non-dimensional parameter groupings derived during the scaling process described below control the final solution; thus, you can become aware of these beforehand.

6.2.1 Scaling

Non-dimensionalization removes the units from an equation by dividing or multiplying by suitable constants. Because this can be a largely arbitrary process, it may not be very productive or particularly informative with respect to determining which terms dominate an equation. The solution to this dilemma lies in choosing meaningful reference quantities (constants), which are called *scales*. Scales are intrinsic to the problem. If one chooses proper scales, then each term in the dimensional equation can be transformed into 1) the product of a constant dimensionless factor, which closely estimates the term's order of magnitude, and 2) a dimensionless term, i.e. unitless second derivative, unitless first derivative, etc. of unit order of magnitude (Lin and Segel, 1974; Bear and Bachmat, 1992). So how does one recognize the scales in a given problem? Generally, they are the intrinsic measures of the maximum variation in the variables; specifically:

i) For an independent *spatial variable*, x, there may be more than one choice. Firstly, if the problem is defined on a specific finite interval, say $0 \leq x \leq L$, then normally one takes the scale, x_{ref}, to be L; alternatively, if there is an inhomogeneous term in the equation(s), then one can take the inverse of the factor that multiplies x in this inhomogeneous term, e.g. in

$$D\frac{d^2C}{dx^2} - v\frac{dC}{dx} - kC + R_o\, exp(-\alpha x) = 0 \tag{6.2}$$

one could take $1/\alpha$ as the length scale. If both L and one or more α-type inhomogeneous parameters are defined, normally one takes L, although there may be a specific advantage in using one of the alternative scales. For example, if most of the interesting diagenesis occurs on the scale of $1/\alpha$ and $L > 1/\alpha$, then $1/\alpha$ is probably the better choice. If L and $1/\alpha$ are absent from a problem, certain combinations of other parameters can supply a length scale, e.g. $(D/k)^{1/2}$, D/v, v/k. If one believes that the problem might be dominated by a balance between diffusion and reaction, then take $(D/k)^{1/2}$; if the balance might be between diffusion and advection, D/v is appropriate, etc. One can always check these assumptions *a posteriori.* Because there may be more than one reaction in a problem, different reactions may dominate on different scales, which may lead one to divide the problem into separate, but connected problems.

ii) For the independent *temporal variable*, t, we face similar choices: if the problem is defined on a finite time interval, i.e. $0 \leq t \leq T$, then take $t_{ref} = T$; otherwise, if there is an inhomogeneous term in the equation(s), or in the boundary conditions, then take the inverse of the factor that multiplies t in this inhomogeneous term. For example, a time-dependent boundary condition can contain a frequency γ,

$$C(0,t) = C_o \cdot \frac{1}{2}\left[sin(\gamma t) + 1\right] \tag{6.3}$$

Then try $1/\gamma$ as the time scale. If T is not specified and time constants are absent, certain combinations of other parameters can supply a time scale, e.g. L^2/D, L/v, 1/k. If it is believed that the problem might be dominated by diffusion, then take L^2/D, etc. (Note that for some pure diffusion problems, no time scale can be assigned.)

iii) The dependent *concentration variable*, C, should be scaled by the maximum value C can attain; for many species, this is the value at the sediment-water interface, as given by the boundary conditions; for others, there may be an equilibrium value, C_s, attained at great depth. A problem arises when C is zero at the sediment-water interface and is produced by conversion from another species, e.g. H_2S from SO_4. In this case, the interfacial value of the producing species may be a good choice. For those species produced by an inhomogeneous reaction, e.g. R_o *exp*(-αx), one may estimate a maximum C as $R_o/(D\,\alpha^2)$.

iv) For *derivatives* of concentration (i.e. dC, ∂C, d^2C, and ∂^2C), use the maxi-mum *change* in C, i.e. designated ΔC. If C is zero at the sediment-water interface and is produced by an inhomogeneous reaction, e.g. R_o *exp*(-αx), one may estimate a maximum change as $\Delta C \approx C(0,t) \pm R_o/(D\,\alpha^2)$.

As an example of scaling analysis, consider the steady-state distribution of a dissolved species that is influenced by diffusion, porewater advection, second-order homogeneous decay, and inhomogeneous production (all at constant porosity), i.e.

$$D\frac{d^2C}{dx^2} - v\frac{dC}{dx} - kC^2 + R_o\, exp(-\alpha x) = 0 \tag{6.4}$$

subject to the boundary conditions,

$$C\,(x{=}0) = C(0) \tag{6.5}$$

$$\left.\frac{dC}{dx}\right|_{x\to\infty} = 0 \tag{6.6}$$

The above is a fairly common type of problem with no set interval length L; the guide given above suggests that $x_{ref} = 1/\alpha$. The maximum value of C is not readily apparent because there is both nonlinear consumption and inhomogeneous production; however, some thought shows that the homogeneous decay and the exponentially decreasing production guarantee that $C \to 0$ as $x \to \infty$. Thus, C_o is a first approximation to the maximum value of C. There could be an intermediate depth maximum in C, so that one could be more exact by adding $R_o/(D\,\alpha^2)$ to C_o, but little is gained actually by doing this. Likewise, let the maximum ΔC be C_o. As a consequence, both dimensionless depth and concentration can be defined as:

$$\xi = \alpha x \tag{6.7}$$

and

$$\Theta = C/C(0) \tag{6.8}$$

and Eq. (6.4) becomes:

$$\frac{d^2\Theta}{d\xi^2} - Pe_\alpha \frac{d\Theta}{d\xi} - Da_2\Theta^2 + R_\alpha \, exp(-\xi) = 0 \tag{6.9}$$

where the dimensionless parameter groupings, Pe_α, Da_2, and R_α, are defined as: $Pe_\alpha \equiv v/(D\alpha) \equiv$ a Peclet number based on $1/\alpha$ as a characteristic length, $Da_2 \equiv kC(0)/(D\alpha^2) \equiv$ a nonlinear Damkohler number for a second-order reaction, and $R_\alpha \equiv R_o/(D\alpha^2) \equiv$ nondimensional maximum rate for zeroth-order reaction. The meaning of these new parameter groupings, or numbers, is quite elementary.

Peclet numbers give the relative importance (ratio) of advective to diffusive transport over a chosen length scale; thus, if $Pe < 1$, then diffusion is dominant over that distance, and if $Pe > 1$, then advection is dominant over the same distance.

Damkohler numbers give the relative importance of reaction to diffusion over the distance contained in Da, so that if $Da < 1$, then diffusion dominates over the effects of the reaction over this distance; and if $Da > 1$, then the effects of reaction dominate over diffusion over the same distance. With more than one homogeneous reaction in a problem, there can be more that one Damkohler number present in its scaled form. Damkohler numbers can also be written to compare advection and reaction (see Domenico, 1977).

Finally, R_α can be interpreted as a measure of the relative importance of inhomogeneous reaction (production/consumption) and diffusion over the chosen reference distance, i.e. in this case $1/\alpha$. For another example of the use of scaling with coupled diagenetic equations, see Boudreau (1990). For another geochemically interesting use of scaling, see Lichtner (1993).

In our example, Eq. (6.9), if $Pe_\alpha << 1$, then the advective term would not contribute much to the balance on a scale defined by $1/\alpha$. It may be advantageous to remove this small advection term, if it is an impediment to exact solution. One can verify that the omitted term is small by solving the reduced equation, and then calculating the size of the excluded term from the answer so obtained. This does not, however, tell us how to correct the solution for the effects of the neglected term. The methodology for calculating this correction is called a *perturbation method.* The details of this technique differ depending on the term being dropped. We will examine only the case in which the undesired term is not the highest derivative in the equation, known as a *regular perturbation.*

One cannot eliminate the highest derivative of a differential equation without great consequences. Careful handling is needed. A hint of the problem this causes is offered by considering the boundary conditions. A second-order ODE, such as Eq. (6.1), satisfies two boundary conditions. Removing the second-order derivative term transforms the problem to a first-order ODE, which satisfies only one boundary condition. Something is radically different. The technique for removing the highest order derivative is called a *singular perturbation.* It is beyond the scope of this book (see instead Cole, 1968, or Lankin and Sanchez, 1970, or Bender and Orszag, 1978). Consequently, we will examine only the elimination of a lower order derivative or reaction term.

6.2.2 Regular Perturbation Method

To explain this method, it's best to use a concrete example. All other problems are done in about the same way. A variation on Eq. (6.4) is worth considering, just to illustrate another way of scaling, i.e.

$$D\frac{d^2C}{dx^2} - v\frac{dC}{dx} - kC^n = 0 \tag{6.10}$$

which is subject to the same the boundary conditions. The first and essential step in a perturbation solution is to scale the equation. There is no explicit length scale to Eq. (6.10) because the equation is valid on the interval $0 \le x < \infty$. However, there are natural scale lengths that remain in the problem related to the interaction of diffusion and reaction, i.e. $L_r \equiv [D/(kC(0)^{n-1})]^{1/2}$, diffusion and advection, i.e. D/v, and advection and reaction, i.e. $v/(kC(0)^{n-1})$. In this case let us choose the diffusion-reaction scale because we will be exploring the common case in which advection is small. Then, the non-dimensional distance and concentration are, respectively,

$$\xi = \left(\frac{kC_o^{n-1}}{D}\right)^{1/2} x \tag{6.11}$$

and

$$\Theta(\xi) = C(x)/C(0) \tag{6.12}$$

With these quantities, the scaled problem becomes

$$\frac{d^2\Theta}{d\xi^2} - Pe_r\frac{d\Theta}{d\xi} - \Theta^n = 0 \tag{6.13}$$

with

$$\Theta(0) = 1 \tag{6.14}$$

$$\left.\frac{d\Theta}{d\xi}\right|_{\xi\to\infty} = 0 \tag{6.15}$$

as the boundary conditions. The Peclet number Pe_r is defined as:

$$Pe_r \equiv \frac{vL_r}{D} \tag{6.16}$$

where, again, L_r is the reaction-diffusion length scale (see above).

Whereas we normally consider Θ (and by inference C) as a function of ξ (i.e. x) only, it is also in a very general sense a function of the parameter Pe_r, even though Pe_r is a constant in any given calculation. The dependence of Θ on Pe_r can be expressed as a power series in Pe_r, i.e.

$$\Theta = \Theta_o + Pe_r\,\Theta_1 + (Pe_r)^2\,\Theta_2 + (Pe_r)^3\,\Theta_3 + ... \tag{6.17}$$

where Θ_o is the solution when $Pe_r = 0$ (i.e. zero advection), and Θ_1, Θ_2, ... are progressively higher order additional solution terms that correct for non-zero Pe_r.

For illustration purposes only, let n = 2 in Eq. (6.13), so that

$$\frac{d^2\Theta}{d\xi^2} - Pe_r\frac{d\Theta}{d\xi} - \Theta^2 = 0 \tag{6.18}$$

with the same boundary conditions. Substitution of Eq. (6.17) into Eq. (6.18) gives

$$\begin{aligned}&\frac{d^2\Theta_o}{d\xi^2} + Pe_r\frac{d^2\Theta_1}{d\xi^2} + (Pe_r)^2\frac{d^2\Theta_2}{d\xi^2} + ...\\ &-Pe_r\frac{d\Theta_o}{d\xi} - (Pe_r)^2\frac{d\Theta_1}{d\xi} - (Pe_r)^3\frac{d\Theta_2}{d\xi} - ....\\ &-\left(\Theta_o + Pe_r\,\Theta_1 + (Pe_r)^2\,\Theta_2 + ...\right)^2 = 0\end{aligned} \tag{6.19}$$

or

$$\begin{aligned}&\left[\frac{d^2\Theta_o}{d\xi^2} - \Theta_o^2\right] + Pe_r\left[\frac{d^2\Theta_1}{d\xi^2} - \frac{d\Theta_o}{d\xi} - 2\,\Theta_o\Theta_1\right]\\ &+ (Pe_r)^2\left[\frac{d^2\Theta_2}{d\xi^2} - \frac{d\Theta_1}{d\xi} - \left(2\Theta_o\Theta_2 + \Theta_1^2\right)\right] + ... = 0\end{aligned} \tag{6.20}$$

For the left-hand side of Eq. (6.20) to always equal zero, each term in square brackets in this equation must itself be equal to zero (given that $Pe_r \neq 0$). This creates a hierarchy of simpler differential equations that can be solved to obtain better and better approximation(s) of the solution that retains the advective term. Specifically, the first set in square brackets gives the lowest order solution, called the *zeroth-order* solution,

$$\left[\frac{d^2\Theta_o}{d\xi^2} - \Theta_o^2\right] = 0 \tag{6.21}$$

with boundary conditions

$$\Theta_o(0) = 1 \tag{6.22}$$

and

$$\left.\frac{d\Theta_o}{d\xi}\right|_{\xi\to\infty} = 0 \tag{6.23}$$

which are the original boundary conditions.

The first-order correction for $Pe_r \neq 0$ can be calculated from the solution of the ODE in the second square-bracketed term in Eq. (6.20), i.e.

$$\left[\frac{d^2\Theta_1}{d\xi^2} - \frac{d\Theta_o}{d\xi} - 2\,\Theta_o\Theta_1\right] = 0 \tag{6.24}$$

with

$$\Theta_1(0) = 0 \tag{6.25}$$

$$\left.\frac{d\Theta_1}{d\xi}\right|_{\xi\to\infty} = 0 \tag{6.26}$$

where Θ_o is already known as the solution of Eq. (6.21). Notice that the boundary condition given by Eq. (6.25) has changed compared with Eq. (6.22) and is now homogeneous. If we can now solve Eq. (6.24) for Θ_1, we would have the next correcting term in the perturbation solution, Eq. (6.7). The best way to solve Eq. (6.24) may be numerical, the results of which could then be fit with a polynomial or spline or exponentials to give the first-order correction. The next higher order correction comes from the ODE for Θ_3. This is obtained by setting the third term in square brackets in Eq. (6.20) to zero. Θ_3 would be subject to boundary conditions identical to Eqs (6.25) and (6.26). This process is continued as needed. (An example of this type of regular perturbation expansion can be found in Sen, 1988.)

6.3 Solution of First-Order ODEs

Having decided which ODE to solve, we must now examine actual methods of solution. Such methods constitute a large literature, which I can hardly begin to explain in this short monograph. Some techniques, outlined below, I have found to be useful when dealing with transport-reaction models. However, my treatment tends to be matter-of-fact, with little theoretical background, and there is space for only a limited number of examples. The reader is advised to consult mathematical texts dedicated to this topic for a more in-depth presentation and analysis, e.g. Ince

(1956), Murphy (1960), Ames (1965), Hildebrand (1976), Boyce and DiPrima (1977), Bender and Orszag (1978), and Zwillinger (1989), along with many others too numerous to name here.

Conservation equations can be in the form of first-order ODEs if no diffusion term is present. The method of solution depends on whether they are linear or not.

6.3.1 Linear Equations

Linear equations have a general solution that applies to all such equations. For example, consider the general first-order ODE,

$$\frac{dC}{dx} + p(x)C = g(x) \tag{6.27}$$

where p(x) and g(x) are defined, continuous functions of x (or equivalently for time by replacing x with t). The concentration in this equation is generic, and it could represent either dissolved or solid concentrations. The general solution to this equation is

$$C = \frac{1}{\mu(x)}\left[\int_x \mu(x')g(x')dx' + c\right] \tag{6.28}$$

where x' is a dummy variable, c is an integration constant, and $\mu(x)$ is an integration factor, which is calculated as

$$\mu(x) = exp\left[\int_x p(x')dx'\right] \tag{6.29}$$

One should not imagine that the integrals in Eqs (6.28) and (6.29) can always be evaluated to give a specific and recognizable function of x, i.e. what is called a *closed-form* solution. Often they cannot be expressed in terms of known functions, in which case one might as well evaluate the ODE numerically (see Chapter 8). However, let us consider a few examples, assuming closed-form solutions are available. Vast tables of integrals exist that may be of considerable help in evaluating Eqs (6.28) and (6.29), e.g. Gradshteyn and Ryzhik (1980).

Example 6-1. Begin by considering the simple and familiar problem of steady-state advection and decay of a solid tracer with constant porosity,

$$w\frac{dB}{dx} + kB = 0 \tag{6.30}$$

with the boundary condition

$$B(0) = B_o \tag{6.31}$$

By comparing this with Eq. (6.27), we deduce that p = k/w and g(x) = 0; thus

$$\mu(x) = exp\left(\int_x \frac{k}{w} dx'\right) = exp\left(\frac{kx}{w}\right) \tag{6.32}$$

and the solution from Eq. (6.28) is

$$B = c\, exp(-kx / w) \tag{6.33}$$

where c is an integration constant. But $B(0) = B_o$, so that $c = B_o$, and

$$B = B_o\, exp(-kx / w) \tag{6.34}$$

which is the well-known solution.

Example 6-2. Now let us modify Eq. (6.30) such that there is also an inhomogeneous production of the tracer given by $g(x) = A\, exp(-\alpha x)$, and

$$w\frac{dB}{dx} = -kB + A\, exp(-\alpha x) \tag{6.35}$$

again with

$$B(0) = B_o \tag{6.36}$$

The integration factor, $\mu(x)$, is still given by Eq. (6.32), so that

$$B = exp\left(-\frac{kx}{w}\right)\left[A\int_x exp\left(\frac{kx'}{w} - \alpha x'\right)dx' + c\right] \tag{6.37}$$

If $\alpha \neq k/w$, then

$$B = \frac{A\, exp(-\alpha x')}{\left(\frac{k}{w} - \alpha\right)} + c\, exp\left(-\frac{kx}{w}\right) \tag{6.38}$$

and, if $\alpha = k/w$, then

$$B = exp(-kx / w)[Ax + c] \tag{6.39}$$

The integration constant c is evaluated with the boundary condition $B(0) = B_o$; i.e. for Eq. (6.38),

$$c = B_o - \frac{A}{\left(\frac{k}{w} - \alpha\right)} \tag{6.40}$$

and, for Eq. (6.39),

$$c = B_o \tag{6.41}$$

Example 6-3. Procedurally, the presence of a porosity gradient represents a slight modification to this treatment. For example, say that instead of Eq. (6.30), we have

$$\frac{d\varphi_s wB}{dx} + \varphi_s kB = 0 \tag{6.42}$$

where φ_s is the solid volume fraction. In this case φ and, consequently, w are known functions of depth, as governed by Eqs (3.66) and (3.68). We also know from Chapter 3 that $d(\varphi w)/dx = 0$ in this case. Thus, Eq. (6.42) can be simplified to

$$w(x)\frac{dB}{dx} + kB = 0 \tag{6.43}$$

The solution of Eq. (6.43) is again given by Eqs (6.28) and (6.29) with

$$p(x) = \frac{k}{w(x)} \tag{6.44}$$

so that

$$B = c\, exp\left(-k\int_x \frac{dx}{w(x)}\right) \tag{6.45}$$

where c is an integration constant, evaluated with the boundary condition. Whether or not the integral in Eq. (6.45) can be evaluated in closed form depends entirely on the mathematical form of w(x). For an example of this type of treatment, see Christensen (1982).

Example 6-4. Finally, consider the equations for total solids and a solid tracer in a sediment layer subject to conveyor-belt mixing without biodiffusion, i.e. Eqs (3.124) through (3.127), which include an integro-differential equation. At steady state and with constant porosity, total solids conservation is given by

$$\frac{dw}{dx} = -K_R(x) \tag{6.46}$$

which defines the burial velocity of the sediment, w(x). If $K_R(x)$ is a constant with depth, K_R, then $\mu = 1$ and

$$w = -\int_x K_R dx + w(0) \tag{6.47}$$

or

$$w = w(0) - K_R x \tag{6.48}$$

where the integration constant is identified as the value of w at x = 0. Equation (6.48) says that the burial velocity decreases linearly through the conveyor-belt zone $0 \le x < L$. The quantity w(0) is not usually known *a priori*, but it can be calculated from Eq. (6.48) by setting x = L, i.e. the depth at which the velocity should be equal to that without biological activity, w_∞; thus,

$$w(0) = w_\infty + K_R L \tag{6.49}$$

The value of w(0) could also have been calculated with the boundary condition given by Eq. (3.126).

For a solid tracer species, the corresponding conservation equation is

$$\frac{dwB}{dx} = -(\lambda + K_R)B \tag{6.50}$$

or, multiplying Eq. (6.46) by B and subtracting it from Eq. (6.49),

$$w(x)\frac{dB}{dx} = -\lambda B \tag{6.51}$$

Here g(x) = 0 and $p(x) = \lambda/w(x) = \lambda/(w(0) - K_R x)$, so that

$$\mu(x) = exp\left[\int_x \frac{\lambda}{(w(0) - K_R x)} dx'\right] \tag{6.52}$$

or, carrying out the integration,

$$\mu(x) = \left(w(0) - K_R x\right)^{-\lambda/K_R} \tag{6.53}$$

so that

$$B = c\left(w(0) - K_R x\right)^{-\lambda/K_R} \tag{6.54}$$

The integration constant c is obtained from the boundary condition, Eq. (3.127), which in this case reads as

$$w(0)B(0) = F_o + K_R \int_0^L B(x)dx \tag{6.55}$$

Thus,

$$c = F_o\left[\left(w(0) - K_R L\right)^{\frac{\lambda}{K_R}-1}\right] \tag{6.56}$$

and the solution is complete.

Example 6-5. The final example of a first-order ODE comes from Eq. (3.123), which describes constant, symmetrical, nonlocal mixing within a sediment (see Boudreau and Imboden, 1987). For steady state, constant porosity and first-order decay, a tracer's distribution is described by

$$w\frac{dB}{dx} = -(k + K_{ex})B + \frac{K_{ex}}{L}\int_0^L B(x)dx \tag{6.57}$$

At first this does not appear to be an ODE because of the integral, but this is a deception, as the integral term in this case is simply equal to a number, i.e. the value of this definite integral. We don't know the value of the integral in advance, but it is simply a constant, say b_o. One can then identify $p(x) = (k+K_{ex})/w$, which is also a constant, and $g(x) = K_{ex}b_o/(wL)$. These produce

$$\mu(x) = exp\left((k + K_{ex})x / w\right) \tag{6.58}$$

and

$$B = c\ exp\left(\frac{(k + K_{ex})x}{w}\right) + \frac{K_{ex}b_o}{L(k + K_{ex})} \tag{6.59}$$

The integration constant c is found by employing the boundary condition at x=0, which, we assume, is still $B(0) = B_o$, so that

$$c = B_o - \frac{K_{ex}b_o}{L(k + K_{ex})} \tag{6.60}$$

and the unknown b_o is obtained by integrating Eq. (6.59),

$$b_o = \frac{\dfrac{wB_o}{(k + K_{ex})}\left(1 - exp\left(-(k + K_{ex})L / w\right)\right)}{\left(1 + \dfrac{wK_{ex}\left(1 - exp\left(-(k + K_{ex})L / w\right)\right)}{L(k + K_{ex})^2} - \dfrac{K_{ex}}{(k + K_{ex})}\right)} \tag{6.61}$$

and the problem is solved. The method can be generalized for symmetric exchange functions that are depth-dependent and of the form $K_{ex}(x,x') = K_1(x)K_2(x')$, i.e. the product of two independent functions. However, the resulting integrals may be too complicated for closed-form (exact) solutions.

6.3.2 Nonlinear Equations

Unlike Eq. (6.28), nonlinear equations do not have general solutions. Each equation must be considered and solved individually; however, there are some broad classes of nonlinear ODEs for which certain methods are often useful (see Murphy, 1960; Boyce and DiPrima, 1977). In my experience, most of the first-order non-linear ODEs that come from diagenetic modelling and that can be integrated analytically are of the type called *separable*. These are equations of the form

$$\frac{dC}{dx} = f(C)g(x) \tag{6.62}$$

subject to $C(0) = C_o$, and where f(C) is only a function of C and g(x) is only a function of x (x could be replaced by t). Such equations can be written as

$$\frac{dC}{f(C)} = g(x)dx \tag{6.63}$$

where $f(C) \neq 0$, so that the left-hand side is solely a function of C, and the right-hand side is only dependent on x (or t). Integration of Eq. (6.63) produces

$$\int_{C_o}^{C} \frac{1}{f(C)} dC = \int_0^x g(x)dx \tag{6.64}$$

However, it is not guaranteed that the integrals in Eq. (6.64) can be calculated in closed form.

Example 6-6. As an example of this type of equation, consider burial with second-order reaction,

$$\frac{dC}{dx} = -\frac{k}{w}C^2 \tag{6.65}$$

so that $f(C) = C^2$ and $g(x) = -k/w$ in Eq. (6.62). Therefore, we get from Eq. (6.63)

$$\int_{C_o}^{C} C^{-2} dC = -\frac{k}{w}\int_0^x dx \tag{6.66}$$

Upon integration and slight rearrangement, one obtains

$$C = \left[\frac{1}{C_o} + \frac{kx}{w}\right]^{-1} \tag{6.67}$$

While types other than separable forms of first-order ODEs can be generated by transport-reaction modelling, it is my experience that if they are not separable, it is usually easier to solve them by numerical methods (Chapter 8).

6.4 Solution of Second-Order ODEs

Like first-order ODEs, the solution of most second-order ODEs depends on their linearity, but the methodology also depends on the x or t functionality of the terms that multiply the derivatives of C, i.e. the *coefficients*, as will be demonstrated in the following sections.

6.4.1 Directly Integratable Equations

It is useful to recognize this subclass of ODEs because their solution is comparatively easy to obtain. Generally, they are equations that have no first-order derivative term (no advection) and the reaction term, if present, depends only on x. If this is the case, then the solution is obtained by two direct integrations, as illustrated below.

Example 6-7. Assume that there is only diffusion in a zone of a constant porosity sediment. A solute's distribution is then given by

$$D'\frac{d^2C}{dx^2} = 0 \tag{6.68}$$

which integrates twice to

$$C(x) = Ax + B \tag{6.69}$$

where A and B are integration constants, resolved with the boundary conditions.

Example 6-8. Consider next a solute species that has a prescribed depth-dependent source, R(x), which takes into account variable porosity, and a depth-dependent diffusivity, D(x), such that

$$\frac{d}{dx}\left(D(x)\frac{dC}{dx}\right) + R(x) = 0 \tag{6.70}$$

where we assume that advection is negligible and that there are no other reactions. This could represent a nutrient species produced by organic matter decay. A first integration of Eq. (6.70) gives

$$D(x)\frac{dC}{dx} + \int_x R(x')dx' + A = 0 \qquad (6.71)$$

where A is an integration constant. This result is a first-order linear ODE. A second integral, i.e. the solution for C(x), is obtained by another direct integration, the practicality of which depends on the exact forms of R(x) and D(x). In any event, the problem has been reduced to solving a first-order ODE which is, in principle, easier.

Example 6-9. While most of the examples in this chapter deal with Cartesian depth as the independent variable, there are cases in which another independent variable may be of interest. An example of this type is provided by the radial micro-environment of Jørgensen (1977) and Jahnke (1985); see Chapter 3. Here the concentration of O_2 (or NO_3) within a small isolated organic-rich particle in a sediment is governed by

$$\frac{D}{r^2}\frac{d}{dr}\left[r^2\frac{dC}{dr}\right] = R_o \qquad (6.72)$$

where D is the molecular diffusion coefficient and R_o is a zeroth-order (constant) rate of O_2 consumption. To solve, one multiplies Eq. (6.72) by r^2, and two direct integrations later one obtains

$$C(r) = \frac{r^2 R_o}{6D} - \frac{A_1}{r} + A_2 \qquad (6.73)$$

where A_1 and A_2 are integration constants that depend on the boundary conditions.

6.4.2 Linear Equations with Constant Coefficients

This category contains most of the classic and familiar steady-state ADR diagenetic equations. They are all of the form

$$D\frac{d^2C}{dx^2} - v\frac{dC}{dx} - kC = g(x) \qquad (6.74)$$

where D, v, and k are generic constants, and g(x) is a function of x, including zero. The g(x) term is referred to as an *inhomogeneous source/sink*or *forcing*, or as a zeroth-order reaction. Notice that one cannot integrate Eq. (6.74) directly.

Homogeneous Equations, g(x) = 0. Like linear first-order ODEs, linear second-order ODEs with constant coefficients have a general solution, i.e.

$$C = A\,exp(ax) + B\,exp(bx) \qquad (6.75)$$

where

$$a \equiv \frac{v - \sqrt{v^2 + 4Dk}}{2D} \tag{6.76}$$

$$b \equiv \frac{v + \sqrt{v^2 + 4Dk}}{2D} \tag{6.77}$$

and A and B are arbitrary integration constants, which must be determined with the boundary conditions (see below). Note that it is always true that

$$v < \sqrt{v^2 + 4Dk} \tag{6.78}$$

as long as v, D, and k are positive constants. This implies that, in Eq. (6.75), the constant a is a negative constant, while b is a positive constant. A proof of Eq. (6.75) is given further below in the context of another problem, but one can verify that Eq. (6.75) is the solution by substituting it into Eq. (6.74).

I stated previously that the solution to ODEs was really a function of non-dimensional parameter groupings, like the Peclet number and the Damkohler number. Where are these numbers in Eqs (6.75) through (6.77)? To discover them, let Eq. (6.74) be valid over some depth L; then, if we set $\xi = x/L$, Eq. (6.75) becomes

$$C = A\,exp(a'\xi) + B\,exp(b'\xi) \tag{6.79}$$

where

$$a' \equiv \frac{1}{2}\left[\frac{vL}{D} - \sqrt{\left(\frac{vL}{D}\right)^2 + \frac{4kL^2}{D}}\right] = \frac{1}{2}\left[Pe - \sqrt{(Pe)^2 + 4Da}\right] \tag{6.80}$$

$$b' \equiv \frac{1}{2}\left[\frac{vL}{D} + \sqrt{\left(\frac{vL}{D}\right)^2 + \frac{4kL^2}{D}}\right] = \frac{1}{2}\left[Pe + \sqrt{(Pe)^2 + 4Da}\right] \tag{6.81}$$

and $Pe \equiv vL/D$, which is a Peclet number, and $Da \equiv kL^2/D$, which is a Damkohler number; thus, we see that the solutions do indeed depend on dimensionless numbers.

Next, to determine the values of the integration constants A and B in Eq. (6.75) or Eq. (6.79), the boundary conditions must be used. Three examples of this procedure are supplied below.

Example 6-10. A second-order ODE needs two boundary conditions, as explained in Chapter 5. If the concentration is known at the sediment-water interface, $x = 0$, then

$$C(x=0) = C_o \tag{6.82}$$

The second condition must describe what happens at the lower boundary of the model. In this example, we assume that the sediment extends to such a great depth that we can safely set the bottom to infinite depth. We want the concentration to remain finite as $x \to \infty$, so

$$\left.\frac{dC}{dx}\right|_{x\to\infty} = 0 \tag{6.83}$$

To evaluate the integration constants, start with Eq. (6.83). The first derivative of the solution, Eq. (6.75), is

$$\frac{dC}{dx} = A\,a\,exp(ax) + B\,b\,exp(bx) \tag{6.84}$$

Now, as $x \to \infty$, $exp(ax) \to 0$ because the exponent ax is a negative number; however, $exp(bx) \to \infty$ because bx is a positive number. For Eq. (6.84) to behave like Eq. (6.83) as $x \to \infty$, it is necessary that $B = 0$.

Subsequently, Eq. (6.82) gives that

$$C_o = A \tag{6.85}$$

and both A and B have been defined.

Example 6-11. Let us change the problem so that the condition at the interface is a boundary-layer condition and the sediment is of finite thickness, $x = L$, where it becomes impermeable. The first condition means that (see Chapter 5)

$$-D\left.\frac{dC}{dx}\right|_0 = \beta(C_{sw} - C(x=0)) \tag{6.86}$$

where β is the mass-transfer coefficient and C_{sw} is the known constant concentration in the overlying waters. For the bottom condition,

$$-D\left.\frac{dC}{dx}\right|_L + vC(L) = 0 \tag{6.87}$$

By substituting Eq. (6.75) into Eq. (6.87), we first obtain that

$$-D\big(A\,a\,exp(aL) + B\,b\,exp(bL)\big) + v\big(A\,exp(aL) + B\,exp(bL)\big) = 0 \tag{6.88a}$$

or

$$B = -\frac{A(v - aD)\,exp((a - b)L)}{(v - bD)} \tag{6.88b}$$

which expresses the unknown constant B in terms of the unknown, but soon to be determined, constant A. Thus, at this stage of the solution,

$$C = A\left[exp(ax) - \frac{(v - aD)\,exp((a - b)L)}{(v - bD)}\,exp(bx)\right] \tag{6.89}$$

and, upon differentiation,

$$\frac{dC}{dx} = A\left[a\,exp(ax) - \frac{b(v - aD)\,exp((a - b)L)}{(v - bD)}\,exp(bx)\right] \tag{6.90}$$

Now substitute Eqs (6.89) and (6.90) into Eq. (6.86),

$$-AD\left[a - \frac{b(v - aD)\,exp((a - b)L)}{(v - bD)}\right] = \beta\left(C_{sw} - A\left[1 - \frac{(v - aD)\,exp((a - b)L)}{(v - bD)}\right]\right) \tag{6.91a}$$

or, rearranging to obtain an explicit expression for A,

$$A = \frac{\beta C_{sw}}{\beta\left[1 - \frac{(v - aD)\,exp((a - b)L)}{(v - bD)}\right] - D\left[a - \frac{b(v - aD)\,exp((a - b)L)}{(v - bD)}\right]} \tag{6.91b}$$

The constant B is non-zero in this case because we need that term to make the flux disappear at $x = L$.

Example 6-12. This final example illustrates the use of continuity conditions between adjacent sediment layers. Let us assume that the diffusivity, D, has a value of D_1 in an upper zone of the sediment, $0 \le x \le L_1$, and a different value, D_2, in the adjacent lower zone, $L_1 < x < \infty$. The diagenetic conservation equations for C are then

$$D_1\frac{d^2C}{dx^2} - v\frac{dC}{dx} - kC = 0 \qquad \text{for } 0 \le x \le L_1 \tag{6.92}$$

and

$$D_2 \frac{d^2C}{dx^2} - v\frac{dC}{dx} - kC = 0 \qquad \text{for } L_1 < x < \infty \tag{6.93}$$

The boundary conditions for this example will be

$$C(x{=}0) = C_o \tag{6.94}$$

$$\left.\begin{aligned} &C(L_1^+) = C(L_1^-) \\ &D_1 \left.\frac{dC}{dx}\right|_{L_1^+} = D_2 \left.\frac{dC}{dx}\right|_{L_1^-} \end{aligned}\right\} \qquad \text{at } x = L_1 \tag{6.95a,b}$$

and

$$\left.\frac{dC}{dx}\right|_{x\to\infty} = 0 \tag{6.96}$$

The solutions to Eqs (6.92) and (6.93) are, respectively,

$$C = A_1\, exp(a_1x) + B_1\, exp(b_1x) \qquad 0 \le x \le L_1 \tag{6.97}$$

and

$$C = A_2\, exp(a_2x) + B_2\, exp(b_2x) \qquad L_1 < x < \infty \tag{6.98}$$

where

$$a_1, b_1 \equiv \frac{v \mp \sqrt{v^2 + 4D_1k}}{2D_1} \tag{6.99}$$

$$a_2, b_2 \equiv \frac{v \mp \sqrt{v^2 + 4D_2k}}{2D_2} \tag{6.100}$$

and A_1, A_2, B_1, and B_2 are all integration constants.

First, apply the boundary conditions, beginning with Eq. (6.96). This condition necessitates that $B_2 = 0$. Next apply Eq. (6.94) to get

$$B_1 = C_o - A_1 \tag{6.101}$$

Therefore, we have in the two zones that

$$C = A_1\big(exp(a_1x) - exp(b_1x)\big) + C_o\, exp(b_1x) \qquad 0 \le x \le L_1 \tag{6.102}$$

and

$$C = A_2\, exp(a_2 x) \qquad L_1 < x < \infty \tag{6.103}$$

From these equations, we can calculate that

$$\frac{dC}{dx} = A_1\big(a_1 exp(a_1 x) - b_1 exp(b_1 x)\big) + C_o\, b_1\, exp(b_1 x) \qquad 0 \le x \le L_1 \tag{6.104}$$

and

$$\frac{dC}{dx} = A_2\, a_2\, exp(a_2 x) \qquad L_1 < x < \infty \tag{6.105}$$

Next, substitute Eqs (6.102) and (6.103) into Eq. (6.95a) to get

$$A_2 = \frac{A_1\big(exp(a_1 L_1) - exp(b_1 L_1)\big) + C_o\, exp(b_1 L_1)}{exp(a_2 L_1)} \tag{6.106}$$

Finally, substitute Eqs (6.104) through (6.106) into Eq. (6.95b) to get

$$A_1 = \frac{(D_2 a_2 - D_1 b_1) C_o\, exp(b_1 L_1)}{\big[D_1\big(a_1 exp(a_1 L_1) - b_1 exp(b_1 L_1)\big) - D_2 a_2\big(exp(a_1 L_1) - exp(b_1 L_1)\big)\big]} \tag{6.107}$$

so that A_1, B_1, and A_2 are now all defined in terms of known parameters. If there are more than two layers in the sediment, one generates a system of linear algebraic equations for the resulting As and Bs. Computer programs that are capable of doing symbolic mathematics (such as Mathematica, Maple and Macsyma) can do these reductions automatically, or one can resort to purely numerical calculations of their values.

Special Cases. Many special cases of Eq. (6.74) appear in the geochemical literature, and these have somewhat simpler solutions. One common form assumes that $v = 0$, i.e. that advection is negligibly small. The solution then becomes

$$C = A\, exp(ax) + B\, exp(b_1 x) \tag{6.108}$$

where

$$a \equiv -\sqrt{\frac{k}{D}} \tag{6.109}$$

$$b_1 \equiv \sqrt{\frac{k}{D}} \tag{6.110}$$

The second form occurs when k = 0; that is to say, a first-order reaction is not present; then

$$C = A + B\,exp(b_2 x) \tag{6.111}$$

where

$$b_2 \equiv \frac{v}{D} \tag{6.112}$$

Inhomogeneous Equations $g(x) \neq 0$. Because this problem is linear, it has a general solution, which is of the form $C = C_h + C_p$ where C_h is the solution to Eq. (6.74) with $g(x) = 0$, and C_p is an added particular solution that specifically accounts for the g(x) term. As before, the homogeneous part of the solution is given by Eq. (6.75), but the integration constants, A and B, cannot be determined until the solution is complete.

The essence of the matter is that we now need to derive C_p. The steps toward obtaining C_p are in principle easy, but in practice, they may be so difficult as to recommend a numerical solution. The method outlined below is often called the method of *undetermined coefficients* (Boyce and DiPrima, 1977). Specifically, assume that

$$C_p = c\, g(x) \tag{6.113}$$

where c is an arbitrary constant; substitute this C_p for C into Eq. (6.74) with $g(x) \neq 0$, and derive an algebraic equation for c. Check that c g(x) is not equal to one of the homogeneous solutions to Eq. (6.74); if it is, then start again with

$$C_p = x^s\, c\, g(x) \tag{6.114}$$

where s is a constant equal to 1 or 2 that will ensure that Eq. (6.114) is different from a homogeneous solution. Finally, put the pieces of the solution together, and use the boundary conditions to derive the values of the integration constants, A and B.

Example 6-13. A simple, but common, inhomogeneous forcing is an exponential, e.g.

$$D\frac{d^2C}{dx^2} - v\frac{dC}{dx} - kC = -R_o\, exp(-\gamma x) \tag{6.115}$$

where R_o is a known constant, and this g(x) term constitutes a source. The boundary conditions are chosen to be

$$C(x=0) = C_o \tag{6.116}$$

$$\left.\frac{dC}{dx}\right|_{x\to\infty} = 0 \tag{6.117}$$

The homogeneous solution, C_h, is given by Eq. (6.75), and the method of undetermined coefficients suggests that C_p is of the form

$$C_p = c\, exp(-\gamma x) \tag{6.118}$$

where c is an unknown constant (i.e. undetermined coefficient). Substitute Eq. (6.118) into Eq. (6.115)

$$D\gamma^2 c\, exp(-\gamma x) + v\gamma c\, exp(-\gamma x) - k c\, exp(-\gamma x) = -R_o\, exp(-\gamma x) \tag{6.119}$$

However, *exp*(-γx) multiplies each term, so we can divide this term out and rearrange to get

$$c = \frac{-R_o}{D\gamma^2 + v\gamma - k} \tag{6.120}$$

The combined (final) solution is then:

$$C = A\, exp(ax) + B\, exp(bx) - \frac{R_o\, exp(-\gamma x)}{D\gamma^2 + v\gamma - k} \tag{6.121}$$

Boundary condition Eq. (6.117) immediately gives that

$$B = 0 \tag{6.122}$$

while boundary condition Eq. (6.116) requires that

$$A = C_o + \frac{R_o}{D\gamma^2 + v\gamma - k} \tag{6.123}$$

Example 6-14. A second example gives the reader a chance to see not only the solution of an inhomogeneous problem, but also the use of boundary conditions at an unknown point (see Chapter 5). The problem is a classic example of sediment-water interaction, that is, the diffusion of O_2 into a sediment where it is consumed by *zeroth-order kinetics* in a finite zone $0 \le x \le L$ (Bouldin, 1968), i.e.

$$D\frac{d^2C}{dx^2} - k' = 0 \tag{6.124}$$

where k' is a zeroth-order consumption rate, $g(x) = k'$. The trick to the problem is that L is not known *a priori*, but must be calculated as part of the problem. As a consequence, there are three boundary conditions,

$$C(x=0) = C_o \tag{6.125}$$

$$C(x=L) = 0 \tag{6.126}$$

and

$$\left.\frac{dC}{dx}\right|_L = 0 \tag{6.127}$$

It is the availability of a third boundary condition that supplies enough information to calculate L.

The solution to Eq. (6.124) is obtained either by direct integration or by assuming that g(x) is a quadratic. Why a quadratic? Because the homogeneous solution, C_h, is a linear function of depth. In accord with Eq. (6.114), C_p must be at least a quadratic. Either way,

$$C(x) = \frac{k'x^2}{2D} + c_1 x + c_2 \tag{6.128}$$

where c_1 and c_2 are integration constants. Now from Eq. (6.125),

$$c_2 = C_o \tag{6.129}$$

and from Eq. (6.126),

$$c_1 = -\frac{k'L}{D} \tag{6.130}$$

Finally, Eq. (6.127) gives that

$$\frac{k'L^2}{2D} + c_1 L + c_2 = 0 \tag{6.131}$$

or, after substitution of Eqs (6.129) and (6.130),

$$L = \sqrt{\frac{2DC_o}{k'}} \tag{6.132}$$

which defines the unknown depth.

6.4.3 Linear Equations with Variable Coefficients

Equations in this category are characterized by any or all of the coefficients D, v, or k in Eq. (6.74) being functions of depth. Examples include diagenetic equations for a solid species when depth-dependent bioturbation, $D = D_B(x)$, is present or when porosity, φ, is a function of depth. Such equations are of the general form:

$$p(x)\frac{d^2C}{dx^2} - q(x)\frac{dC}{dx} - r(x)C = g(x) \tag{6.133}$$

where p(x), q(x), or r(x) can be functions of x. These can be interpreted as generalized diffusivity, advective velocity, and reactivity, respectively.

Equations with variable coefficients were studied extensively in the 19th and early 20th centuries. Their solutions are the bases for many of the *special functions* found in mathematics, e.g. Bessel, Legendre, Hypergeometric, and Confluent Hypergeometric functions (see Abramowitz and Stegun, 1964; Lebedev, 1972; Hildebrand, 1976; Spanier and Oldham, 1987). Today, we often resort to numerical methods, but I give here a short exposé of the analytical method for their solution, because some equations are still conveniently solved this way, and numerical methods need to be tested against known solutions. (The reader less interested in the underlying details of series solution method can skip directly to Section 6.4.6.)

6.4.4 Series Solution about an Ordinary Point

A depth x is said to be a *regular* or *ordinary point* in the mathematical sense if, for the coefficients in Eq. (6.133), $p(x) \neq 0$, $q(x) \neq \infty$, and $r(x) \neq \infty$ at that point. Any depth where these conditions are not true is called a *singular* point. At an ordinary point one can seek a solution to Eq. (6.133) in the form of a simple power series (see Chapter 2). If x = 0 is a regular point, then the desired solution has the form:

$$C(x) = a_o + a_1x + a_2x^2 + a_3x^3 + a_4x^4 + \ldots \tag{6.134}$$

This solution is generally valid between x = 0 and the first singular point that exists in the depth range of interest (if any). In addition, let me be quite honest in saying that the application of Eq. (6.134) as a solution to Eq. (6.133) does not necessarily lead to a useful result from a computational point of view. Nevertheless, when it does lead to a workable solution, it can be faster and easier to calculate than a numerical method, and it offers insights that are absent in the numerical treatment.

Example 6-15. As a first example, let us deal with Eq. (6.74). This is a very special case of Eq. (6.133) where p(x), q(x), and r(x) are constants and g(x) = 0; hence, all depths are regular points. This demonstrates the origin of the exponential solutions to this equation. Furthermore, to make the mathematics a bit more transparent, set v = 0, so that we are trying to find the solution of the equation

$$D\frac{d^2C}{dx^2} - kC = 0 \tag{6.135}$$

Differentiating Eq. (6.134) twice, one obtains

$$\frac{d^2C}{dx^2} = 2a_2 + 2\cdot 3a_3x + 3\cdot 4a_4x^2 + 4\cdot 5a_5x^3 + 5\cdot 6a_6x^4 \ldots \tag{6.136}$$

Substituting Eqs (6.134) and (6.136) into Eq. (6.135) produces

$$\begin{aligned} &D\left(2a_2 + 2\cdot 3a_3x + 3\cdot 4a_4x^2 + 4\cdot 5a_5x^3 + 5\cdot 6a_6x^4 \ldots\right) \\ &\qquad - k\left(a_o + a_1x + a_2x^2 + a_3x^3 + a_4x^4 + \ldots\right) = 0 \end{aligned} \tag{6.137}$$

or, with a bit of rearrangement,

$$\begin{aligned} &(2a_2D - ka_o) + (2\cdot 3a_3D - ka_1)x + (3\cdot 4a_4D - ka_2)x^2 \\ &\qquad + (4\cdot 5a_5D - ka_3)x^3 + (5\cdot 6a_6D - ka_4)x^4 + \ldots = 0 \end{aligned} \tag{6.138}$$

In order that Eq. (6.138) be identically zero for all values of x, every coefficient in Eq. (6.138) must be zero simultaneously. This creates the sequence of algebraic equations,

$$a_2 = \frac{k}{1\cdot 2D}a_o \tag{6.139}$$

$$a_3 = \frac{k}{2\cdot 3D}a_1 \tag{6.140}$$

$$a_4 = \frac{k}{3\cdot 4D}a_2 = \frac{k^2}{1\cdot 2\cdot 3\cdot 4D^2}a_o \tag{6.141}$$

$$a_5 = \frac{k}{4\cdot 5D}a_3 = \frac{k^2}{2\cdot 3\cdot 4\cdot 5D^2}a_1 \tag{6.142}$$

etc. This sequence defines the infinite series of coefficients of the power series; all of which leads to the result:

$$C(x) = a_o + a_1 x + \frac{a_o k}{1 \cdot 2D} x^2 + \frac{a_1 k}{2 \cdot 3D} x^3 + \frac{a_o k^2}{1 \cdot 2 \cdot 3 \cdot 4D^2} x^4 + \frac{a_1 k^2}{1 \cdot 2 \cdot 3 \cdot 4 \cdot 5D^2} x^5 + \ldots \tag{6.143}$$

which still contains two arbitrary constants, a_o and a_1; these can be defined via the two boundary conditions on Eq. (6.135), which have not been stated.

We could leave Eq. (6.143) in this state, but a more interesting form of this solution exists. Start by defining

$$a_1 \equiv a_1' \sqrt{\frac{k}{D}} \tag{6.144}$$

where a_1' is a different arbitrary constant. Substitution of this new constant and rearrangement of Eq. (6.143) lead to:

$$C(x) = a_o \left(1 + \frac{1}{2!}\left(\frac{k}{D}\right)x^2 + \frac{1}{4!}\left(\frac{k}{D}\right)^2 x^4 + \ldots \right) + a_1' \left(\left(\frac{k}{D}\right)^{1/2} x + \frac{1}{3!}\left(\frac{k}{D}\right)^{3/2} x^3 + \frac{1}{5!}\left(\frac{k}{D}\right)^{5/2} x^5 + \ldots \right) \tag{6.145}$$

where ! indicates factorials; i.e. 2! = 1·2, 3! = 1·2·3, 4! = 1·2·3·4, etc. Finally, let

$$a_o \equiv A_o + A_1 \tag{6.146}$$

and

$$a_1' \equiv A_o - A_1 \tag{6.147}$$

where A_o and A_1 are again two other arbitrary replacement constants. Introduction of these two relationships into Eq. (6.145) brings us to:

$$C(x) = A_o \left(1 + \left(\frac{k}{D}\right)^{1/2} x + \frac{1}{2!}\left(\frac{k}{D}\right)x^2 + \frac{1}{3!}\left(\frac{k}{D}\right)^{3/2} x^3 + \frac{1}{4!}\left(\frac{k}{D}\right)^2 x^4 + \ldots \right) + A_1 \left(1 - \left(\frac{k}{D}\right)^{1/2} x + \frac{1}{2!}\left(\frac{k}{D}\right)x^2 - \frac{1}{3!}\left(\frac{k}{D}\right)^{3/2} x^3 + \frac{1}{4!}\left(\frac{k}{D}\right)^2 x^2 - \ldots \right) \tag{6.148}$$

But, from the section on Taylor series in Chapter 2, we have

$$exp\left(x\sqrt{\frac{k}{D}}\right) = 1 + \left(\frac{k}{D}\right)^{1/2} x + \frac{1}{2!}\left(\frac{k}{D}\right)x^2 + \frac{1}{3!}\left(\frac{k}{D}\right)^{3/2} x^3 + \ldots \tag{6.149}$$

and

$$exp\left(-x\sqrt{\frac{k}{D}}\right) = 1 - \left(\frac{k}{D}\right)^{1/2} x + \frac{1}{2!}\left(\frac{k}{D}\right)x^2 - \frac{1}{3!}\left(\frac{k}{D}\right)^{3/2} x^3 + \ldots \tag{6.150}$$

Thus the first term in Eq. (6.148) is a growing exponential and the second is a decay exponential, which recovers the solution given for Eq. (6.108).

Example 6-16. The second example of this category is related to Example 6-4 above, which described the distribution of a decaying tracer in a sediment layer, $0 \leq x \leq L$, affected by head-down deposit-feeding (the biological conveyor-belt). The only processes in that model were advection and decay; what if constant biodiffusive mixing is also present? At constant porosity, the steady-state conservation equation for the tracer is then

$$D_B \frac{d^2C}{dx^2} - (w_\infty + K_R(L - x))\frac{dC}{dx} - kC = 0 \tag{6.151}$$

where D_B is the biodiffusion coefficient, w_∞ is the net burial velocity below the conveyor-belt zone, K_R is the rate constant for sediment removal by deposit feeding, and k is the decay constant. In this example $q(x) = - w_\infty - K_R(L-x)$, which remains finite over the depth range of interest, i.e. $0 \leq x < L$.

Equation (6.151) is most conveniently solved through use of Eq. (6.134) if a small change is made in the dependent variable. Specifically, let

$$\xi \equiv \frac{w_\infty + K_R(L - x)}{K_R L} \tag{6.152}$$

Then, Eq. (6.151) becomes

$$\frac{D_B}{L^2}\frac{d^2C}{d\xi^2} + K_R \xi \frac{dC}{d\xi} - kC = 0 \tag{6.153}$$

or, after multiplying by L^2/D_B,

$$\frac{d^2C}{d\xi^2} + \text{Pe}\, \xi \frac{dC}{d\xi} - \text{Da}\, C = 0 \tag{6.154}$$

where

$$Pe \equiv \frac{K_R L^2}{D_B} \tag{6.155}$$

$$Da \equiv \frac{kL^2}{D_B} \tag{6.156}$$

The first of these dimensionless numbers is a Peclet number for the maximum downward velocity, $K_R L$, created by head-down deposit-feeding (remember that to a solid species this removal looks like a sink). The second is a Damkohler number for decay process. This equation bears some resemblance to the Hermite equation (Boyce and DiPrima, 1977, p. 159), but the signs of the central and final terms are reversed, and there is a general coefficient Pe before the second term rather than a 2.

Assuming Eq. (6.134) is the solution, then we have

$$\frac{dC}{d\xi} = a_1 + 2a_2\xi + 3a_3\xi^2 + 4a_4\xi^3 + 5a_5\xi^4 + 6a_6\xi^5 + \ldots \tag{6.157}$$

and

$$\frac{d^2C}{d\xi^2} = 1\cdot 2a_2 + 2\cdot 3a_3\xi + 3\cdot 4a_4\xi^2 + 4\cdot 5a_5\xi^3 + 5\cdot 6a_6\xi^4 + \ldots \tag{6.158}$$

which, along with a version of Eq. (6.134) with ξ substituted for x, can be plugged into Eq. (6.154) to obtain

$$\begin{aligned}&(1\cdot 2)a_2 + (2\cdot 3)a_3\xi + (3\cdot 4)a_4\xi^2 + (4\cdot 5)a_5\xi^3 + \ldots\\ &\qquad + Pe\left(a_1\xi + 2a_2\xi^2 + 3a_3\xi^3 + 4a_4\xi^4 + \ldots\right)\\ &\qquad - Da\left(a_o + a_1\xi + a_2\xi^2 + a_3\xi^3 + a_4\xi^4 + \ldots\right) = 0\end{aligned} \tag{6.159}$$

As in Example 6-15, we can group same powers of ξ in Eq. (6.159) and equate the resulting coefficient groupings to zero. This process leads to the following equations for the coefficients:

$$a_2 = \frac{Da}{1\cdot 2}a_o \tag{6.160}$$

$$a_3 = \frac{(Da - Pe)}{2\cdot 3}a_1 \tag{6.161}$$

$$a_4 = \frac{(Da - 2Pe)}{3\cdot 4}a_2 \tag{6.162}$$

$$a_5 = \frac{(Da - 3Pe)}{4 \cdot 5} a_3 \tag{6.163}$$

and, in general,

$$a_{n+2} = \frac{(Da - n\,Pe)}{(n+1) \cdot (n+2)} a_n \tag{6.164}$$

Observe that all even-numbered coefficients are solely related to a_o and that odd-numbered coefficients are related only to a_1. This allows the resulting series to be divided into two sub-series, i.e.

$$\begin{aligned} C(\xi) = a_o \left\{ 1 + \frac{Da}{2!} \xi^2 + \frac{Da(Da - 2Pe)}{4!} \xi^4 + \ldots \right\} \\ + a_1 \left\{ \xi + \frac{(Da - Pe)}{3!} \xi^3 + \frac{(Da - Pe)(Da - 3Pe)}{5!} \xi^5 + \ldots \right\} \end{aligned} \tag{6.165}$$

which is the desired solution. As best as I can determine, the two series in brackets in Eq. (6.165) don't have official names, but they could perhaps be called modified Hermite functions. The solution can now be expressed in terms of the original depth variable x by substitution of Eq. (6.152) into Eq. (6.165).

One cannot imagine that the easy separation of solutions into two sub-series, such as in Eqs (6.148) and (6.165), is the norm. Usually, the two series are jumbled together in such a way that they are essentially inseparate, particularly if p(x), q(x), and/or r(x) are nonlinear functions of depth (e.g. Christensen, 1982).

6.4.5 Series Solution about a Regular Singularity

There are those variable-coefficient diagenetic equations for which there is at least one depth x_o where $p(x) = 0$ or $q(x) = \infty$ or $r(x) = \infty$, and some examples are described in the next section. The solution of these equations can be obtained in the form of a series about this singular point, if it is not "too singular". Not too singular means that the following limits must exist:

$$\underset{x \to x_o}{\text{limit}} \left[(x - x_o) \frac{q(x)}{p(x)} \right] < \infty \tag{6.166}$$

and

$$\underset{x \to x_o}{\text{lim it}} \left[(x - x_o)^2 \frac{r(x)}{p(x)} \right] < \infty \tag{6.167}$$

In this case, x_o is called a *regular singular point* of the equation. (Sounds a bit like an oxymoron?!). In most cases, $x_o = 0$ is the point in question, and this will certainly be true of all the cases discussed below. If the depth of interest is an *irregular singular point*, i.e. Eqs (6.166) and (6.167) are not true, the solution to this type of ODE as a series is beyond the scope of this book, and I suggest that the reader explore a numerical method. (A determined researcher/student might wish to consult Ince, 1956, or Bender and Orszag, 1978.)

At a regular singular point, a solution to Eq. (6.133) can be sought in the form of a *Frobenius series* (see Hildebrand, 1976; Boyce and DiPrima, 1977), i.e.

$$C(x) = x^{s}\left\{a_o + a_1 x + a_2 x^2 + a_3 x^3 + a_4 x^4 + \ldots\right\} \tag{6.168}$$

which resembles Eq. (6.134), but it is multiplied by the extra term in x raised to the power s. The power s has an undetermined value that must be evaluated as part of the solution to the problem.

The solution technique progresses in a manner similar to a regular power series. Equation (6.168) has as its first and second derivatives

$$\frac{dC}{dx} = s a_o x^{s-1} + (s+1)a_1 x^{s} + (s+2)a_2 x^{s+1} + (s+3)a_3 x^{s+2} + \ldots \tag{6.169}$$

and

$$\frac{d^2C}{dx^2} = (s-1)s\,a_o x^{s-2} + s(s+1)a_1 x^{s-1} + (s+1)(s+2)a_2 x^{s} + \ldots \tag{6.170}$$

and these are substituted into Eq. (6.133). Coefficients of like powers of x are then grouped and equated to zero. This gives a sequence of equations that define the series coefficients in terms of a_o.

The equation generated for the lowest power of x defines two values of s. If the values of s are not identical or do not differ by a positive integer, the separate substitution of each of the values of s into the equations for the coefficients, a_n, generates a unique series. The sum of these two series is the solution to Eq. (6.133) when p(x), q(x), and r(x) satisfy Eqs (6.166) and (6.167).

In the exceptional case in which the calculated s values are identical or differ by a positive integer, only the series generated with the larger s value is useful. Nevertheless, a second solution can be generated from the first. If the s values are equal and $C_1(x)$ designates the series obtained with this value, then the second series solution is given by

$$C_2(x) = C_1(x)\,ln|x| + |x|^{s}\sum_{n=1}^{\infty} b_n x^n \tag{6.171}$$

If the s values differ by a positive integer and $C_1(x)$ is the series obtained with the larger s values, say s_1, then the second series is given by

$$C_2(x) = b_o C_1(x) \ln|x| + |x|^{s_1} \left\{ 1 + \sum_{n=1}^{\infty} b_n x^n \right\} \tag{6.172}$$

where the constants b_o, b_1, etc. are determined by substitution into Eq. (6.133).

In practice one rarely, if ever, goes through the steps of using these series. Mathematicians and scientists long ago studied a large number of forms for Eq. (6.133) that were likely to need this treatment, and they codified and tabulated their solutions. This is the source of the functions associated with such names as Bessel, Legendre, etc. Because they've done all this work for us already, a much simpler approach is to see if the model equation is one of the studied equations or if it can be transformed into one. Then the solution is waiting for us, and this is the topic of the next section.

6.4.6 Solution By Variable Transformation and the "Look-Up" Technique

Perhaps it reflects an underlying unity in nature, or perhaps it is just our limited imagination, but either way, many of the same second-order ODEs result from models of fundamentally different phenomena. If someone else has already studied and solved an equation that is mathematically identical to the one you wish to solve, then yours has the same solution, and the previous solution is at your disposal. If you can recognize this equivalence, or create it by some simple variable transformation, then you can employ this most efficient solution method (Zwillinger, 1989).

Actually, this is the way most professional modellers deal with such equations. (Moreover, if a model equation is not related to one of the standard equations, it's often best to go directly to numerical methods if computation of a solution is the main goal.) Fortunately, there are a number of compendia of the standard equations and of the transformations that take related equations to these canonical forms (see Kamke, 1948; Murphy, 1960; Abramowitz and Stegun, 1964; Zwillinger, 1989).

For this method to be successful, it is usually necessary to perform two steps:

i) Make the equation nondimensional, although it need not be scaled; mathematicians don't report their work in dimensional form, so this is a necessary step in the "look-up" procedure.

ii) Alter the independent variable so that only simple powers of the transformed independent variable appear as coefficients in the equations; this may sound a bit obscure, but the examples below make this point more obvious. Zwillinger (1989) in his Chapter 1B gives a modern overview of the use of transformations.

The solution of the resulting ODEs often involves *special functions* or *higher transcendental functions* (see Abramowitz and Stegun, 1964; Lebedev, 1965; Erdélyi, 1953; Spanier and Oldham, 1987). These may be unfamiliar, but they are well studied, and there are algorithms (computer codes) for calculating their values (see below and the Appendix).

Example 6-17. Depth-variable porosity poses surprising difficulties to the solution of diagenetic equations. Under suitable assumptions, an analytical solution with variable porosity is possible, and this example illustrates such a case.

Consider a solute subject to molecular diffusion, first-order decay, nonlocal irrigative exchange, and production from a constant source in the solids, all occurring in a sediment with variable porosity. The tortuosity is assumed to obey Archie's Law, with m=3 (see Table 4.11). This model could describe either a radio-tracer such as radon, or a nutrient from zeroth-order organic matter decomposition, if the decay constant is set to zero. Conservation for this solute is then

$$\frac{d}{dx}\left[\varphi^3 D\frac{dC}{dx}\right] - \varphi(\lambda+\alpha)C + (1-\varphi)P = 0 \qquad (6.173)$$

or, upon expansion and dividing by φ,

$$\varphi^2 D\frac{d^2C}{dx^2} + 3\varphi D\frac{d\varphi}{dx}\frac{dC}{dx} - (\lambda+\alpha)C + \frac{(1-\varphi)}{\varphi}P = 0 \qquad (6.174)$$

where λ is the decay constant, α is the irrigation constant, and φ is the depth-dependent porosity. This equation is valid only for $0 \le x \le L$.

The usual treatment of depth-dependent porosity assumes that

$$\varphi(x) = \varphi_\infty + \left(\varphi(0) - \varphi_\infty\right) exp\left(-\gamma x\right) \qquad (6.175)$$

where φ_∞ is the final porosity at great depth in the sediment, $\varphi(0)$ is the porosity at the sediment-water interface, and γ is an empirical constant. A useful analytical solution with Eq. (6.175) substituted into Eq. (6.174) is not possible, as best as I can determine. Yet, use of Eq. (6.175) is not an ordained requirement. Successful modelling is in part learning the art of reasonable compromise†. If φ were a linear function of depth over a fixed interval, then we might stand a better chance of finding a solution. In fact, this is not an outrageous simplification. The first two terms of a Maclaurin series (Chapter 2) expansion of the exponential in Eq. (6.175) give that

$$\varphi(x) \approx \varphi_\infty + \left(\varphi(0) - \varphi_\infty\right)(1-\gamma x) \qquad (6.176)$$

which is linear over the depth interval $0 \le x \le L$.

---

† Whoever said that "politics is the art of the possible" never met an applied mathematician.

Secondly, even if Eq. (6.176) is not a good approximation of Eq. (6.175) over the entire depth domain of the model, the domain can be divided into a limited number of zones in which a linear approximation is reasonably accurate. Consequently, look for a solution to Eq. (6.174) with

$$\varphi(x) = a + \frac{bx}{L} \tag{6.177}$$

where a and b are empirical constants obtained from fitting the porosity data, and L is the thickness of the zone over which Eq. (6.177) is valid.

With Eq. (6.177), Eq. (6.174) becomes

$$\left(a + \frac{bx}{L}\right)^2 \frac{d^2C}{dx^2} + \frac{3b}{L}\left(a + \frac{bx}{L}\right)\frac{dC}{dx} - \frac{(\lambda + \alpha)}{D}C + \frac{\left(1 - a - \frac{bx}{L}\right)}{\left(a + \frac{bx}{L}\right)}\frac{P}{D} = 0 \tag{6.178}$$

Now the two steps cited above need to be taken, and these can be done simultaneously. Define a new independent variable,

$$\zeta \equiv a + \frac{bx}{L} \tag{6.179}$$

from which there results

$$\frac{b}{L}\frac{d}{d\zeta} = \frac{d}{dx} \tag{6.180}$$

and

$$\frac{b^2}{L^2}\frac{d^2}{d\zeta^2} = \frac{d^2}{dx^2} \tag{6.181}$$

Equation (6.179) means that C is being expressed as a function of $\varphi(x)$, rather than x itself.

Substitution of Eqs (6.179) through (6.181) into Eq. (6.178) and multiplication by L^2/b^2 produces

$$\zeta^2 \frac{d^2C}{d\zeta^2} + 3\zeta\frac{dC}{d\zeta} - Da'C + \frac{(1-\zeta)}{\zeta}P' = 0 \tag{6.182}$$

where

$$Da' \equiv \frac{(\lambda+\alpha)L^2}{Db^2} \tag{6.183}$$

which is a Damkohler number for combined decay and irrigation, and

$$P' \equiv \frac{PL^2}{Db^2} \tag{6.184}$$

which is the nondimensional production. Equation (6.182) is nondimensional and the coefficients are the simplest possible powers of ζ. How did I know to do this? Practice and familiarity with what others have previously done, for which there is absolutely no substitute.

Equation (6.182) is a linear inhomogeneous ODE. Its solution is made up of two parts: the solution to the homogeneous part, C_h, without the term containing P', and the particular solution, C_p, that accounts for the P' term. It is the homogeneous part, i.e.

$$\zeta^2 \frac{d^2C_h}{d\zeta^2} + 3\zeta \frac{dC_h}{d\zeta} - Da'C_h = 0 \tag{6.185}$$

which we solve by the "look-up" method. Specifically, I have had access to Murphy (1960) while writing this book. Flipping through the tables for linear second-order ODEs, I find on p. 341 of that book a form (his Eq. 311) completely consistent with my Eq. (6.185). Checking the section referred to in that text, one discovers that Eq. (6.185) is an Euler equation (see also Boyce and DiPrima, 1977, p. 174). Its solution is

$$C_h(\zeta) = c_1 \zeta^{r_1} + c_2 \zeta^{r_2} \tag{6.186}$$

where

$$r_1, r_2 \equiv -1 \pm \sqrt{1+Da'} \tag{6.187}$$

and c_1 and c_2 are integration constants that must be evaluated with the boundary conditions.

Next we need to solve for the inhomogeneous solution, and unfortunately, this part cannot be done by the "look-up" technique. We need to employ the method of *undetermined coefficients*, as discussed previously. Given the form of the inhomogeneous part of Eq. (6.182), let us look for a solution of the form

$$C_p(\zeta) = \frac{c_3}{\zeta} + c_4 \tag{6.188}$$

So substitute Eq. (6.188) into Eq. (6.182) to obtain

$$\frac{1}{\zeta}\left(P' - c_3(1 + Da')\right) - \left(Da' c_4 + P'\right) = 0 \qquad (6.189)$$

which is possible for all values of ζ only if, first,

$$c_3 = \frac{P'}{1 + Da'} \qquad (6.190)$$

and, secondly,

$$c_4 = -\frac{P'}{Da'} \qquad (6.191)$$

The full solution is then the sum of the homogeneous and particular solutions,

$$C(\zeta) = c_1 \zeta^{r_1} + c_2 \zeta^{r_2} + \frac{P'}{(1 + Da')\zeta} - \frac{P'}{Da'} \qquad (6.192)$$

or, in terms of the original depth variable x,

$$C(x) = c_1 \left(a + \frac{bx}{L}\right)^{r_1} + c_2 \left(a + \frac{bx}{L}\right)^{r_2} + \frac{P'}{(1 + Da')\left(a + \frac{bx}{L}\right)} - \frac{P'}{Da'} \qquad (6.193)$$

Example 6-18. Let us now consider the problem of a bioturbated solid species undergoing first-order decay in a sediment with linear depth-dependent porosity. The species could be a radio-tracer, organic matter, or an organic contaminant with oxidant-independent decomposition. If we ignore advection, the conservation equation for this species is:

$$\frac{d}{dx}\left[\varphi_s D_B \frac{dB}{dx}\right] - \varphi_s \lambda B = 0 \qquad (6.194a)$$

or, upon expansion,

$$\varphi_s D_B \frac{d^2B}{dx^2} + D_B \frac{d\varphi_s}{dx}\frac{dB}{dx} - \varphi_s \lambda B = 0 \qquad (6.194b)$$

where $\varphi_s \equiv (1\text{-}\varphi) \equiv$ the solid volume fraction.

If solid volume fraction can be taken to be a linear function of depth in the zone where Eq. (6.194) applies, say $0 \leq x \leq L$, then

$$(a+\frac{bx}{L})D_B\frac{d^2B}{dx^2}+D_B\frac{b}{L}\frac{dB}{dx}-(a+\frac{bx}{L})\lambda B=0 \tag{6.195}$$

where the constants a and b in this equation are not the same as in Example 6-17. Nevertheless, the linear porosity function suggests defining the new independent variable ζ, as in Eq. (6.179); consequently, Eq. (6.195) becomes,

$$\zeta\frac{d^2B}{d\zeta^2}+\frac{dB}{d\zeta}-\zeta Da'B=0 \tag{6.196}$$

where Da´ is also defined as before ($\alpha = 0$).

A search in Murphy (1960, p. 329) reveals his Eq. 199, which is of the form of Eq. (6.196). His suggestion is simply to introduce another, slightly modified, independent variable

$$\xi=\zeta\sqrt{Da'} \tag{6.197}$$

and to multiply the resulting equation by ξ, which transforms Eq. (6.196) into

$$\xi^2\frac{d^2B}{d\xi^2}+\xi\frac{dB}{d\xi}-\xi^2 B=0 \tag{6.198}$$

A second consultation with this table identifies Eq. (6.198) as the modified Bessel equation of order 0 (Murphy, 1960, p. 339, his Eq. 275). This equation has as its solution

$$B(\xi)=c_1\,I_o(\xi)+c_2\,K_o(\xi) \tag{6.199}$$

where $I_o(\xi) \equiv$ *hyperbolic* or *modified Bessel function* of the first kind of order zero. This function has the appearance of an increasing exponential (Fig. 6.1). $K_o(\xi) \equiv$ modified Bessel function of the second kind of order 0. $K_o(\xi)$ decays as $\xi \to \infty$ and is unbounded as $\xi = 0$ (Fig. 6.1). $K_o(\xi)$ is also known as Weber's function, Macdonald's function, and the Basset function.

These two modified Bessel functions are defined by the following series (e.g. Lebedev, 1972, p. 108 and 110; Spanier and Oldham, 1987, p. 483 and 502):

$$I_o(\xi)=\sum_{n=0}^{\infty}\frac{(\xi/2)^{2n}}{(n!)^2} \tag{6.200}$$

and

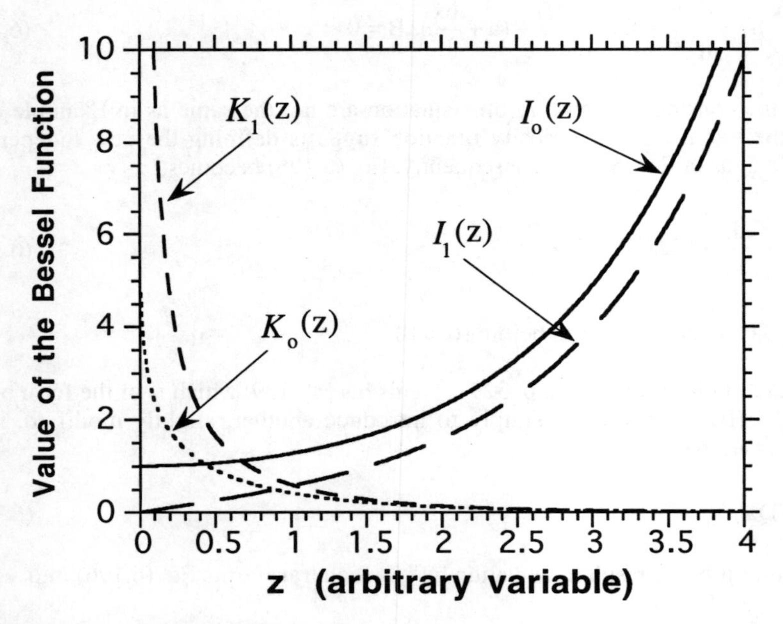

Figure 6.1. Plots of the modified Bessel functions $I_o(z)$, $I_1(z)$, $K_o(z)$, and $K_1(z)$ as functions of the arbitrary input variable z.

$$K_o(\xi) = -\sum_{n=0}^{\infty} \frac{(\xi/2)^{2n}}{(n!)^2}\left[\log\left(\frac{\xi}{2}\right) - \psi(n+1)\right] \tag{6.201}$$

where ψ( ) is the logarithmic derivative of the Gamma function, which also bears the names of Digamma and Psi functions (e.g. Lebedev, 1972, p. 6; Spanier and Oldham, 1987, p. 425),

$$\psi(n+1) = -\gamma + \sum_{k=1}^{n} \frac{1}{k} \tag{6.202}$$

and γ = Euler's constant = 0.57721566. Computer methods to calculate these special functions are discussed below in Section 6.4.7.

In terms of the original depth variable, we have the solution

$$B(x) = c_1\, I_o\left(\left[a + \frac{bx}{L}\right]\sqrt{Da'}\right) + c_2\, K_o\left(\left[a + \frac{bx}{L}\right]\sqrt{Da'}\right) \tag{6.203}$$

This solution in no way resembles the homogeneous part of Eq. (6.193), which also confirms that different porosity functionalities lead to different ODEs and different analytical solutions.

Example 6-19. Another common source for depth-dependent coefficients is a *depth-dependent biodiffusion.* It is well known that the intensity of this process must decrease and eventually disappear with depth in a sediment. This means that the biodiffusion coefficient is a decreasing function of depth. Boudreau (1986a) presents an extensive investigation of the models for this phenomenon, and two of these problems are worth repeating here as examples.

The first problem of this type describes the distribution of a tracer subject to first-order decay, burial, and linearly decreasing biodiffusion in the finite depth-range $0 \le x \le L$ (the bioturbated zone) of a constant porosity sediment,

$$\frac{d}{dx}\left[D_B\left(1-\frac{x}{L}\right)\frac{dB}{dx}\right] - w\frac{dB}{dx} - \lambda B = 0 \tag{6.204}$$

or, upon expansion,

$$D_B\left(1-\frac{x}{L}\right)\frac{d^2B}{dx^2} - \left(w+\frac{D_B}{L}\right)\frac{dB}{dx} - \lambda B = 0 \tag{6.205}$$

This bears some mathematical resemblance to Eq. (6.195) of the last example, and we will discover that this resemblance is more than superficial.

As in the other two examples, a new independent variable can be defined

$$\zeta \equiv 1-\frac{x}{L} \tag{6.206}$$

such that

$$-\frac{1}{L}\frac{d}{d\zeta} = \frac{d}{dx} \tag{6.207}$$

and

$$\frac{1}{L^2}\frac{d^2}{d\zeta^2} = \frac{d^2}{dx^2} \tag{6.208}$$

Substituting these last three relations into Eq. (6.205) and multiplying by L^2/D_B, one obtains

$$\zeta\frac{d^2B}{d\zeta^2}+(1+Pe)\frac{dB}{d\zeta}-Da\,B=0 \tag{6.209}$$

where $Pe \equiv wL/D_B$ (Peclet number) and $Da \equiv \lambda L^2/D_B$ (Damkohler number).

This last equation is now in a form suitable for searching in the tables, and this one is found in Abramowitz and Stegun (1964, p. 362 and 377) as a special form of the modified Bessel equation, i.e. their Eq. 9.1.53 with $z = \zeta^{1/2}$ $p = Pe/2$, $q = 1/2$, $\nu = Pe$, and $\lambda^2 = 4\,Da$. This source also indicates that its solution is

$$B(\zeta)=\left[2\sqrt{Da\,\zeta}\right]^{-Pe}\left\{c_1 I_{Pe}(2\sqrt{Da\,\zeta})+c_2 I_{-Pe}(2\sqrt{Da\,\zeta})\right\} \tag{6.210}$$

if Pe is not an integer, and

$$B(\zeta)=\left[2\sqrt{Da\,\zeta}\right]^{-Pe}\left\{c_1 I_{Pe}(2\sqrt{Da\,\zeta})+c_2 K_{Pe}(2\sqrt{Da\,\zeta})\right\} \tag{6.211}$$

if Pe is an integer, and where c_1 and c_2 are integration constants. In these equations $I_{Pe}(\,)$ and $K_{Pe}(\,)$ are again *modified Bessel functions*, but this time they are of order Pe (or -Pe). However, the second term of each of these solutions is unbounded at $\zeta = 0$ (i.e. $x = L$), which is not physically realizable. Therefore, $c_2 = 0$ in both Eqs (6.210) and (6.211), and the solution becomes

$$B(\zeta)=c_1\left[2\sqrt{Da\,\zeta}\right]^{-Pe} I_{Pe}(2\sqrt{Da\,\zeta}) \tag{6.212}$$

or, in terms of dimensional depth,

$$B(x)=c_1\left[2\sqrt{Da\left(1-\frac{x}{L}\right)}\right]^{-Pe} I_{Pe}\left(2\sqrt{Da\left(1-\frac{x}{L}\right)}\right) \tag{6.213}$$

The integration constant c_1 is determined by the boundary condition at $x = 0$.

Example 6-20. A diagenetic equation need not have polynomial coefficients for the transform method to work. As a final example of this genre, consider the bioturbation and burial of a linearly decaying tracer when mixing decreases exponentially with depth. At constant porosity, the appropriate diagenetic equation is

$$D_B\frac{d}{dx}\left[exp(-\sigma x)\frac{dB}{dx}\right]-w\frac{dB}{dx}-kB=0 \tag{6.214}$$

over the domain $0 \le x \le \infty$ and where D_B, w, k, and σ are all constants.

The trick here is to see if Eq. (6.214) can be transformed into an equation with polynomial coefficients. The exponential function can be considered to be a mapping that takes the original depth domain, $0 \le x \le \infty$, to a new domain {0,1}, if its argument is negative, and {1,∞} if the argument is positive. These properties suggest defining a new independent variable, either as $\zeta = exp(-\sigma x)$ or $\zeta = exp(\sigma x)$. Both will work, but the latter gets to the solution with fewer steps; therefore, set a new variable

$$\zeta \equiv \frac{exp(\sigma x)}{Pe_\sigma} \tag{6.215}$$

where

$$Pe_\sigma \equiv \frac{w}{D_B \sigma} \tag{6.216}$$

The factor Pe_σ does not need to be here, but trial and error will show that the derivation is simpler if it's put in. In addition,

$$\frac{d}{dx} = \sigma \frac{exp(\sigma x)}{Pe_\sigma} \frac{d}{d\zeta} = \sigma \zeta \frac{d}{d\zeta} \tag{6.217}$$

and

$$\frac{d^2}{dx^2} = \sigma^2 \left(\zeta^2 \frac{d^2}{d\zeta^2} + \zeta \frac{d}{d\zeta} \right) \tag{6.218}$$

Now substitute Eqs (6.215) through (6.218) into Eq. (6.214) to obtain

$$\zeta \frac{d^2 B}{d\zeta^2} - \zeta \frac{dB}{d\zeta} - Da_\sigma B = 0 \tag{6.219}$$

where $Da_\sigma \equiv k/(w\sigma)$, a type of Damkohler number. (Remember that a large Da_σ value means that reaction dominates over advection on the scale defined by $1/\sigma$.)

If we consult the literature on special equations, we can discover the confluent hypergeometric equation (Murphy, 1960, p. 331, #211; Abramowitz and Stegun, 1964, Chapter 13; Zwillinger, 1989, p. 124), i.e.

$$\zeta \frac{d^2 B}{d\zeta^2} - (b - \zeta) \frac{dB}{d\zeta} - Da_\sigma B = 0 \tag{6.220}$$

where b is a constant. A comparison readily reveals that Eq. (6.219) is a special case of Eq. (6.220) produced by b = 0. The solution to Eq. (6.219) is then:

$$B(\zeta) = c_1\, U(Da_\sigma;0;\zeta) + c_2\, exp(\zeta)\, U(-Da_\sigma;0;\zeta) \tag{6.221}$$

where $U(\mathrm{Da}_\sigma,0;\zeta)$ and $U(-\mathrm{Da}_\sigma,0;\zeta)$ are two *Tricomi* or *confluent hypergeometric functions* (Fig. 6.2).

For the special case in which b = 0, a Tricomi function can be expressed as (Spanier and Oldham, 1987, p. 473)

$$U(\mathrm{Da};0;\zeta) = \frac{1}{\Gamma(1+\mathrm{Da})} \sum_{n=0}^{\infty} \frac{(\mathrm{Da})_n x^n}{(n!)^2} \{1 + n\, ln(\zeta) + n\,\psi(\mathrm{Da}+n) - 2n\,\psi(n)\} \tag{6.222}$$

where $\Gamma(\,)$ indicates a Gamma function, i.e.

$$\Gamma(z) = \int_0^\infty s^{z-1}\, exp(-s)\, ds \tag{6.223}$$

where s in Eq. (6.223) is a dummy variable. In addition, $\psi(\,)$ in Eq. (6.222) represents the Digamma function (for a non-integer),

$$\psi(z) = -\gamma + \int_0^\infty \frac{exp(-s) - exp(-zs)}{1 - exp(-s)} ds \tag{6.224}$$

where γ is in this case Euler's constant. Finally, $(\mathrm{Da})_n$ in Eq. (6.222) is shorthand for

$$(\mathrm{Da})_n = \mathrm{Da}\,(\mathrm{Da}+1)(\mathrm{Da}+2)(\mathrm{Da}+3)\cdots(\mathrm{Da}+n) \tag{6.225}$$

which is sometimes called a *Pochhammer polynomial* (Spanier and Oldham, 1987, p. 149). (Note that $(\mathrm{Da})_o = 1$, by definition.) The second Tricomi function would be calculated simply by replacing Da with (-Da).

The first term in Eq. (6.221) decays with depth, while the second increases without bound with depth (see Spanier and Oldham, 1987, p. 475, #48:7:2). Consequently, for a solution on the domain of $0 \le x \le \infty$, c_2 must be equal to zero, and the second Tricomi function is removed from the solution. Thus,

$$C(x) = c_1 U(\mathrm{Da}_\sigma; 0; exp(\sigma x)/\mathrm{Pe}_\sigma) \tag{6.226}$$

Note that this Tricomi function has the property that, in the limit $x \to \infty$,

$$\lim_{x \to \mathrm{large}} \left[U(\mathrm{Da}_\sigma; 0; exp(\sigma x)/\mathrm{Pe}_\sigma)\right] \approx exp(-kx/w) \tag{6.227}$$

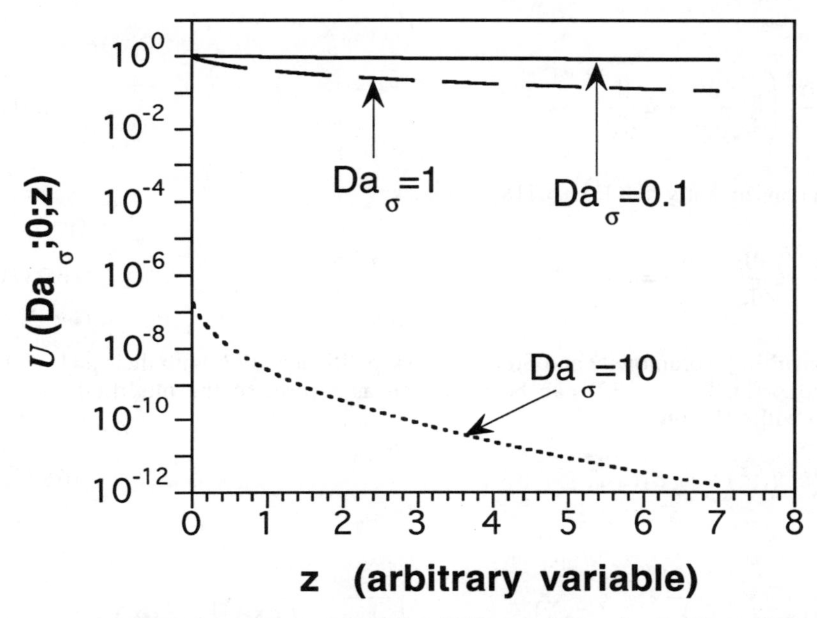

Figure 6.2. Plot of the Tricomi function, $U(\mathrm{Da}_\sigma,0,z)$, for three values of Da_σ. Note the extreme dependence on the value of Da_σ.

which is the solution to the advection-decay equation, as expected.

The Tricomi function is quite difficult to calculate for large values of Da_σ (say $\geq$ 10) and ζ, because of round-off and other errors when computing such small values of the function. A large value of Da_σ implies that the tracer has essentially decayed before advection becomes an important term in its balance. This fact suggests an alternative to Eq. (6.214) with w = 0, first advanced by Nozaki (1977),

$$\frac{d^2B}{dx^2} - \sigma\frac{dB}{dx} - \frac{\lambda}{D_B}\,exp(\sigma x)\,B = 0 \tag{6.228}$$

By introducing the new variable,

$$\xi \equiv 2\sqrt{\frac{k}{D_B\sigma}}\,exp\left(\frac{\sigma x}{2}\right) \tag{6.229}$$

so that

$$\frac{d}{dx} = \frac{\sigma}{2} \xi \frac{d}{d\xi} \tag{6.230}$$

$$\frac{d^2}{dx^2} = \frac{\sigma^2}{4} \left(\xi^2 \frac{d^2}{d\xi^2} + \xi \frac{d}{d\xi} \right) \tag{6.231}$$

We find consequently that Eq. (6.218) becomes:

$$\xi^2 \frac{d^2 B}{d\xi^2} - \xi \frac{dB}{d\xi} - \xi^2 B = 0 \tag{6.232}$$

By consulting Abramowitz and Stegun (1964, p. 362 and 377, with their p=1, q=1, ν=1, and λ=1), Eq. (6.232) can be identified as a form of the modified Bessel equation with solution

$$B(\xi) = [c_1' \, I_1(\xi) + c_2' \, K_1(\xi)] \xi \tag{6.233}$$

or

$$B(x) = \left[c_1 \, I_1\left(2\sqrt{\frac{k}{D_B \sigma}} \, exp\left(\frac{\sigma x}{2}\right) \right) + c_2 \, K_1\left(2\sqrt{\frac{k}{D_B \sigma}} \, exp\left(\frac{\sigma x}{2}\right) \right) \right] exp\left(\frac{\sigma x}{2}\right) \tag{6.234}$$

where c_1 and c_2 are two new integration constants, and $I_1(\ )$ and $K_1(\ )$ are the *modified Bessel functions* of the first and second kinds, respectively, of order 1 (see Fig. 6.1). Note that, because the solution should remain finite as $x \to \infty$, $c_1 = 0$. Modified Bessel functions are easier to calculate than is the Tricomi function, but personally I prefer to do this problem numerically (see Chapter 8).

6.4.7 Calculation of Special Functions

The higher transcendental functions are perhaps unfamiliar to many readers, and they are not normally available as call-up functions on most computer systems. Furthermore, the formal definitions given above are not always the best method to calculate values of these functions; so their calculation requires the availability of specific *algorithms* (specific computational formulas or computer codes).

Public-domain algorithms can be found in variety of sources. Abramowitz and Stegun (1964) present computational formulas for a limited set of functions, which include the $I_o(x)$, $I_1(x)$, $K_o(x)$, and $K_1(x)$. Spanier and Oldham (1987) give relatively simple "pseudo-codes" for the calculation of most special functions, including $I_\nu(x)$ and $K_\nu(x)$. Pseudo-code is a set of generic instructions that can be implemented in almost any computer language (FORTRAN, PASCAL, C, etc.).

Included in that list are the modified Bessel functions and the Tricomi function. Cody (1983) provides a FORTRAN code for the modified Bessel function of arbitrary order. Luke (1977) offers a set of FORTRAN codes for some higher functions, but the beginner may find them difficult to understand. Finally, Press et al. (1992) also offer algorithms and computer codes for some special functions, but the reader is reminded that these are copyrighted. Well-tested public-domain FORTRAN codes for many special functions can be found in the large repository at the NETLIB site on the INTERNET. The Appendix explains how these codes may be obtained from either NETLIB or the author.

6.5 Nonlinear Second-Order ODEs

These equations are of the general form:

$$\frac{d}{dx}\left[D(C)\frac{dC}{dx}\right]-\frac{dv(C)C}{dx}-f(C)=g(x) \qquad (6.235)$$

and are nonlinear because at least one of the following conditions is true: i) D(C) is a function of the (generic) concentration C, e.g. biodiffusion as a function of organic matter concentration; ii) burial, v(C), is a function of C, e.g. burial is dependent on calcite or opal content when these minerals dissolve with depth in a sediment (Schink and Guinasso, 1980); and iii) most commonly, the reaction kinetics, f(C), are nonlinear functions of C, including nonlinear coupling with other species.

The one truth about such equations is that there is no general solution, such as Eq. (6.75) for linear equations. If a given equation has an analytical solution (and that's a big "if"), then this solution is unique in that it applies only to that equation. Most such equations don't have analytical solutions, and numerical methods are then the only recourse (Chapter 8). Those equations that admit to analytical solutions normally don't contain an advective term and inhomogeneous (depth-dependent) source/sink terms; that is to say, they are of the simpler form,

$$\frac{d}{dx}\left[D(C)\frac{dC}{dx}\right]\pm f(C)=0 \qquad (6.236)$$

A first integral (i.e. a conversion to a first-order ODE) may then be possible by multiplying the equation by 2 dC/dx (Ames, 1965). If need be, a little rearrangement may facilitate this process. The second integration for concentration follows the methods discussed for nonlinear first-order ODEs. A few examples will illustrate this methodology.

Example 6-21. For solids such as calcite and aragonite ($CaCO_3$ polymorphs), the rate of dissolution is often modelled as a function of the difference between the ion concentration-product at equilibrium, i.e. $(ICP)_{eq} = (Ca^{2+})_{eq}(CO^{3=})_{eq}$, and that which actually exists in the porewater (see Schink and Guinasso, 1977a,b; Keir,

1982)[§]. A primitive model for $CO_3^=$, without forcing from organic matter decomposition, ignoring the dissolved acid-base reactions, and assuming constant excess solid $CaCO_3$ and constant porosity, is given by

$$D'\frac{d^2CO_3^=}{dx^2} + k'\left[(Ca^{2+})_{eq}(CO_3^{2-})_{eq} - (Ca^{2+})(CO_3^{2-})\right]^n = 0 \qquad (6.237)$$

where k' is an apparent rate constant and n is the order of reaction, assumed to be greater than 1, but not necessarily an integer. Note that $k' = (1-\varphi)k_h\bar{s}B/\varphi$, where φ is the porosity, k_h is the heterogeneous rate constant, $\bar{s}$ is the specific surface area, and B is the mass of dissolvable $CaCO_3$. The dissolved calcium concentration can frequently be regarded as a constant in porewaters (Schink and Guinasso, 1977a,b) because of its relatively high concentration in seawater-like solutions. With this assumption, Eq. (6.237) can be written as

$$D'\frac{d^2C}{dx^2} - kC^n = 0 \qquad (6.238)$$

where $C \equiv (CO_3^=)_{eq} - (CO_3^=)$, $n > 1$, and k is a new constant equal to $k'[(Ca^{2+})_{eq}]^n$.

Appropriate boundary conditions for Eq. (6.238) are

$$C(x=0) = C(0) \qquad (6.239)$$

$$\left.\frac{dC}{dx}\right|_{x\to\infty} = 0 \qquad (6.240)$$

To solve, multiply each term in Eq. (6.238) by 2 dC/dx,

$$2D'\frac{dC}{dx}\frac{d^2C}{dx^2} - 2\frac{dC}{dx}kC^n = 0 \qquad (6.241)$$

then recognize that

$$\frac{d}{dx}\left(\frac{dC}{dx}\right)^2 = 2\frac{dC}{dx}\frac{d^2C}{dx^2} \qquad (6.242)$$

so that Eq. (6.241) can be written as

$$\frac{d}{dx}\left(\frac{dC}{dx}\right)^2 - \frac{2k}{D'}C^n\frac{dC}{dx} = 0 \qquad (6.243)$$

[§] I am not trying to defend this approach as a good model of carbonate diagenesis, but simply saying how the resulting equations are solved as an example.

Equation (6.243) can be integrated directly, using boundary condition Eq. (6.240) to obtain

$$\frac{dC}{dx} = -\left(\frac{2k}{(n+1)D'}\right)^{1/2} C^m \tag{6.244}$$

with $m \equiv (n+1)/2$. The negative root is chosen because the gradient should decrease with depth. Fortunately, this ODE is separable, i.e.

$$\frac{dC}{C^m} = -\left(\frac{2k}{(n+1)D'}\right)^{1/2} dx \tag{6.245}$$

Finally, integrate both sides of this last equation using Eq. (6.239) and reorganize slightly, to obtain

$$C = \left[\left(\frac{n-1}{2}\right)\left(\frac{2k}{(n+1)D}\right)^{1/2} x + c_o\right]^{\frac{1}{1-m}} \tag{6.246}$$

where

$$c_o = C(0)^{1-m} \tag{6.247}$$

If the $CaCO_3$ concentration, B, is a known quantity, then concentration profiles can be generated by substituting this value, as well as those of the other parameters, into Eq. (6.246). More often, however, we wish to predict B from a prescribed input flux of $CaCO_3$, F_B. In this case, Eq. (6.246) is coupled to the conservation equation for B, which is an integral conservation equation of the form of Eq. (3.83), i.e.

$$F_B - (1-\varphi)wB + \varphi D' \left.\frac{dC}{dx}\right|_{x=0} = 0 \tag{6.248}$$

The derivative term in Eq. (6.248) can be evaluated with Eq. (6.244). On the other hand, the burial velocity w must be a function of the $CaCO_3$ preserved in the sediment (Schink and Guinasso, 1980). This relation can be derived from the integral conservation for the inert material (clay) accumulating in the sediment and the static conservation of solid volume, i.e.

$$F_M - (1-\varphi)wM = 0 \tag{6.249}$$

and

$$\frac{B}{\rho_B} + \frac{M}{\rho_M} = 1 \tag{6.250}$$

respectively, where M is the concentration of inert material, and ρ_B and ρ_M are the intrinsic densities of $CaCO_3$ and clay, respectively. Together, these equations give

$$w = \frac{F_M}{(1-\varphi)\rho_M\left(1-\frac{B}{\rho_B}\right)} \tag{6.251}$$

Substituting Eqs (6.244) and (6.251) into Eq. (6.248) produces

$$F_B - \frac{F_M B}{\rho_M\left(1-\frac{B}{\rho_B}\right)} - \varphi D'\left(\frac{2(1-\varphi)k_h\bar{s}B\left(Ca^{2+}\right)_{eq}^n}{\varphi(n+1)D'}\right)^{1/2} C(0)^m = 0 \tag{6.252}$$

This nonlinear algebraic equation must now be solved for B, which is relatively easy to do numerically (see Chapter 8).

Example 6-22. To illustrate that the solution to nonlinear ODEs is equation dependent, we consider the same problem as Eq. (6.238) with Monod kinetics,

$$D'\frac{d^2C}{dx^2} - k\frac{C}{K_M + C} = 0 \tag{6.253}$$

where k is a rate constant and K_M is the saturation constant. Equation (6.253) might describe the diffusive penetration of a reactive organic pollutant into a sediment. The difference with Eq. (6.238) is the form of the kinetics. Following the scheme outlined in the previous problem, we multiply by 2 dC/dx and integrate to get

$$\left(\frac{dC}{dx}\right)^2 = \frac{2k}{D'}\int\frac{C}{K_M + C}dC \tag{6.254}$$

To evaluate the integral in this last equation, we can either look it up in a table of integrals (see, for example, CRC 1979 or Gradshteyn and Ryzhik, 1980), or we can work it out manually, which is quite simple in this case. Define a new (temporary) dependent variable,

$$Y = K_M + C \tag{6.255}$$

then

$$\int\frac{C}{K_M + C}dC = \int\frac{Y - K_M}{Y}dY = \int dY - K_M\int Y^{-1}dY \tag{6.256}$$

and

$$\int dY - K_M \int Y^{-1} dY = Y - K_M \, ln(Y) + c_o \tag{6.257}$$

where c_o is an integration constant. Equation (6.254) then becomes

$$\frac{dC}{dx} = \left[\frac{2k}{D'} \{ Y - K_M \, ln(Y) + c_o \} \right]^{1/2} \tag{6.258}$$

or, in terms of the original concentration variable,

$$\frac{dC}{dx} = \left[\frac{2k}{D'} \{ (K_M + C) - K_M \, ln(K_M + C) + c_o \} \right]^{1/2} \tag{6.259}$$

If we implement boundary condition (6.240), i.e. no flux at depth, then we get

$$c_o = -K_M + K_M \, ln(K_M) \tag{6.260}$$

and Eq. (6.259) now reads

$$\frac{dC}{dx} = \left[\frac{2k}{D'} \left\{ C + K_M \, ln\left(\frac{K_M}{K_M + C} \right) \right\} \right]^{1/2} \tag{6.261}$$

Try as you might, you won't find a second integral (solution) of this last ODE in terms of analytical functions, even though it is formally separable! Not all integrals can be evaluated in closed form. This does not mean that all the work to this point has been for naught; on the contrary, Eq. (6.261) is much easier to integrate numerically than the original Eq. (6.249). Boudreau (1990b) employs just this type of result to advantage in modelling silica diagenesis in non-mixed sediments.

Example 6-23. Coupled nonlinear equations can be solved in the same way, if they can be reduced to a single equation of the form of Eq. (6.238). In particular, Wong and Grosch (1978) analytically solve a model of silica diagenesis by this technique. They begin with the equations

$$D' \frac{d^2C}{dx^2} + \varepsilon k \left(C_{eq} - C \right) B = 0 \tag{6.262}$$

$$D_B \frac{d^2B}{dx^2} - k \left(C_{eq} - C \right) B = 0 \tag{6.263}$$

for dissolved silica and solid opal, respectively. Note that ε is simply a unit conversion constant, if needed. Boudreau (1990a) discusses fully the geochemical limitations of this particular description, but these need not detain us here.

To make the mathematical development simpler to follow, it is advisable to introduce the new dependent variable for dissolved silica given by

$$\Theta = C_{eq} - C \tag{6.264}$$

Thus, Eqs (6.262) and (6.263) become

$$D' \frac{d^2\Theta}{dx^2} - \varepsilon k \Theta B = 0 \tag{6.265}$$

and

$$D_B \frac{d^2B}{dx^2} - k \Theta B = 0 \tag{6.266}$$

At this stage, Eq. (6.266) can be multiplied by ε and subtracted from Eq. (6.265) to obtain an equation linking Θ and B,

$$D' \frac{d^2\Theta}{dx^2} - \varepsilon D_B \frac{d^2B}{dx^2} = 0 \tag{6.267}$$

or, upon integrating once,

$$D' \frac{d\Theta}{dx} - \varepsilon D_B \frac{dB}{dx} = c_1 \tag{6.268}$$

and twice,

$$D'\Theta - \varepsilon D_B B = c_1 x + c_2 \tag{6.269}$$

where c_1 and c_2 are integration constants.

Normally, the sediment is taken to be semi-infinite; i.e. it occupies the region $0 \le x < \infty$. If the concentrations of silica and opal are to remain finite as $x \to \infty$, then it is necessary that $c_1 = 0$ in Eqs (6.268) and (6.269). If we now write Eq. (6.269) for B as a function of Θ and substitute into Eq. (6.265), we arrive at

$$D' \frac{d^2\Theta}{dx^2} - \frac{k}{D_B} \Theta (D'\Theta - c_2) = 0 \tag{6.270}$$

which is a single equation of the form of Eq. (6.238), and it can be solved by the method of Example 6-21. The details can be found in Wong and Grosch (1978), who explain the calculation of the integration constant c_2, which they call an eigenvalue, as a function of the boundary conditions and the amount of opal preserved in the sediment.

Example 6-24. This last example illustrates what can result when nonlinearity appears in the diffusion coefficient. This example looks at the effects of concentration-dependent biodiffusion on the distribution of organic matter. Specifically, if the intensity of biodiffusion is related to the reactive (labile) organic matter concentration, G, then, as a first approximation we can assume that

$$D_B(G) = D^* G \tag{6.271}$$

where D^* is a constant with units of squared length per unit concentration per unit of time. This relationship has conceptual appeal. The smaller the amount of organic matter, the less intense the bioturbation, a result which is not contradicted by observation, although one may want to argue about the exact form of Eq. (6.271).

Assuming first-order decay, constant porosity, and advection, then the conservation equation for the labile organic matter is

$$\frac{d}{dx}\left[G\frac{dG}{dx}\right] - \frac{k}{D^*}G = 0 \tag{6.272}$$

How might we go about solving such an equation? There is nothing quite like it in Murphy (1960), for example. As with many nonlinear equations, there is a mathematical trick that becomes more obvious after sufficient practice. The definition of a derivative of a function tells us that

$$G\frac{dG}{dx} = \frac{1}{2}\frac{d(G^2)}{dx} \tag{6.273}$$

This means that Eq. (6.272) can be written as

$$\frac{d^2(G^2)}{dx^2} - \frac{2k}{D^*}G = 0 \tag{6.274}$$

Now define a new dependent variable,

$$Y \equiv G^2 \tag{6.275}$$

and introduce it into Eq. (6.274),

$$\frac{d^2Y}{dx^2} - \frac{2k}{D^*}Y^{1/2} = 0 \tag{6.276}$$

This equation balances the diffusion of Y with consumption by *fractional order kinetics*, i.e. Eq. (6.238) with $n<1$ (see Frank-Kamenetskii, 1969). The emergence of fractional kinetics is meaningful. Remember that the solution to diffusion with first-order kinetics on a semi-infinite depth interval, Eq. (6.108), is an exponential that decreases as $x \to \infty$. In contrast, Example 6-14 shows that diffusion with zeroth-order kinetics creates a situation in which the diffusing species disappears at a finite depth x=L, which must be calculated as part of the solution. Fractional kinetics, links these two types of behavior. As the order of reaction goes from 0 to 1, the depth of disappearance will increase, for set values of the parameters and interfacial concentration (x=0). Thus, even though the original equation for G did not have fractional kinetics, the interaction between linear kinetics and a biodiffusion coefficient that depends linearly on G generates similar behavior.

As long as the order is less than one, the disappearance depth (x=L) will also be an unknown point. This means that in order to solve Eq. (6.276) properly, we must specify three boundary conditions:

$$G(x=0) = G_o \tag{6.277}$$

$$G(x=L) = 0 \tag{6.278}$$

$$\left.\frac{dG}{dx}\right|_L = 0 \tag{6.279}$$

where, again, L is unknown. (These equations are easily converted to boundary conditions in Y.)

Equation (6.276) is solved by the method for solving Eq. (6.236); i.e. multiply by 2 dY/dx and integrate directly to obtain

$$\left(\frac{dY}{dx}\right)^2 = \frac{8k}{3D^*}Y^{3/2} + c_1 \tag{6.280}$$

where c_1 is an integration constant; however, we know that at some depth L, Eqs (6.278) and (6.279) hold true, so that $c_1 = 0$ based on the values at that depth.

Equation (6.280) is separable, and a second integration produces

$$4Y^{1/4} = -\sqrt{\frac{8k}{3D^*}}\,x + c_2 \tag{6.281}$$

The negative reflects the decrease in concentration with depth. Application of Eq. (6.277) informs us that

$$c_2 = 4Y_o^{1/4} = 4C_o^{1/2} \tag{6.282}$$

so that

$$Y = \left[-\frac{1}{4}\sqrt{\frac{8k}{3D^*}}\ x + Y_o^{1/4} \right]^4 \qquad (6.283)$$

or

$$G = \left[-\frac{1}{4}\sqrt{\frac{8k}{3D^*}}\ x + G_o^{1/2} \right]^2 \qquad (6.284)$$

To calculate the depth L, apply Eq. (6.278) to Eq. (6.283), i.e.

$$L = 4\left(\frac{3D^*}{8k} \right)^{1/2} Y_o^{1/4} \qquad (6.285)$$

or

$$L = 4\sqrt{\frac{3D^* G_o}{8k}} \qquad (6.286)$$

which is always finite for finite G_o. Notice the resemblance to Eq. (6.132) with its square-root concentration dependence.

6.6 Solution of Steady-State DAEs

To remind the reader, DAEs are differential-algebraic equations (see Chapter 3). They are generated when local equilibrium conditions govern one or more reactions affecting the species of interest. Such equations may have analytical solutions in at least two circumstances. First, if the equilibrium relation(s) is (are) linear, then the problem can always be reduced to solving one or more ODEs. Still, this does not guarantee that the resulting ODE has an analytical solution. This case is discussed in detail by Berner (1980).

Secondly, an "exact" solution may be possible, even with nonlinear equilibria, if the ODE(s) is (are) linear, advection is negligible, and all reactions are inhomogeneous sources and sinks (i.e. they are dependent only on x). Olander (1960), Otto and Quinn (1971), Chang and Rochelle (1982), and Meldon et al. (1982) provide examples of such problems from the field of chemical engineering. Boudreau (1987, 1991) and Boudreau and Canfield (1988, 1993) make use of this fact in their models for porewater pH, and Raiswell et al. (1993) in their model of localized pyrite formation. The final example of this chapter illustrates a simple model that follows this procedure.

Example 6-25. The situation under consideration is the dissolved carbonate system, i.e. $CO_2(aq)$, HCO_3^-, and $CO_3^=$, with the inhomogeneous production of either $CO_2(aq)$ from oxic organic matter decay or HCO_3^- from sulfate reduction. These dissolved species are interconvertible through the reversible reaction

$$CO_2(aq) + CO_3^= + H_2O \leftrightarrow 2\,HCO_3^- \tag{6.287}$$

which is assumed to be at equilibrium. To greatly simplify this example, we also assume that there is no $CaCO_3$ dissolution during oxic decay and no precipitation of this solid during sulfate reduction.

For a steady state with constant porosity and negligible advection, the three conservation equations for these dissolved carbonate species are

$$D_{CO2}\frac{d^2(CO_2)}{dx^2} - r_e + f(x) = 0 \tag{6.288}$$

$$D_{HCO3}\frac{d^2(HCO_3)}{dx^2} + 2r_e + g(x) = 0 \tag{6.289}$$

$$D_{CO3}\frac{d^2(CO_3)}{dx^2} - r_e = 0 \tag{6.290}$$

for $CO_2(aq)$, HCO_3^-, and $CO_3^=$, respectively, and where r_e is the net rate of reaction (6.287), f(x) is the source of $CO_2(aq)$ during oxic decomposition, and g(x) is bicarbonate production by sulfate reduction.

In the standard approach to the problem, the rate term r_e would be resolved by introducing some sort of kinetic expression. Yet we know that this reaction is fast and reversible; consequently, in the DAE approach, we treat r_e as an unknown and eliminate it from the equations. Manipulating these equations to remove r_e leads to

$$D_{CO2}\frac{d^2(CO_2)}{dx^2} + D_{HCO3}\frac{d^2(HCO_3)}{dx^2} + D_{CO3}\frac{d^2(CO_3)}{dx^2} = -(f(x) + g(x)) \tag{6.291}$$

and

$$D_{HCO3}\frac{d^2(HCO_3)}{dx^2} + 2D_{CO3}\frac{d^2(CO_3)}{dx^2} = -g(x) \tag{6.292}$$

Equation (6.291) can be interpreted as a statement of conservation of (weighted) total CO_2, while Eq. (6.292) describes charge balance, or weighted alkalinity in the system. A third possible combination of these equations is not linearly independent of these two new equations.

There is then a problem in that there are now only two equations, i.e. Eqs (6.291) and (6.292), for three dissolved species. To eliminate this indeterminacy, we can add the equilibrium relation for reaction (6.287), i.e.

$$K_{eq}\,(CO_2)(CO_3) = (HCO_3)^2 \tag{6.293}$$

To solve this new set of equations, note that Eqs (6.291) and (6.292) are linear, and they can be integrated directly twice to produce two linear algebraic equations:

$$D_{CO2}(CO_2) + D_{HCO3}(HCO_3) + D_{CO3}(CO_3) = -F(x) - G(x) + a_1 x + a_2 \tag{6.294}$$

from Eq. (6.291), and

$$D_{HCO3}(HCO_3) + 2D_{CO3}(CO_3) = -G(x) + a_3 x + a_4 \tag{6.295}$$

from Eq. (6.292), where

$$F(x) = \int_0^x \int_{x'}^L f(x'')dx'' \tag{6.296}$$

$$G(x) = \int_0^x \int_{x'}^L g(x'')dx'' \tag{6.297}$$

and a_1, a_2, a_3, and a_4 are integration constants. The depth x=L is the base of the model. Normally, gradients are made to disappear at this depth, and if $L = \infty$, then a_1 and a_3 are both zero.

Equations (6.293) through (6.295) constitute a system of three algebraic equations for three unknown concentrations as functions of depth. One can solve for any of the unknown concentrations, but the result is a nonlinear equation; e.g. for HCO_3^-,

$$\begin{aligned}\frac{2D_{CO2}D_{CO3}(HCO_3)^2}{K_{eq}\left[a_3x + a_4 - F(x) - D_{HCO3}(HCO_3)\right]} + D_{HCO3}(HCO_3)\\ + \frac{1}{2}\left[a_3x + a_4 - D_{HCO3}(HCO_3)\right] = -\frac{1}{2}F(x) - G(x) + a_1x + a_2\end{aligned} \tag{6.298}$$

To calculate the HCO_3^- concentration, one must find the value of (HCO_3) that makes the left-hand side of this equation equal to the right-hand side. This is called an *implicit solution*, and the process of finding the correct (HCO_3) is called finding the *root* of this equation. Finding the root can be done numerically (see Chapter 8).

7 ANALYTICAL SOLUTIONS FOR TIME-DEPENDENT AND/OR MULTI-DIMENSIONAL DIAGENETIC EQUATIONS

"Are five nights warmer than one night, then?" Alice ventured to ask.

"Five times as warm, of course."

"But they should be five times as cold, by the same rule -"

"Just so!" cried the Red Queen.

(Lewis Carroll, *Through the Looking-Glass,* 1865)

7.1 Introduction

Time-dependent and/or multi-dimensional diagenetic equations are in the form of partial differential equations (PDE). Such equations can be classified into three broad categories:

i) *Parabolic Equations* involve one derivative of time and at least one second derivative in a space dimension, e.g. Fick's second law:

$$\frac{\partial C}{\partial t} = D\frac{\partial^2 C}{\partial x^2} \tag{7.1}$$

ii) *Elliptic Equations* are time independent and involve second derivatives in at least two spatial dimensions, e.g. steady-state diffusion in a 2-D rectangular domain:

$$D\left[\frac{\partial^2 C}{\partial x^2} + \frac{\partial^2 C}{\partial y^2}\right] = 0 \tag{7.2}$$

iii) *Hyperbolic Equations* contain the same order of space and time derivatives, e.g. the 1-D transient advection equation:

$$\frac{\partial C}{\partial t} + w\frac{\partial C}{\partial x} = 0 \tag{7.3}$$

The presence or absence of additional advection (i.e. first derivative) or reaction terms in any of these equations is irrelevant to this classification, although it may affect their analytic solvability. Nevertheless, this classification is more than mathematical pigeonholing, because the optimum method of solution for a given PDE is partially a function of its type. All three types of equations appear as diagenetic models (e.g. Lerman and Weiler, 1970; Lerman and Jones, 1973; Guinasso and Schink, 1975; Lasaga and Holland, 1976; Matisoff, 1980; Aller, 1977, 1980, 1982; Boudreau, 1986a,b; Boudreau and Imboden, 1987; Wheatcroft et al., 1990.)

In addition, PDEs may be *linear* or *nonlinear* for the same reasons as ODEs (see Chapter 2). Linearity is also a prime determinant in the choice of solution method. In fact, nonlinear PDEs are sufficiently difficult to solve exactly that the reader is referred to the books by Ames (1965, 1968), Özisik (1980), Zauderer (1983), and Zwillinger (1992), if an analytical solution is absolutely necessary. Otherwise, numerical methods can deal with nonlinearity in PDEs in a relatively straightforward manner (see Chapter 8).

There are four commonly applied methods for the solution of *linear* PDEs from transport-reaction models: a) similarity variables; b) separation of variables and Fourier series; c) integral transforms, particularly Laplace transforms; and d) Green's functions. Even from this limited selection, only the first three are discussed here in any detail. Green's functions are addressed only as they pertain to the use of the convolution integral (see below).

7.2 Transformations and Simplifications

The simpler the form of PDE, the easier it is to apply these methods. Consequently, if it is possible to simplify the PDE, there may be a significant reduction in the effort needed to gain a solution. A well-known example of this type of simplification applies to the *linear Advection-Diffusion-Reaction* equation (ADR). This equation can be transformed into the simpler *Diffusion* equation (Carslaw and Jaeger, 1959). Specifically, one starts with

$$\frac{\partial C}{\partial t} = D\frac{\partial^2 C}{\partial x^2} - v\frac{\partial C}{\partial x} - kC + R(x) \qquad (7.4)$$

where D, v, and k are the generic diffusion coefficient, advective velocity, and first-order rate constant, and where R(x) is an inhomogeneous source/sink term that depends only on x.

The first simplification is to eliminate R(x) from the PDE; let

$$C(x,t) = U(x,t) + V(x) \qquad (7.5)$$

where V(x) is a steady-state component which is the solution to the steady-state diagenetic equation (i.e. the ODE),

$$D\frac{d^2V}{dx^2} - v\frac{dV}{dx} - kV + R(x) = 0 \tag{7.6}$$

Consequently, U(x,t) can be interpreted as the temporal deviation from the steady state given by V(x). Substitution of Eq. (7.5) into Eq. (7.4) leads to

$$\frac{\partial(U+V)}{\partial t} = D\frac{\partial^2(U+V)}{\partial x^2} - v\frac{\partial(U+V)}{\partial x} - k(U+V) + R(x) \tag{7.7a}$$

or

$$\frac{\partial U}{\partial t} - \left[D\frac{\partial^2 U}{\partial x^2} - v\frac{\partial U}{\partial x} - kU\right] = D\frac{\partial^2 V}{\partial x^2} - v\frac{\partial V}{\partial x} - kV + R(x) \tag{7.7b}$$

because $\partial V/\partial t = 0$, by definition. The right-hand side of Eq. (7.7b) is equal to zero by Eq. (7.6), so that

$$\frac{\partial U}{\partial t} = D\frac{\partial^2 U}{\partial x^2} - v\frac{\partial U}{\partial x} - kU \tag{7.8}$$

which removes the inhomogeneous R(x) term.

Next, assume that U(x,t) can be written in terms of a new variable,

$$U(x,t) = W(x,t)\, exp\left(\frac{vx}{2D} - \frac{v^2t}{4D} - kt\right) \tag{7.9}$$

Substitution of Eq. (7.9) into Eq. (7.8) produces (see Özisik, 1980, p. 74)

$$\frac{\partial W}{\partial t} = D\frac{\partial^2 W}{\partial x^2} \tag{7.10}$$

which is the Diffusion Equation. Remember that these same transformations must also be made to the boundary and initial conditions. This may create a new problem with respect to any time-dependent boundary condition, and the resulting problem may need to be solved via Duhamel's Theorem (see Section 7.5). Furthermore, the transform U = W *exp*(-kt), included in Eq. (7.9), can be used to eliminate the linear reaction term in transport equations in all the standard coordinate systems.

The solution to the original problem, Eq. (7.4), will be the sum of the solutions to Eqs (7.6) and (7.10), taking into account Eq. (7.9). This fact is an important feature of the solution to linear PDEs; that is to say, their solution can be expressed as the sum of solutions to other linear PDEs. This is called *superposition*, and it will reappear frequently in the presentation given below.

Another transformation, found in Crank (1975, p. 104), simplifies the treatment of time-dependent diffusivity, and it has proven to be useful in diagenetic studies (e.g. Aller, 1977). Crank considers expressly a problem of the form

$$\frac{\partial C}{\partial t} = D(t)\frac{\partial^2 C}{\partial x^2} \tag{7.11}$$

where x is depth, t is time, C is a generic concentration, and D(t) is a time-dependent generic diffusion coefficient. D(t) may represent either molecular diffusion or bio-diffusion, in which case the time dependence results from temperature variations, e.g. caused by seasonal effects. Crank (1975) invokes the transformation

$$d\tau = D(t)dt \tag{7.12}$$

or, explicitly,

$$\tau = \int_0^t D(t')dt' \tag{7.13}$$

where τ is a new time variable. If D(t) is simple enough, this integral can be evaluated to give τ as a function of t or, just as importantly, t as a function of τ. Substitution of Eq. (7.12) into Eq. (7.11) converts the latter to

$$\frac{\partial C}{\partial \tau} = \frac{\partial^2 C}{\partial x^2} \tag{7.14}$$

which is much easier to solve.

Aller (1977) recognized that this procedure could easily be extended to treat diagenetic equations of the form

$$\frac{\partial C}{\partial t} = D_o A_1(t)\frac{\partial^2 C}{\partial x^2} + R(x)A_2(t) \tag{7.15}$$

where D_o is a mean diffusivity, $A_1(t)$ and $A_2(t)$ are activation energy functions for diffusion and reaction, respectively, and R(x) is a reference or mean rate of reaction at depth x. As an example of the activation function, Aller (1977) gives

$$A_1(t) = A_o exp\left(-\frac{E_a}{B + P\,cos(\omega t)}\right) \tag{7.16}$$

where A_o, B, and P are empirical constants, E_a is the activation energy, and ω is the frequency of the variations. Other forms are certainly possible (see Basha and El-Habel, 1993).

If one applies the transform Eq. (7.12) to Eq. (7.15), then one obtains

$$\frac{\partial C}{\partial \tau} = D_o \frac{\partial^2 C}{\partial x^2} + R(x)A(\tau) \tag{7.17}$$

where $A(\tau) \equiv A_2(\tau)/A_1(\tau)$. In the special case when $A(\tau)$ is equal to 1, i.e. $A_1(\tau) = A_2(\tau)$, the problem is reduced to a constant coefficient equation. Otherwise, one can use Duhamel's Theorem to arrive at a solution Eq. (7.17), as given in Section 7.5.

Finally, a special transformation is available for radially symmetric spherical problems, such as modelling concretions and micro-environments. First, the transport-reaction equation

$$\frac{\partial C}{\partial t} = D\left(\frac{\partial^2 C}{\partial r^2} + \frac{2}{r}\frac{\partial C}{\partial r}\right) - kC \tag{7.18}$$

can be reduced to the pure diffusion equation by defining $U(r,t) = C(r,t)\, exp(-kt)$, which removes the first-order reaction. So far, this is available to equations in all coordinate systems; however, now set

$$W(r,t) = U(r,t)\cdot r \tag{7.19}$$

which produces

$$\frac{\partial W}{\partial t} = D\frac{\partial^2 W}{\partial r^2} \tag{7.20}$$

The reader should try this out. This last result is the simple 1-D diffusion equation, for which a host of solutions are known. Further information on these transformations can be found in Carslaw and Jaeger (1959), Crank (1975), and Özisik (1980).

7.3 Similarity Variables

Only a relatively small number of PDEs are amenable to solution by similarity variables, and most are parabolic (see Zwillinger, 1989, for details). Yet, some are so famous that this book would be remiss in not discussing at least one example. The basic premise is to find a single new independent variable that is a combination of the original independent variables and that transforms the original PDE into an ODE. This is done because ODEs are simpler to solve (in theory, anyway).

The requirements for this method to work are: i) a semi-infinite space domain, i.e. $0 \le x < \infty$, and ii) the collapse of two of the boundary conditions into one upon substitution of the similarity variable; i.e. the ODE must have one less condition.

Example 7-1 (The Error Function). One important example of the use of a similarity variable to solve a PDE is pure diffusion in a semi-infinite sediment, i.e.

$$\frac{\partial C}{\partial t} = D\frac{\partial^2 C}{\partial x^2} \tag{7.21}$$

with the initial condition

$$C(x,0) = 0 \tag{7.22}$$

and boundary conditions

$$C(0,t) = C_o \tag{7.23}$$

$$C(\infty,t) = 0 \tag{7.24}$$

There exists for this problem the following similarity variable (see Zwillinger, 1989, p. 375, on how to derive it):

$$\eta = \frac{x}{\sqrt{4Dt}} \tag{7.25}$$

(Please do not confuse η with the viscosity in Chapter 4.) The "chain rule" for differentiation can be used to derive

$$\frac{\partial C}{\partial t} = -\frac{\eta}{2t}\frac{dC}{d\eta} \tag{7.26}$$

$$\frac{\partial^2 C}{\partial x^2} = \frac{\eta^2}{x^2}\frac{d^2 C}{d\eta^2} \tag{7.27}$$

With these results, Eq. (7.21) becomes the ODE

$$\frac{d^2 C}{d\eta^2} + 2\eta\frac{dC}{d\eta} = 0 \tag{7.28}$$

with boundary conditions

$$C(\eta=0) = C_o \tag{7.29}$$

$$C(\eta\to\infty) = 0 \tag{7.30}$$

Note the collapse of the initial condition and boundary condition at infinity into one.

A first direct integration of Eq. (7.28) gives

$$\frac{dC}{d\eta} = a\,exp\left(-\eta^2\right) \tag{7.31}$$

where a is an integration constant. This is a separable equation, and a second integration gives

$$C = a\int_0^{\eta} exp\left(-\eta^2\right)d\eta + b \tag{7.32}$$

where b is also an integration constant. Using boundary condition Eq. (7.29), we get that

$$b = C_o \tag{7.33}$$

and boundary condition Eq. (7.30) gives

$$a = \frac{-C_o}{\int_0^{\infty} exp\left(-\eta^2\right)d\eta} \tag{7.34}$$

Now the integral in the denominator of Eq. (7.34) can be evaluated to obtain (Gradshteyn and Ryzhik, 1980, p. 306)

$$\int_0^{\infty} exp\left(-\eta^2\right)d\eta = \frac{\sqrt{\pi}}{2} \tag{7.35}$$

so that the solution, Eq. (7.32), becomes

$$C = C_o\left[1 - \frac{2}{\sqrt{\pi}}\int_0^{\eta} exp\left(-\eta^2\right)d\eta\right] \tag{7.36}$$

To some readers, this will look familiar. The integral in the brackets is not expressible in terms of simple functions; however, its values are easily tabulated (numerically), and it is now known as the *Error function*, $erf(\eta)$,

$$erf(\eta) = erf\left(\frac{x}{\sqrt{4Dt}}\right) = \frac{2}{\sqrt{\pi}}\int_0^{\eta} exp\left(-\eta^2\right)d\eta \tag{7.37}$$

It is one of the more famous of the special or higher transcendental functions (see Fig. 7.1). The *Complementary Error function* is defined as

$$erfc(\eta) = 1 - erf(\eta) \tag{7.38}$$

Thus, the Error function is defined by the solution of an ODE. The properties of the Error function are such that (see Fig. 7.1)

$$erf(0) = 0 \tag{7.39}$$

$$erf(\infty) = 1 \tag{7.40}$$

$$erf(-\eta) = -\, erf(\eta) \tag{7.41}$$

$$erf(-\infty) = -1 \tag{7.42}$$

Algorithms for calculating $erf(\eta)$ and $erfc(\eta)$ can be found in Abramowitz and Stegun (1964, Chapt. 7), van Genuchten and Alves (1982), Spanier and Oldham (1987, Chapt. 40), Press et al. (1992), and in the "specfun" directory of the NETLIB (see also the Appendix of this volume).

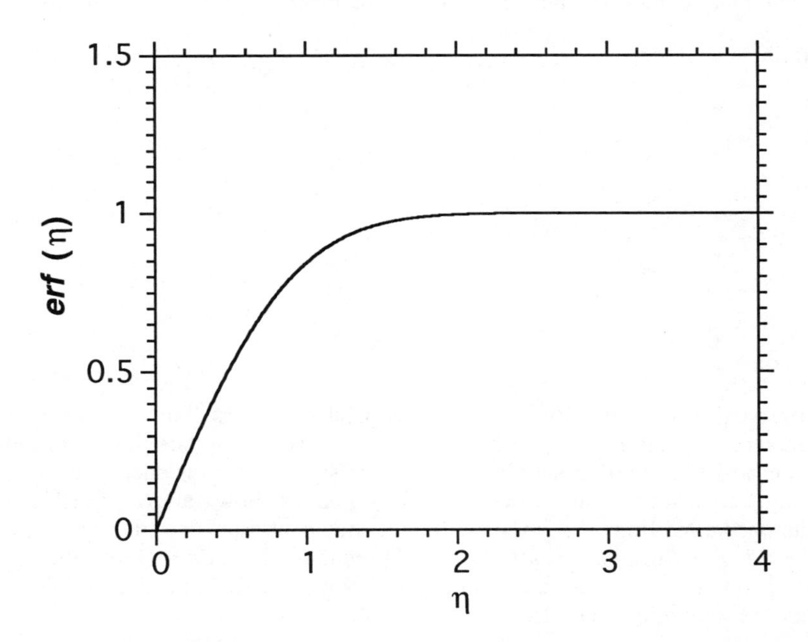

Figure 7.1. Plot of the Error function, $erf(\eta)$, for positive values of an arbitrary real argument η (as defined in Eq. (7.25)).

7.4 Separation of Variables

Separation of variables (SoV) is arguably the most commonly employed solution method for PDEs, and it forms the backbone of texts by Carslaw and Jaeger (1959), Crank (1975), Özisik (1980), van Genuchten (1981), and van Genuchten and Alves (1982). These authors present so many solutions obtained via this technique that these books are customarily used simply to look up solutions for problems of interest. Nevertheless, as extensive as these compendia are, problems arise in diagenesis that cannot be found in these sources. Therefore, both to understand the derivation of the solutions that already exist and to be able to deal with new unsolved problems, the reader is here presented with a review of the SoV method.

The idea behind this method of solution is deceptively simple. Is it possible to divide a PDE into parts that depend only on one of the independent variables? If this can be done, then one only needs to solve each separated ODE independently to obtain the complete solution. For this method to work, the problem must be *homogeneous*; i.e.

i) there can be no source/sink terms, R(x,t), that depend on either x or t, or both, in the differential equation; and

ii) the boundary conditions must be of one of the following forms (Chapter 5):

$$C = 0 \tag{7.43}$$

or

$$\frac{\partial C}{\partial x} = 0 \tag{7.44}$$

or

$$vC - D\frac{\partial C}{\partial x} = 0 \tag{7.45}$$

This requisite homogeneity may sound restrictive at first, but most problems I have encountered can be transformed or split up to accommodate this requirement. The superposition of solutions makes all this possible. For instance, removing an inho-mogeneous source/sink, R(x), is fairly simple, as shown above. Furthermore, the boundary conditions can be made homogeneous at the same time by choosing V(x) in Eq. (7.6) to also satisfy the inhomogeneous boundary conditions in the original problem. To illustrate the division of a problem into a separable state, consider the following example.

Example 7-2. The time-dependent conservation equation for a species subject to diffusion, advection, first-order reaction, and an inhomogeneous source/sink at constant porosity is

$$\frac{\partial C}{\partial t} = D\frac{\partial^2 C}{\partial x^2} - v\frac{\partial C}{\partial x} - kC + R(x) \quad (7.46)$$

where D, v, and k are all generic constants. For boundary conditions, assume that the concentration is known at the sediment-water interface,

$$C(0,t) = C_o \quad (7.47)$$

and that the diffusive flux is known (prescribed) at some depth x = L,

$$-D\frac{\partial C}{\partial x}\bigg|_L = F \quad (7.48)$$

where F is a constant. In addition, the initial condition is some arbitrary, but known, distribution of the species with depth,

$$C(x,0) = f(x) \quad (7.49)$$

To reduce the problem to a homogeneous form, again set

$$C(x,t) = U(x,t) + V(x) \quad (7.50)$$

where V(x) is a steady-state component which is the solution to the ODE,

$$D\frac{d^2V}{dx^2} - v\frac{dV}{dx} - kV + R(x) = 0 \quad (7.51)$$

with boundary conditions

$$V(0) = C_o \quad (7.52)$$

$$-D\frac{dV}{dx}\bigg|_L = F \quad (7.53)$$

We already know that substitution of Eq. (7.50) into Eq. (7.46) makes the conservation equation homogeneous (see Section 7.2). For the boundary conditions, substitution of Eq. (7.50) into Eq. (7.47) gives

$$U(0,t) + V(0) = C_o \quad (7.54)$$

but, from Eq. (7.52), $V(0) = C_o$, so that

$$U(0,t) = 0 \quad (7.55)$$

and the first boundary condition has become homogeneous. For the second boundary condition, substitution produces

$$-D\frac{\partial U}{\partial x}\bigg|_L - D\frac{dV}{dx}\bigg|_L = F \tag{7.56}$$

but Eq. (7.53) reduces this to

$$-D\frac{\partial U}{\partial x}\bigg|_L = 0 \tag{7.57}$$

which is the second desired homogeneous boundary condition. Similarly, many other inhomogeneous problems can be transformed to one or more steady-state ODEs plus one or more homogeneous PDEs.

The SoV method itself is also best explained by considering some actual examples. To this end, consider the following.

7.4.1 SoV for 1-D Diffusion in a Finite Sediment Layer

Example 7-3. Once we have a homogeneous diffusion problem, how do you separate the equation? My first choice to illustrate SoV is based on the classic of all such diagenetic problems, Guinasso and Schink's (1975) model for bioturbation of a solid tracer that is introduced as a unit impulse into a sediment. I will assume that advection can be neglected for the sake of simplicity, and I direct the reader to Guinasso and Schink's (1975) paper for details of its inclusion. The initial problem is then stated as

$$\frac{\partial C}{\partial t} = D\frac{\partial^2 C}{\partial x^2} - kC \tag{7.58}$$

where D is a constant. The initial condition for Guinasso and Schink's original problem was a Delta function (see Chapter 2) at the surface, but we can generalize the problem to any initial distribution, f(x); therefore, let

$$C(x,0) = f(x) \tag{7.59}$$

The boundary conditions are

$$\frac{\partial C}{\partial x}\bigg|_0 = 0 \tag{7.60}$$

$$\frac{\partial C}{\partial x}\bigg|_L = 0 \tag{7.61}$$

where x = L is the base of the bioturbated zone. Equations (7.60) and (7.61) mean that there is no species (tracer) loss from either the top or the base of the mixed zone by bioturbation. As there is also no advection, this model simply redistributes the initial tracer distribution with time.

Equation (7.58) was previously solved as a heat-conduction problem by Carslaw and Jaeger (1959), but the first solution as a diagenetic transport-reaction problem, nevertheless, lies with Guinasso and Schink (1975). The decay process can be removed from this problem by introducing the transformation,

$$C(x,t) = U(x,t)\,exp(-kt) \tag{7.62}$$

Equations (7.58) through (7.61) become, respectively,

$$\frac{\partial U}{\partial t} = D\frac{\partial^2 U}{\partial x^2} \tag{7.63}$$

subject to

$$U(x,0) = f(x) \tag{7.64}$$

$$\left.\frac{\partial U}{\partial x}\right|_0 = 0 \tag{7.65}$$

$$\left.\frac{\partial U}{\partial x}\right|_L = 0 \tag{7.66}$$

We are fortunate in this case that the original boundary conditions were homogeneous; otherwise, Eq. (7.62) would have introduced time-dependent right-hand sides, which would necessitate the use of Duhamel's Theorem for solution (see further below).

Step #1: Assume that U(x,t) can be written as the product of a term T(t) that depends only on time, t, and a term X(x) that depends only on depth, x, i.e.

$$U(x,t) = T(t)\cdot X(x) \tag{7.67}$$

(Don't confuse T(t) in Eq. (7.67) with temperature.) Now substitute this into Eq. (7.63),

$$X\frac{\partial T}{\partial t} = D\,T\frac{\partial^2 X}{\partial x^2} \tag{7.68a}$$

or

$$\frac{1}{T}\frac{\partial T}{\partial t} = \frac{D}{X}\frac{\partial^2 X}{\partial x^2} \tag{7.68b}$$

All the left-hand side of Eq. (7.68b) is a function of t only, while the right-hand side is a function of x only, i.e. we have isolated the dependencies in the problem into two separate but equal parts. Two arbitrary functions of two independent variables can be equal to each other, for all t and x, if and only if they are both equal to a constant. This constant to which both sides of Eq. (7.68b) are equal is called the *separation constant*, say $-\mu^2$. (The choice of the negative and the square is explained below.) The introduction of this constant generates two ODEs, one in time and one in space, i.e.

$$\frac{dT}{dt} = -\mu^2 T \tag{7.69}$$

and

$$\frac{d^2X}{dx^2} + \beta^2 X = 0 \tag{7.70}$$

where $\beta^2 = \mu^2/D$ for convenience. (Don't confuse μ with a viscosity and β for a mass-transfer coefficient.)

Eventually we will need to calculate the value(s) of μ^2. But why is the separation constant negative? Looking Eq. (7.69), one recognizes it as the decay equation. If the separation constant were not chosen to be negative, T(t) would grow to infinity with time. Similarly, why square μ? Taking the square of a real number always produces a positive value, ensuring that $-\mu^2$ is always negative.

Step #2: The solution to Eq. (7.69) is

$$T(t) = A\,exp\left(-\mu^2 t\right) \tag{7.71}$$

However, we also need the solution to Eq. (7.70). Notice that it has a positive sign before the X term, unlike the diffusion-reaction equations we have so far examined. If a power series solution such as Eq. (6.134) is substituted into this equation, one eventually deduces that

$$X(x) = a\,cos(\beta x) + b\,sin(\beta x) \tag{7.72}$$

where a and b are integration constants (see Boyce and DiPrima, 1977).

Substituting Eq. (7.72) into the boundary condition at x=0, i.e. Eq. (7.65),

$$\left.\frac{dX}{dx}\right|_0 = b\beta = 0 \tag{7.73}$$

so that b = 0 to avoid a trivial solution that does not describe the problem. At the other boundary,

$$\left.\frac{dX}{dx}\right|_L = a\beta\, sin(\beta L) = 0 \tag{7.74}$$

We cannot take a = 0 in Eq. (7.74) or both integration constants would be zero, and this would lead to the result U(x,t) = 0, which is either trivial or incorrect. Therefore, it is necessary that

$$sin(\beta L) = 0 \tag{7.75}$$

As $L \neq 0$, the value of β must be such that Eq. (7.75) is true. As the function *sin*(x) is oscillatory, there are an infinite number of values of β for which Eq. (7.75) is true. Specifically, in this case,

$$\beta = \frac{n\pi}{L} \qquad n = 0, 1, 2, 3, 4, 5, \ldots \tag{7.76}$$

Thus, let an arbitrary allowed value be β_n; then

$$\mu_n^2 = D\left(\frac{n\pi}{L}\right)^2 \tag{7.77}$$

and the separation(s) constant is now defined. For each β_n there will also be a solution component,

$$X_n(x) = a_n\, cos(\beta_n x) \tag{7.78}$$

In Eq. (7.73), $X_n(x)$ is often called an *Eigenfunction*, and the corresponding β_n is called an *Eigenvalue.*

Step #3: The solution for U(x,t) is obtained by summing all the components $X_n(x)$,

$$U(x,t) = \sum_{n=0}^{\infty} c_n\, cos(\beta_n x)\, e^{-\mu_n^2 t} \tag{7.79}$$

where $c_n = a_n \cdot A$, as A does not need to be calculated explicitly. This sum reflects the superposition principle, mentioned earlier; each element of this sum contributes a component to the full problem, represented by U(x,t) and, thereby, C(x,t).

Step #4: Finally, we need to be able to calculate the c_n. To do so, recognize that Eq. (7.79) is a *Fourier Series* (see Chapter 2) that represents the function U(x,t). There are some particular properties of Fourier Series that allow us to calculate c_n. The procedure that allows us to take advantage of these properties is to apply Eq. (7.79) at the initial condition, t = 0, i.e. Eq. (7.64). It is then necessary that

$$f(x) = \sum_{n=0}^{\infty} c_n \, cos(\frac{n\pi x}{L}) \tag{7.80}$$

Multiply both sides of Eq. (7.80) by cos(mπx/L):

$$f(x)\, cos(\frac{m\pi x}{L}) = \sum_{n=0}^{\infty} c_n \, cos(\frac{n\pi x}{L}) cos(\frac{m\pi x}{L}) \tag{7.81}$$

where m is an arbitrary positive integer. Then integrate Eq. (7.81) with respect to depth, x, from 0 to L:

$$\int_0^L f(x)\, cos(\frac{m\pi x}{L}) dx = \sum_{n=0}^{\infty} c_n \int_0^L cos(\frac{n\pi x}{L}) cos(\frac{m\pi x}{L}) dx \tag{7.82}$$

The integral on the right-hand side of Eq. (7.82) can be evaluated using trigonometric identities (e.g. Hildebrand, 1976). This procedure tells us that, if $m \neq n$,

$$\int_0^L cos(\frac{n\pi x}{L}) cos(\frac{m\pi x}{L}) dx = \frac{1}{2}\int_0^L \left[cos\left(\frac{(n-m)\pi x}{L}\right) + cos\left(\frac{(n+m)\pi x}{L}\right)\right] dx$$

$$= -\frac{1}{2}[(n-m) sin((n-m)\pi) + (n+m) sin((n+m)\pi)] \tag{7.83}$$

But as (n-m) is an integer i and (n+m) is another integer j, a basic property of the *sin* function gives that

$$sin(i\pi) = sin(j\pi) = 0 \tag{7.84}$$

so that

$$\int_0^L cos(\frac{n\pi x}{L}) cos(\frac{m\pi x}{L}) dx = 0 \qquad (m \neq n) \tag{7.85}$$

On the other hand, if n = m,

$$\int_0^L cos^2(\frac{n\pi x}{L}) dx = \frac{1}{2}\int_0^L \left[1 + cos\left(\frac{2n\pi x}{L}\right)\right] dx = \frac{L}{2} - \left[\frac{L}{2n\pi} sin\left(\frac{2n\pi x}{L}\right)\right]_0^L = \frac{L}{2} \tag{7.86}$$

In summary then,

$$\int_0^L cos(\frac{n\pi x}{L})cos(\frac{m\pi x}{L})dx = \begin{cases} \frac{L}{2} & m = n \\ 0 & m \neq n \end{cases} \quad (7.87)$$

This is the so-called *orthogonality relationship* for the product of two cosines. With Eq. (7.87), Eq. (7.82) reduces to

$$c_n = \frac{2}{L}\int_0^L f(x)\, cos(\frac{n\pi x}{L})dx \quad (7.88)$$

Calculation of values for c_n requires the specification of f(x) and evaluation of the integral in Eq. (7.88), perhaps numerically (see Chapter 8).

Step #5. The solution given by Eq. (7.79) then becomes complete with substitution of Eq. (7.88), i.e.

$$U(x,t) = \sum_{n=0}^{\infty} \frac{2}{L}\int_0^L f(x')\, cos(\frac{n\pi x'}{L})dx'\, cos(\beta_n x)\, exp\left(-\mu_n^2 t\right) \quad (7.89)$$

General 1-D Problems. Özisik (1980) has done an excellent job of codifying the SoV solutions to Eq. (7.63) under all possible homogeneous boundary conditions, along with their associated orthogonality relationships and eigenvalues (repeated in Tables 7.1A and 7.1B). The interested reader might wish to consult Mikhailov and Özisik (1984) as well. For such problems, Eqs (7.71) and (7.72) are always the general solutions for the temporal and spatial parts, respectively. The solutions in these tables include that for the original Guinasso and Schink (1975) problem that included advection. These are very powerful and complete tables.

Orthogonality is defined for all such problems by

$$\int_0^L X(\beta_m, x)X(\beta_n, x)dx = \begin{cases} N(\beta_m) & m = n \\ 0 & m \neq n \end{cases} \quad (7.90)$$

where $X(\beta_m,x)$ is the spatial solution with eigenvalue β_m, and where $N(\beta_m)$ is called a *norm* and depends on the boundary conditions employed. Values of the inverse of the norm can also be found in Table 7.1B.

The general solution to a homogeneous diffusion problem can be written as

$$U(x,t) = \sum_{n=1}^{\infty} \frac{1}{N(\beta_n)}\int_0^L f(x')\, X(\beta_n, x')dx'\, X(\beta_n x)\, exp\left(-\mu_n^2 t\right) \quad (7.91)$$

where $X(\beta_n,x)$ is given in Table 7.1A, $N(\beta_m)$ is from Table 7.1B, and f(x) is the initial condition.

Example 7-4. Spatially Variable Coefficients. When the coefficients of the PDE, i.e. D, v, and/or k, are depth dependent, SoV is still possible, but it leads to solutions in terms of *higher transcendental functions*. Most such problems cannot be found in the standard texts, and are easier to deal with numerically. However, as an illustration of the SoV method, consider the case of a radioactive tracer in a surficial mixed-zone with linearly decreasing biodiffusivity (Boudreau, 1986a). At constant porosity, conservation of a decaying tracer is given by

$$\frac{\partial B}{\partial t} = D_B \frac{\partial}{\partial x}\left[(1-\frac{x}{L})\frac{\partial B}{\partial x}\right] - w\frac{\partial B}{\partial x} - k\,B \tag{7.92}$$

where D_B is a constant, and with surface boundary condition

$$w\,B(0,t) - D_B \left.\frac{\partial B}{\partial x}\right|_0 = 0 \tag{7.93}$$

In addition, B(x,t) must remain finite as x approaches L, i.e.

$$B(L,t) \rightarrow \text{finite} \tag{7.94}$$

Equation (7.94) is a type of Dirichlet boundary condition that allows us to reject any solution that blows up at x=L. As an initial condition, let

$$B(x,0) = f(x) \tag{7.95}$$

As a first step, introduce the new depth and time variables

$$\xi \equiv 1 - \frac{x}{L} \tag{7.96}$$

$$\tau \equiv \frac{D_B t}{L^2} \tag{7.97}$$

so that Eq. (7.92) becomes

$$\frac{\partial B}{\partial \tau} = \frac{\partial}{\partial \xi}\left[\xi \frac{\partial B}{\partial \xi}\right] + Pe\frac{\partial B}{\partial \xi} - Da\,B \tag{7.98}$$

where Pe is the Peclet number (wL/D_B) and Da is the Damkohler number (kL^2/D_B). (Please don't confuse the dimensionless time τ with the transformed time τ of Eq. (7.13). They are not the same.)

Table 7.1A. Solutions, $X(\beta_m,x)$, for 1-D Diffusion in a Finite Layer of Sediment, $0 \le x \le L$, as a Function of the Boundary Conditions. (Source Özisik, 1980; reproduced with the kind permission of John Wiley and Sons, Inc.)

Case No.	Boundary Condition at x = 0	Boundary Condition at x = L	$X(\beta_n,x)$ in Eqs (7.90) and (7.91)
1	$X=0$	$X=0$	$sin(\beta_n x)$
2	$X=0$	$\frac{dX}{dx}=0$	$sin(\beta_n x)$
3*	$X=0$	$\frac{dX}{dx}+h_2X=0$	$sin(\beta_n x)$
4	$\frac{dX}{dx}=0$	$X=0$	$cos(\beta_n x)$
5	$\frac{dX}{dx}=0$	$\frac{dX}{dx}=0$	$cos(\beta_n x)^{\dagger}$
6	$\frac{dX}{dx}=0$	$\frac{dX}{dx}+h_2X=0$	$cos(\beta_n x)$
7*	$-\frac{dX}{dx}+h_1X=0$	$X=0$	$sin(\beta_n[L-x])$
8	$-\frac{dX}{dx}+h_1X=0$	$\frac{dX}{dx}=0$	$cos(\beta_n[L-x])$
9	$-\frac{dX}{dx}+h_1X=0$	$\frac{dX}{dx}+h_2X=0$	$\beta_n\, cos(\beta_n x) + h_1\, sin(\beta_n x)$

* h_1 and h_2 can represent the advective velocity divided by D, or a mass-transfer coefficient divided by D, or a heterogeneous rate constant divided by D.

† $\beta_n = 0$ is a valid eigenvalue in this case.

Table 7.1B. Norms, $N(\beta_n)$, and Eigenvalues, β_n, for 1-D Diffusion in a Finite Layer of Sediment, $0 \le x \le L$, as a Function of the Boundary Conditions. (From Özisik, 1980; reproduced with the kind permission of John Wiley and Sons, Inc.).

Case No.	$1/N(\beta_n)$	Equation for the eigenvalues, β_n (Eigenvalues are the roots)
1	$2/L$	$\mathit{sin}(\beta_n L) = 0$
2	$2/L$	$\mathit{cos}(\beta_n L) = 0$
3	$2\left[\dfrac{\beta_n^2 + h_2^2}{L\left(\beta_n^2 + h_2^2\right) + h_2}\right]$	$\beta_n \, \mathit{cotan}(\beta_n L) = - h_2$
4	$2/L$	$\mathit{cos}(\beta_n L) = 0$
5*	$\begin{cases} 2/L & \beta_n \neq 0 \\ 1/L & \beta_n = 0 \end{cases}$	$\mathit{sin}(\beta_n L) = 0$
6	$2\left[\dfrac{\beta_n^2 + h_2^2}{L\left(\beta_n^2 + h_2^2\right) + h_2}\right]$	$\beta_n \, \mathit{tan}(\beta_n L) = h_2$
7	$2\left[\dfrac{\beta_n^2 + h_1^2}{L\left(\beta_n^2 + h_1^2\right) + h_1}\right]$	$\beta_n \, \mathit{cotan}(\beta_n L) = - h_1$
8	$2\left[\dfrac{\beta_n^2 + h_1^2}{L\left(\beta_n^2 + h_1^2\right) + h_1}\right]$	$\beta_n \, \mathit{tan}(\beta_n L) = h_1$
9	$2\left[\left(\beta_n^2 + h_1^2\right)\left(L + \dfrac{h_2}{\beta_n^2 + h_2^2}\right) + h_1\right]^{-1}$	$\mathit{tan}(\beta_n L) = \dfrac{\beta_n\left(h_1 + h_2\right)}{\beta_n^2 - h_1 h_2}$

* Zero is a legitimate eigenvalue in this case.

Separation involves writing $B(x,t) = B(\xi,\tau)$ as

$$B(\xi,\tau) = X(\xi)T(\tau) \tag{7.99}$$

so that Eq. (7.98) becomes

$$\frac{1}{T}\frac{\partial T}{\partial \tau} = \frac{1}{X}\left\{\frac{\partial}{\partial \xi}\left[\xi\frac{\partial X}{\partial \xi}\right] + \mathrm{Pe}\frac{\partial X}{\partial \xi} - \mathrm{Da}\,X\right\} \tag{7.100}$$

As the left-hand side of this last equation is a function of τ only, and the right-hand side of ξ only, we can set them equal to a separation constant, say $-\lambda^2$, so that

$$\frac{dT}{d\tau} = -\lambda^2 T \tag{7.101}$$

and

$$\frac{d}{d\xi}\left[\xi\frac{dX}{d\xi}\right] + \mathrm{Pe}\frac{dX}{d\xi} + (\lambda^2 - \mathrm{Da})X = 0 \tag{7.102}$$

Solution of Eq. (7.101) is again

$$T(\tau) = A\,exp\left(-\lambda^2 t\right) \tag{7.103}$$

where A is an integration constant that can be set to unity without any loss.

By analogy to Example 6-18 of Chapter 6, we might suspect that the solution to Eq. (7.102) is related to the modified Bessel function, noting however, the difference in sign. In fact, this turns out to be the case which is best illustrated as a two-step procedure. First set

$$X(\xi) = \xi^{-\mathrm{Pe}/2}\Omega(\xi) \tag{7.104}$$

with which Eq. (7.102) becomes

$$\xi^2\frac{d^2\Omega}{d\xi^2} + \xi\frac{d\Omega}{d\xi} + (\kappa^2\xi - \frac{\mathrm{Pe}^2}{16})\Omega = 0 \tag{7.105}$$

where $\kappa^2 = \lambda^2 - \mathrm{Da}$.

The second transformation comes from a change of the independent variable,

$$\zeta = \sqrt{\xi} \tag{7.106}$$

which transforms Eq. (7.105) into a special case of Bessel's equation,

$$\zeta^2 \frac{d^2\Omega}{d\zeta^2} + \zeta \frac{d\Omega}{d\xi} + (\mu^2\zeta^2 - \mathrm{Pe}^2)\Omega = 0 \tag{7.107}$$

where, in this example, $\mu = 2\kappa$. The general solution to Eq. (7.107) is

$$\Omega(\zeta) = A_n\, J_{\mathrm{Pe}}(\mu\zeta) + B_n J_{-\mathrm{Pe}}(\mu\zeta) \tag{7.108}$$

in which $J_{\mathrm{Pe}}(\ )$ is the *Bessel function* of the first kind of order Pe, and A_n and B_n are integration constants.

The Bessel function is another of the special functions, and it can be expressed in terms for a power series as

$$J_\nu(x) = \left(\frac{x}{2}\right)^\nu \sum_{n=1}^{\infty} \frac{(-x^2/4)^n}{\Gamma(1+n+\nu)n!} \tag{7.109}$$

where $\Gamma(\ )$ is the *Gamma function* (Spanier and Oldham, 1987, Chapt. 43) and ν is the order, i.e. Pe and -Pe in Eq. (7.108).

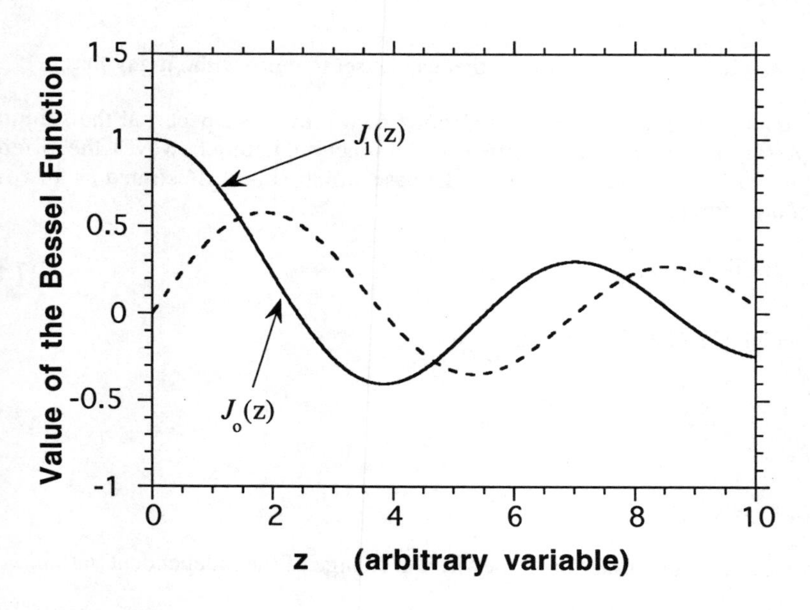

Figure 7.2. Plot of the Bessel function of the first kind, $J_\nu(z)$, with order ν equal to 0 and 1 and for an arbitrary variable z.

Algorithms and formulas for calculating this new Bessel function can be found in Abramowitz and Stegun (1964), Campbell (1979, 1981), Spanier and Oldham (1987), and Press et al. (1992), and public-domain FORTRAN codes can be obtained for its calculation, as explained in the Appendix of this book. Figure 7.2 illustrates the behavior of $J_\nu(z)$ for $\nu = 0$ and 1. This Bessel function is oscillatory, with the amplitude of the oscillation decreasing as $z \to \infty$.

For a negative non-integer order, $J_{-Pe}(\zeta)$ is unbounded at $\zeta = 0$; thus,

$$B_n = 0 \tag{7.110}$$

which fulfills boundary condition (7.94). Boundary condition (7.93) defines the separation constant. At $\zeta = 1$ (x=0),

$$Pe\ \Omega(1) + \left.\frac{d\Omega}{d\zeta}\right|_1 = 0 \tag{7.111}$$

or, substituting Eqs (7.108) and (7.110),

$$A_n\left[Pe\ J_{Pe}(\mu) + \mu \left.\frac{d\,J_{Pe}(\mu\zeta)}{d\mu\zeta}\right|_1\right] = 0 \tag{7.112}$$

However, Abramowitz and Stegun (1964, #9.1.27) provide the identity

$$\frac{d\,J_{Pe}(z)}{dz} = J_{Pe-1}(z) - \frac{Pe}{z} J_{Pe}(z) \tag{7.113}$$

where z is an arbitrary independent variable, which reduces Eq. (7.112) to

$$A_n\ J_{Pe-1}(\mu) = 0 \tag{7.114}$$

Setting A_n to zero leads to a trivial solution, so it is necessary that

$$J_{Pe-1}(\mu) = 0 \tag{7.115}$$

Figure 7.2 shows that this Bessel function is oscillatory; thus, there are an infinite number of values of μ that satisfy Eq. (7.115). These are the eigenvalues, and a unique value of A_n is also associated with each μ_n. The solution is then an infinite sum of Bessel functions, called a Fourier-Bessel series (Hildebrand, 1976).

To calculate the A_n values, we must employ the initial condition and the orthogonality properties of the Bessel function; thus, at $\tau = 0$,

$$\sum_{n=1}^{\infty} A_n\ J_{Pe}(\mu_n\zeta) = \zeta^{Pe}\ f(\zeta) \tag{7.116}$$

To apply orthogonality, multiply each side of Eq. (7.116) by $\zeta\, J_{Pe}(\mu_m\zeta)$, and integrate from 0 to 1, i.e. the dimensionless sediment domain,

$$\sum_{n=1}^{\infty} A_n \int_0^1 \zeta\, J_{Pe}(\mu_m\zeta)\, J_{Pe}(\mu_n\zeta)\, d\zeta = \int_0^1 \zeta^{Pe+1}\, J_{Pe}(\mu_m\zeta)\, f(\zeta)\, d\zeta \qquad (7.117)$$

Hildebrand (1976, p. 226-230) shows that the integral on the left-hand side of Eq. (7.117) has the following values:

$$\int_0^1 \zeta\, J_{Pe}(\mu_m\zeta)\, J_{Pe}(\mu_n\zeta)\, d\zeta = \begin{cases} \frac{1}{2}\left[J_{Pe}(\mu_n)\right]^2 & \text{if } m = n \\ 0 & \text{if } m \neq n \end{cases} \qquad (7.118)$$

which is orthogonality when boundary condition (7.93) is employed; thus,

$$A_n = \frac{2}{\left[J_{Pe}(\mu_n)\right]^2} \int_0^1 \zeta^{Pe+1}\, J_{Pe}(\mu_m\zeta)\, f(\zeta)\, d\zeta \qquad (7.119)$$

which defines the constants. The final solution is

$$B(x,t) = \left(1 - \frac{x}{L}\right)^{-Pe/2} \sum_{n=1}^{\infty} A_n\, J_{Pe}\left(\mu_n \sqrt{1 - \frac{x}{L}}\right) exp\left(-\lambda_n^2 D_B t / L^2\right) \qquad (7.120)$$

where $\lambda^2 = \mu^2/4 + Da$.

In addition to this model, Martin (1989) presents the solution to a diffusion model when the diffusion coefficient decays linearly to a non-zero constant with depth, and Yates (1992) examines the same problem with an exponential decrease. These models might serve as reasonable descriptions of wave-induced hydrodynamic dispersion with omnipresent molecular diffusion in sediments.

7.4.2 SoV for 1-D Diffusion in a Semi-Infinite Sediment

With some modification, the SoV method can be used to solve diffusion problems for a semi-infinite sediment, although it might be preferable to employ the *Laplace Transform* method, as explained below. Let us start with a homogeneous 1-D diffusion problem defined for $0 \le x \le \infty$, say

$$\frac{\partial C}{\partial t} = D \frac{\partial^2 C}{\partial x^2} \qquad (7.121)$$

Separation of variables, $C(x,t) = X(x)T(t)$, again produces two ODEs, one for the time-dependent part,

$$\frac{dT}{dt} = -\mu^2 T \tag{7.122}$$

and one for the spatial part,

$$\frac{d^2X}{dx^2} + \beta^2 X = 0 \tag{7.123}$$

where again $\beta^2 = \mu^2/D$. Solution of Eq. (7.122) is not affected by the spatial domain and is Eq. (7.71). Solution of Eq. (7.123) is again

$$X(x) = a\,cos(\beta x) + b\,sin(\beta x) \tag{7.124}$$

where a and b are arbitrary constants. The boundary condition at x = 0 will define one of these constants, as codified by Özisik (1980) and given in Table 7.2.

There is, however, no second condition to define discrete values of the separation constant. In fact, the solution is defined for any value of β between 0 and ∞; thus, the solution is not an infinite sum of individual solutions, but instead, it is an integral of all the solutions corresponding to all the values of β (see Ozisik, 1980, p. 39-43):

$$C(x,t) = \int_0^\infty \frac{1}{N(\beta)} X(\beta,x)\, e^{-D\beta^2 t} \int_0^\infty X(\beta,x') f(x') dx' d\beta \tag{7.125}$$

where the norms, N(β), and solution components, X(β,x), can be found in Table 7.2.

Table 7.2. Integration Constants, Solutions and Norms for 1-D Diffusion in a Semi-Infinite Layer of Sediment, 0≤x≤∞. (From Özisik, 1980; reproduced with the kind permission of John Wiley and Sons, Inc.)

Case No.	Boundary Condition at x=0	Integration Constant	X(β,x) of Eq. (7.125)	1/N(β)
1	$X = 0$	$a = 0$	$sin(\beta x)$	$2/\pi$
2	$\frac{dX}{dx} = 0$	$b = 0$	$cos(\beta x)$	$2/\pi$
3*	$-\frac{dX}{dx} + h_1 X = 0$	$a\,h_1 = \beta\, b$	$\beta\, cos(\beta x) + h_1 sin(\beta x)$	$\frac{2}{[\pi(b^2 + h_1^2)]}$

* for the meaning of h_1 see footnote in Table 7.1.

Example 7-5. To illustrate the use of Eq. (7.125), consider the time-dependent distribution of a solid radioactive tracer, initially absent in a sediment, but supplied after some arbitrary time (labeled zero) at a constant rate. Furthermore, assume that the porosity and biodiffusion are constant and that advection can be ignored, i.e. $v^2/(D_B k) << 1$. Mathematically, the problem is given by

$$\frac{\partial B}{\partial t} = D_B \frac{\partial^2 B}{\partial x^2} - kB \tag{7.126}$$

with

$$-D_B \left.\frac{\partial B}{\partial x}\right|_0 = F_o \tag{7.127}$$

$$\left.\frac{\partial B}{\partial x}\right|_{x \to \infty} = 0 \tag{7.128}$$

$$B(x,0) = 0 \tag{7.129}$$

Despite the fact that boundary condition (7.127) is quite common in diagenetic problems, it is not homogeneous; thus, before the general solution provided by Eq. (7.125) can be used, the problem must be made homogeneous. The most effective way to do this is to write B as the sum of a steady-state problem, B_{ss}, which satisfies Eqs (7.126) through (7.128) as time approaches infinity, and a transient part, $B_t(x,t)$,

$$B(x,t) = B_{ss}(x) + B_t(x,t)\, exp(-kt) \tag{7.130}$$

where

$$B_{ss}(x) = B_o exp\left(-x\sqrt{\frac{k}{D_B}}\right) \tag{7.131}$$

and

$$B_o = F_o / \sqrt{D_B k} \tag{7.132}$$

The reduced transient part of the problem has become

$$\frac{\partial B_t}{\partial t} = D_B \frac{\partial^2 B_t}{\partial x^2} \tag{7.133}$$

subject to

$$\left.\frac{\partial B_t}{\partial x}\right|_{x=0} = 0 \tag{7.134}$$

$$\left.\frac{\partial B_t}{\partial x}\right|_{x\to\infty} = 0 \tag{7.135}$$

$$B_t(x,0) = -B_o exp\left(-x\sqrt{\frac{k}{D_B}}\right) \tag{7.136}$$

The transient problem can be solved with Eq. (7.125), i.e. $B_t(x,t) = C(x,t)$, and the information provided by Table 7.2, Case 2,

$$B_t(x,t) = -\frac{2B_o}{\pi}\int_0^\infty cos(\beta x)\, exp\left(-D_B\beta^2 t\right)\int_0^\infty cos(\beta x')\, exp\left(-x\sqrt{\frac{k}{D_B}}\right)dx'\, d\beta \tag{7.137}$$

The interior integral in Eq. (7.137) is tabulated (e.g. CRC, 1979, # 538) to give

$$B_t(x,t) = \frac{2B_o\sqrt{k}}{\pi\sqrt{D_B}}\int_0^\infty \frac{cos(\beta x) exp\left(-D_B\beta^2 t\right)}{\left(\frac{k}{D_B}+\beta^2\right)}d\beta \tag{7.138}$$

The remaining integral is also tabulated in Gradshteyn and Ryzhik (1980, #3.954, 2),

$$B_t(x,t) = -\frac{B_o\, exp(kt)}{2}\left[2cosh\left(-x\sqrt{\frac{k}{D_B}}\right) - exp\left(\sqrt{\frac{kx^2}{D_B}}\right) erf\left(\sqrt{kt} - \frac{x}{2\sqrt{D_B t}}\right) - exp\left(-\sqrt{\frac{kx^2}{D_B}}\right) erf\left(\sqrt{kt} + \frac{x}{2\sqrt{D_B t}}\right)\right] \tag{7.139}$$

This example illustrates, moreover, that modellers should be familiar with the use of integral tables if they hope to arrive at solutions in terms of more familiar functions.

7.4.3 SoV for Multi-Dimensional Steady-State Problems

The usefulness of SoV is not limited to parabolic (transient diffusion) problems. It can be employed with 2-D and 3-D steady-state problems, i.e. elliptic equations. Many such solutions are available in Carslaw and Jaeger (1959), Crank (1975), Özisik (1980), and Nowak et al. (1987). On the other hand, situations that demand a multi-dimensional Cartesian description are so far rare in diagenetic studies; consequently, the application of the SoV method to multiple dimensions is illustrated by considering a problem in the cylindrical coordinate system, i.e. Aller's tube model, and another in the spherical coordinate system.

Example 7-6 (Aller's Tube Model). Suppose we are interested in silica diagenesis in one of Aller's average annuli of sediment (see Chapter 3). At constant porosity and with linear kinetics, the dissolved silica distribution (field) is described by:

$$D'\frac{\partial^2 C}{\partial x^2}+\frac{D'}{r}\frac{\partial}{\partial r}\left(r\frac{\partial C}{\partial r}\right)+k(C_s-C)=0 \quad (7.140)$$

where k is a dissolution rate constant and C_s is the saturation concentration. The boundary conditions are:

$$C(0,r)=C_o \quad (7.141)$$

$$C(x,r_1)=C_o \quad (7.142)$$

$$\left.\frac{\partial C}{\partial r}\right|_{r_2}=0 \quad (7.143)$$

$$\left.\frac{\partial C}{\partial x}\right|_{L}=0 \quad (7.144)$$

This last equation indicates that opal dissolution occurs wholly within the irrigated zone, which is acceptable for illustrative purposes.

Step #1. We seek a solution by SoV, but Eq. (7.140) and some of its associated boundary conditions are all inhomogeneous. To make Eq. (7.140) homogeneous, start by defining a new concentration,

$$S= C_s - C \quad (7.145)$$

so that Eq. (7.140) becomes the homogeneous PDE,

$$D'\frac{\partial^2 S}{\partial x^2}+\frac{D'}{r}\frac{\partial}{\partial r}\left(r\frac{\partial S}{\partial r}\right)-kS=0 \quad (7.146)$$

Equation (7.146) is subject to the boundary conditions

$$S(0,r)=S_o \quad (7.147)$$

$$S(x,r_1)=S_o \quad (7.148)$$

$$\left.\frac{\partial S}{\partial r}\right|_{r_2}=0 \quad (7.149)$$

$$\left.\frac{\partial S}{\partial x}\right|_L = 0 \tag{7.150}$$

Fortunately, we do not need to make both Eqs (7.147) and (7.148) homogeneous. In a steady-state 2-D problem, boundary conditions in both directions do not need to be homogeneous. Only the direction used to calculate the eigenvalues must be homogeneous. The choice of the direction for eigenvalue calculations is entirely arbitrary; however, taking the x-direction leads to a few less calculations. Thus, we make the x-direction boundary conditions homogeneous by setting

$$S(x,r) = U(x,r) + V(x) \tag{7.151}$$

where V(x) satisfies the equation

$$D'\frac{d^2V}{dx^2} - kV = 0 \tag{7.152}$$

subject to the boundary conditions

$$V(0) = S_o \tag{7.153}$$

$$\left.\frac{dV}{dx}\right|_L = 0 \tag{7.154}$$

Therefore, from Chapter 6,

$$V(x) = c_1\, exp\left(-(k/D')^{1/2}x\right) + c_2\, exp\left((k/D')^{1/2}x\right) \tag{7.155}$$

where

$$c_1 = \frac{S_o\, exp\left((k/D')^{1/2}L\right)}{\left[exp\left(-(k/D')^{1/2}L\right) + exp\left((k/D')^{1/2}L\right)\right]} \tag{7.156}$$

$$c_2 = S_o - c_1 \tag{7.157}$$

All this transforms Eq. (7.146) into

$$D'\frac{\partial^2 U}{\partial x^2} + \frac{D'}{r}\frac{\partial}{\partial r}\left(r\frac{\partial U}{\partial r}\right) - kU = 0 \tag{7.158}$$

with

$$U(0,r) = 0 \tag{7.159}$$

$$U(x,r_1) = S_o - V(x) \equiv f(x) \tag{7.160}$$

$$\left.\frac{\partial U}{\partial r}\right|_{r_2} = 0 \tag{7.161}$$

$$\left.\frac{\partial U}{\partial x}\right|_{L} = 0 \tag{7.162}$$

which is now a sufficiently homogeneous problem for SoV.

Step #2. Separate Eq. (7.158) by assuming that

$$U(x,r) = X(x)R(r) \tag{7.163}$$

where $X(x)$ is a function of x only and $R(r)$ is a function of r only. Substitute this into Eq. (7.158) to obtain

$$\frac{1}{X}\frac{d^2X}{dx^2} = -\frac{1}{rR}\frac{d}{dr}\left(r\frac{dR}{dr}\right) + \frac{k}{D'} = -\beta^2 \tag{7.164}$$

where $-\beta^2$ is the (negative) separation constant.

Step #3. Now solve the separated equations, starting with the x-direction equation

$$\frac{d^2X}{dx^2} + \beta^2 X = 0 \tag{7.165}$$

which has as its solution

$$X(x) = a\,cos(\beta x) + b\,sin(\beta x) \tag{7.166}$$

Boundary condition Eq. (7.159) immediately gives that a = 0, while boundary condition Eq. (7.162) gives that

$$b\,\beta\,cos(\beta L) = 0 \tag{7.167}$$

Setting b = 0 would lead to a trivial solution, so let β be the roots of $cos(\beta L)$, i.e.

$$\beta = \frac{n\pi}{2L} \qquad n = 0, 1, 2, 3, 4, \ldots \tag{7.168}$$

which are the desired eigenvalues. Since there are an infinite number of β values possible, there are an infinite number of solution components of the form,

$$X_n(x) = b_n \sin(\beta_n x) \tag{7.169}$$

Step #4. Now, solve the r-direction equation,

$$\frac{d^2R}{dr^2} + \frac{1}{r}\frac{dR}{dr} - (\frac{k}{D'} + \beta_n^2)R = 0 \tag{7.170}$$

This is the modified Bessel equation of zero order, as discussed in Chapter 6. The solution to Eq. (7.170) is

$$R(r) = p_n\, I_o(\mu_n r) + q_n\, K_o(\mu_n r) \tag{7.171}$$

where p_n and q_n are integration constants and

$$\mu_n = \sqrt{\frac{k}{D'} + \beta_n^2} \tag{7.172}$$

for convenience, and $I_o(\ )$ and $K_o(\ )$ are the modified Bessel functions of first and second types, respectively (see Fig. 6.1). Note that

$$\frac{dR}{dr} = p_n\, \mu_n\, I_1(\mu_n r) - q_n\, \mu_n\, K_1(\mu_n r) \tag{7.173}$$

which allows us to evaluate the constant q_n with boundary condition Eq. (7.161),

$$q_n = \frac{p_n\, I_1(\mu_n r_2)}{K_1(\mu_n r_2)} \tag{7.174}$$

Putting the two pieces of the solution together, we have that

$$U(x, r) = \sum_{n=0}^{\infty} c_n \left[K_1(\mu_n r_2)\, I_o(\mu_n r) - I_1(\mu_n r_2)\, K_o(\mu_n r)\right] \sin(\beta_n x) \tag{7.175}$$

where $c_n \equiv p_n\, b_n / K_1(\mu_n r_2)$.

Step #5. Calculate each constant c_n with boundary condition Eq. (7.160); i.e.

$$f(x) = \sum_{n=0}^{\infty} c'_n \sin(\beta_n x) \tag{7.176}$$

where, for convenience,

$$c'_n \equiv c_n \left[K_1(\mu_n r_2)\, I_o(\mu_n r_1) - I_1(\mu_n r_2)\, K_o(\mu_n r_1)\right] \tag{7.177}$$

One can utilize orthogonality (see Eq. (7.90) and Table 7.1B) to obtain:

$$c'_n = \frac{2}{L}\int_0^L f(x)\mathit{sin}(\beta_n x)dx \tag{7.178}$$

which completes the last piece of the solution.

Example 7-7. This second example of a steady-state multi-dimensional diagenetic problem is taken from Boudreau and Taylor (1989). The model in question describes the distribution of a decaying solid species (e.g. organic matter, radio-isotope, etc.) in the vicinity of an isolated impenetrable object at the sediment-water interface. This object is represented as a half-sphere in the sediment, and it could be an analog for a rock, nodule, shell, or something man-made (see Fig. 7.3).

Let us assume that the porosity and biodiffusion are constants and the same in all directions (isotropy), that there is no advection, and that the problem is symmetric in all planes parallel to the sediment-water interface; i.e. only variations in the r and ϑ need be considered. The conservation equation for the example species is then (see Eq. (3.113)),

$$D_B\left[\frac{1}{r^2}\frac{\partial}{\partial r}\left(r^2\frac{\partial B}{\partial r}\right)+\frac{1}{r^2 \mathit{sin}\vartheta}\frac{\partial}{\partial \vartheta}\left(\mathit{sin}\vartheta\frac{\partial B}{\partial \vartheta}\right)\right]-kB=0 \tag{7.179}$$

for the sediment domain $r_o \leq r < \infty$ and $0 \leq \vartheta \leq \pi$, where r_o is the radius of the object. The conservation equation for a decaying solute species about an impermeable object would be very similar.

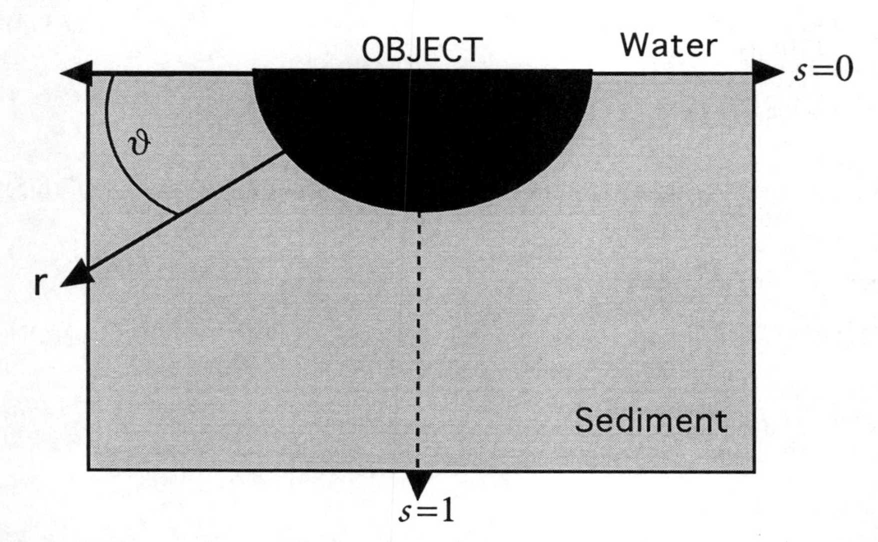

Figure 7.3. A schematic diagram of the isolated half-sphere impermeable object being modelled in Example 7-6.

Before we can state the boundary conditions for this equation, this problem and Eq. (7.179) are more commonly and conveniently stated in a different, but equivalent, form. In particular, define the new variables

$$s \equiv cos(\vartheta) \tag{7.180}$$

$$r \equiv \frac{\mathrm{r}}{\mathrm{r_o}} \tag{7.181}$$

With these new variables, Eq. (7.179) becomes

$$\frac{1}{r^2}\frac{\partial^2 \mathrm{B}}{\partial r^2} + \frac{2}{r}\frac{\partial \mathrm{B}}{\partial r} + \frac{1}{r^2}\frac{\partial}{\partial s}\left((1-s^2)\frac{\partial \mathrm{B}}{\partial s}\right) = \kappa^2 \mathrm{B} \tag{7.182}$$

where

$$\kappa^2 \equiv \frac{\mathrm{kr_o^2}}{\mathrm{D_B}} \tag{7.183}$$

which is a type of Damkohler number.

Because this example involves a solid species, let us assume that the portion of the sediment-water interface that is not blocked by the object is subject to a constant flux of the species,

$$\left.\frac{1}{r}\frac{\partial \mathrm{B}}{\partial s}\right|_0 = -\mathcal{F} \tag{7.184}$$

where $\mathcal{F} \equiv \mathrm{r_o F/D_B}$, and F is the dimensional flux. In addition, the object is impenetrable, so no flux is allowed across its surface,

$$\left.\frac{\partial \mathrm{B}}{\partial r}\right|_1 = 0 \tag{7.185}$$

Finally, Eq. (7.182) is singular at $s=1$, and the concentration must remain bounded as $r \to \infty$. Consequently, we will remove all solution terms that become unbounded at $s=1$ and $r \to \infty$.

As with the other examples encountered so far, it is best to divide this current problem into two parts, one that is valid far from the object (and represents the unperturbed system), $\mathrm{B}_\infty(r,s)$, and one that characterizes the disturbance of the concentration field (i.e. 3-D distribution) due to the presence of the object, $\mathrm{B_o}(r,s)$, i.e.

$$\mathrm{B}(r,s) \equiv \mathrm{B}_\infty(r,s) + \mathrm{B_o}(r,s) \tag{7.186}$$

The far-field solution $B_\infty(r,s)$ is really the solution to the 1-D problem,

$$D_B \frac{d^2B_\infty}{dx^2} - kB_\infty = 0 \tag{7.187}$$

subject to a prescribed flux, F, at the sediment-water interface (x=0) and bounded as $x \to \infty$; that is to say,

$$B_\infty(x) = \frac{F}{\sqrt{D_B k}} exp\left(-x\sqrt{\frac{k}{D_B}}\right) \tag{7.188}$$

Using Eqs (7.180) and (7.181) and the equivalencies provided in Chapter 3, i.e.

$$r = \sqrt{x^2 + y^2 + z^2} \tag{7.189}$$

$$\vartheta = arctan\left(\frac{\sqrt{y^2 + z^2}}{x}\right) \tag{7.190}$$

Eq. (7.188) becomes, in our spherical coordinate system,

$$B_\infty(r,s) = \frac{\mathcal{F}}{\kappa} exp(-\kappa r s) \tag{7.191}$$

Given Eq. (7.191), $B_o(r,s)$ is then the solution to the equation

$$\frac{\partial^2 B_o}{\partial r^2} + 2r\frac{\partial B_o}{\partial r} + \frac{\partial}{\partial s}\left((1-s^2)\frac{\partial B_o}{\partial s}\right) = \kappa^2 r^2 B_o \tag{7.192}$$

subject to,

$$\left.\frac{\partial B_o}{\partial s}\right|_0 = 0 \tag{7.193}$$

$$\left.\frac{\partial B_o}{\partial r}\right|_1 = \mathcal{F} exp(-\kappa s) \tag{7.194}$$

and the requirements of finite concentration.

This problem is now homogeneous in the s-direction, and it can be solved by SoV by assuming that

$$B_o(r,s) = R(r)S(s) \tag{7.195}$$

where $R(r)$ is a function of r only, and $S(s)$ is a function of s only. (Please, do not confuse R with a reaction term. Only so many symbols available, so that some need to be recycled.) Separation leads to the two equations,

$$\frac{d^2R}{dr^2} + 2r\frac{dR}{dr} - \left(\kappa^2 r^2 B_o + \lambda\right)R = 0 \tag{7.196}$$

and

$$\frac{d}{ds}\left((1-s^2)\frac{dS}{ds}\right) + \lambda S = 0 \tag{7.197}$$

where λ is a separation constant.

Equation (7.197) is *Legendre's differential equation* (e.g. see Hildebrand, 1976, p. 230; Spanier and Oldham, 1987, their Chapter 21). It has finite solutions at s=1 if, and only if, λ is a positive integer. For convenience it is best to write λ as

$$\lambda = n(n+1) \qquad n = 0,1,2,3... \tag{7.198}$$

The solutions to Eq. (7.197) are then the so-called *Legendre Polynomials* $P_n(\mu)$. An explicit general formula for these polynomials is

$$P_n(\mu) \equiv \sum_j (-1)^{j/2}\frac{(2n-j-1)!!}{j!!(n-j)!}s^{n-j} \tag{7.199}$$

where the double factorial (Spanier and Oldham, 1987) is defined as:

$$n!! \equiv \begin{cases} 1 & n = -1,0 \\ n(n-2)(n-4)...5\cdot3\cdot1 & n = 1,3,5,... \\ n(n-2)(n-4)...6\cdot4\cdot2 & n = 2,4,6,... \end{cases} \tag{7.200}$$

The first five of these Legendre polynomials are:

$$P_0(s) = 1 \tag{7.201}$$

$$P_1(s) = s \tag{7.202}$$

$$P_2(s) = \frac{3s^2-1}{2} \tag{7.203}$$

$$P_3(s) = \frac{5s^3-3s}{2} \tag{7.204}$$

$$P_4(s) = \frac{35s^4 - 30s^2 + 3}{8} \tag{7.205}$$

More of these can be found in Abramowitz and Stegun (1972) and Spanier and Oldham (1987). To satisfy Eq. (7.193), n can only be even or zero; thus, n is further restricted to the values of 0,2,4,6,... in our solution.

We now need to solve the r-direction equation,

$$\frac{d^2R}{dr^2} + 2r\frac{dR}{dr} - \left(\kappa^2 r^2 B_o + n(n+1)\right)R = 0 \tag{7.206}$$

This looks as if it might be a modified Bessel equation, and in fact, if we look at Eq. 9.1.53 in Abramowitz and Stegun (1972), the solution to Eq. (7.201) is

$$R(r) = r^{-1/2}\left[a\, I_\nu(\kappa r) + b\, K_\nu(\kappa r)\right] \tag{7.207}$$

where a and b are integration constants, $I_\nu(\kappa r)$ and $K_\nu(\kappa r)$ are modified spherical Bessel functions of the third kind (see Chapter 6), and

$$\nu = n + \frac{1}{2} \tag{7.208}$$

In order for this solution to be bounded at infinity, it is necessary that a = 0.

We now have both terms of the solution, and superposition of all the solution components corresponding to the infinite number of n values leads to

$$B_o(r,s) = \frac{1}{\sqrt{r}} \sum_{n=0}^{\infty} b_n K_{n+1/2}(\kappa r) P_n(s) \tag{7.209}$$

We are now left with the problem of calculating the values of b_n.

This last step is accomplished by satisfying Eq. (7.194). Differentiating Eq. (7.209), we obtain

$$\left.\frac{\partial B_o}{\partial r}\right|_{r=1} = -\frac{1}{2} \sum_{n=0}^{\infty} b_n\, \gamma_n\, P_n(s) \tag{7.210}$$

where n = 0,2,4,... and

$$\gamma_n \equiv K_{n+1/2}(\kappa) + \kappa\left(K_{n-1/2}(\kappa) + K_{n+3/2}(\kappa)\right) \tag{7.211}$$

noting that for n=0, $K_{-1/2}(\kappa) = K_{1/2}(\kappa)$. Combining Eqs (7.210) and (7.194) produces

$$-\frac{1}{2}\sum_{n=0}^{\infty} b_n \gamma_n P_n(s) = \mathcal{F}\, exp(-\kappa s) \tag{7.212}$$

As in the other SoV examples, we now multiply Eq. (7.212) by the solution component (eigenfunction) $P_m(s)$ and integrate over the range in s values, i.e. 0 to 1,

$$-\frac{1}{2}\sum_{n=0}^{\infty} b_n \gamma_n \int_0^1 P_n(s)P_m(s)\,ds = \mathcal{F}\int_0^1 exp(-\kappa s)\, P_m(s)\,ds \tag{7.213}$$

Next we employ the orthogonality relation for the Legendre polynomials (Hildebrand, 1976; Özisik, 1980, p. 158),

$$\int_0^1 P_m(s)P_n(s)\,ds = \begin{cases} \dfrac{1}{2n+1} & m = n \\ 0 & m \neq n \end{cases} \tag{7.214}$$

for m and n equal to zero or even integers. Using this relation in Eq. (7.213), one finds that,

$$b_n = -\frac{2(2n+1)\mathcal{F}}{\gamma_n}\int_0^1 exp(-\kappa s)\, P_n(s)\,ds \tag{7.215}$$

The complete solution is then,

$$B(r,s) = \frac{\mathcal{F}}{\kappa} exp(-\kappa r s) + \frac{1}{\sqrt{r}}\sum_{n=0}^{\infty} b_n K_{n+1/2}(\kappa r) P_n(s) \tag{7.216}$$

or, in terms of the original variables,

$$B(r,\vartheta) = \frac{F}{\sqrt{D_B k}}\, exp\left(-r\, cos(\vartheta)\sqrt{\frac{k}{D_B}}\right) + \sqrt{\frac{r_o}{r}}\sum_{n=0}^{\infty} b_n K_{n+1/2}\left(r\sqrt{\frac{k}{D_B}}\right) P_n(cos(\vartheta)) \tag{7.217}$$

with the b_n values from Eq. (7.215).

Multidimensional problems are not common in the diagenetic literature. An investigator contemplating such a problem is first advised to consult Carslaw and Jaeger (1959) and Özisik (1980), who provide concise repertoires of SoV solutions to the diffusion equation in the Cartesian, cylindrical, and spherical coordinate systems.

7.5 Duhamel's Theorem

It might appear, at first, that the SoV method cannot handle inhomogeneities caused by time-dependence in the boundary conditions or in the reaction terms. In and of itself, it cannot. However, *Duhamel's Theorem* provides us with a way to extend solutions of problems with time-independent reaction terms and/or boundary conditions to those in which there is a time-dependence (see Özisik, 1980, Chapter 5, for an extended discussion).

For a problem in depth, x, and time, t, Duhamel's Theorem deals with equations of the form,

$$\frac{\partial C}{\partial t} = D\frac{\partial^2 C}{\partial x^2} \pm g(x,t) \tag{7.218}$$

with an initial condition,

$$C(x,0) = h(x) \tag{7.219}$$

and with the general boundary condition at x = 0,

$$-\varepsilon D\frac{\partial C}{\partial x} + \gamma v C = f(t) \tag{7.220}$$

where ε and γ are parameters of value either 0 or 1, and f(t) is an arbitrary function of time. By setting $\varepsilon = 0$, Eq. (7.220) defines the boundary concentration as a function of time at x = 0; with ε and $\gamma = 1$, we get a prescribed total flux as a function of time; finally with $\varepsilon = 1$ and $\gamma = 0$, we get a diffusive flux as a function of time.

Now let $C(x,t,\tau)$ be the solution to Eq. (7.220) when the time variable t in g(x,t) and f(t) is replaced by a parameter, say τ, so that these terms become the time-independent functions $g(x,\tau)$ and $f(\tau)$, respectively. (Again a symbol must be re-used, i.e. τ, but it is the conventional symbol in this context.) The value of τ can be chosen for convenience, e.g. f(t) =1. $C(x,t,\tau)$ can then be obtained by SoV. Duhamel's Theorem next tells us that the solution to the actual time-dependent problem, C(x,t), is given by (Özisik, 1980)

$$C(x,t) = \frac{\partial}{\partial t}\int_0^t C(x,t-\tau,\tau)d\tau \tag{7.221}$$

where $C(x,t-\tau,\tau)$ indicates that we have replaced t by $t-\tau$ in $C(x,t,\tau)$ before integrating. If we wish to switch the order of integration and differentiation in Eq. (7.221), then this formula becomes (Özisik, 1980)

$$C(x,t) = h(x) + \int_0^t \frac{\partial C(x,t-\tau,\tau)}{\partial t}d\tau \tag{7.222}$$

In Example 7-5, the input flux at the sediment-water interface was a constant, F_o. If, however, it is a function of time, say F(t), then the solution to this new problem is obtained via Duhamel's Theorem by identifying the solution B(x,t), defined by Eq. (7.139), with $C(x,t,\tau)$. The resulting integral would be rather complicated, and it may or may not be expressible in closed form. Alternatively, numerical integration is possible. Aller (1977) provides another example of the use of Duhamel's Theorem when there is a time-dependent source in the governing equation; that example also leads to a very complicated integral.

7.6 The Convolution Integral and Delta Function Inputs

Another way to deal with time-dependent input boundary conditions was introduced to diagenetic modellers by Guinasso and Schink (1975). They consider the fundamental mixing problem for a radioactive tracer, i.e.

$$\frac{\partial B^*}{\partial t} = D_B \frac{\partial^2 B^*}{\partial x^2} - w \frac{\partial B^*}{\partial x} - k B^* \tag{7.223}$$

subject to

$$wB^*(0) - D_B \left. \frac{\partial B^*}{\partial x} \right|_0 = 0 \tag{7.224}$$

$$\left. \frac{\partial B^*}{\partial x} \right|_L = 0 \tag{7.225}$$

$$B^*(x,0) = \delta(x) \tag{7.226}$$

where $\delta(x)$ is the Dirac Delta function (see Chapter 2), which can be interpreted as the deposition of a unit amount of tracer at the sediment-water interface at time t=0, and the asterisk indicates the concentration that results from this Delta function input.

Solution of Eqs (7.223) through (7.226) can be obtained by first applying the transformation given by Eq. (7.9), and then employing SoV (Guinasso and Schink, 1975):

$$B^*(x,t) = exp\left(\frac{wx}{2D_B} - \frac{w^2t}{4D_B} - kt\right) \times \sum_{n=1}^{\infty} \frac{1}{2} + \frac{w}{2D_B\alpha_n^2 L} + \frac{w^2}{8D_B^2\alpha_n^2}$$
$$\times\left(cos(\alpha_n x) + \frac{w}{2D_B\alpha_n} sin(\alpha_n x)\right) exp\left(-D_B\alpha_n^2 t\right) \tag{7.227}$$

where the eigenvalues, α_n, are the roots of

$$tan(\alpha_n L) - \frac{4D_B\alpha_n w}{4D_B^2\alpha_n^2 - w^2} = 0 \tag{7.228}$$

Now what is the solution, B(x,t), if the input boundary condition, Eq. (7.224), is replaced by

$$wB(0) - D_B \frac{\partial B}{\partial x}\bigg|_0 = F(t) \tag{7.229}$$

where F(t) is an arbitrary function of time? This is where the convolution integral immediately tells us that (Özisik, 1980)

$$B(x,t) = \int_{-\infty}^{t} F(\tau) B^*(x, t-\tau)\, d\tau \tag{7.230}$$

where τ is a dummy variable in this case. The integral in Eq. (7.230) informs us that the solution for an arbitrary time-dependent input is equal to the net (integrated) effects of continuous Delta function inputs, weighted to match the real input at each time. Again, this integral, like that from Duhamel's Theorem, may be difficult to evaluate analytically; however, numerical integration is always a possibility (see Chapter 8).

7.7 Laplace Transforms

Laplace transforms are particularly well suited to the solution of the 1-D diagenetic equation (i.e. the ADR equation) for semi-infinite sediment layers (see Carslaw and Jaeger, 1959; Hershey, 1973; Özisik, 1980; van Genuchten, 1981; van Genuchten and Alves, 1982). The application of the transform is a fairly simple procedure, while the solution of the resulting equations ("taking the inverse") can be quite complicated. In fact, formal inversion of a Laplace transformed equation necessitates knowledge of so-called complex† analysis and integration in the complex (imaginary) plane. Such topics are far beyond the scope of this book.

† Complex refers to calculus involving the square-root of negative one, i.e. the so-called imaginary numbers.

Fortunately, tables have been prepared for the inversion of a large class of functions normally associated with diffusion-type problems (e.g. Erdélyi, 1954; Özisik, 1980; van Genuchten and Alves, 1982). In fact, the Laplace transform method is arguably most effective as a look-up technique, and this is the way it is presented here.

The Laplace transform with respect to time of a mathematical function, say F(t), is defined as

$$\mathcal{L}[F(t)] \equiv \int_0^\infty e^{-st} F(t)\,dt \equiv f(s) \tag{7.231}$$

where f(s) is a new function of the transform variable s. The inverse Laplace transform is formally written as

$$\mathcal{L}^{-1}[f(s)] \equiv \mathcal{L}^{-1}\left[\mathcal{L}[F(t)]\right] \equiv F(t) \tag{7.232}$$

This definition implies that, if we have a table of Laplace transforms with various F(t) to f(s) pairs, then reading the table backwards gives the inverse of the transformed function.

Example 7-8. It can be easily verified by doing each integral that

$$\mathcal{L}[1] = \frac{1}{s} \tag{7.233}$$

$$\mathcal{L}[t] = \frac{1}{s^2} \tag{7.234}$$

$$\mathcal{L}\left[exp(\pm at)\right] = \frac{1}{s \mp a} \tag{7.235}$$

where a is a constant, and

$$\mathcal{L}\left[\delta(t)\right] = 1 \tag{7.236}$$

where δ(t) is the Dirac Delta function (see Chapter 2).

Example 7-9. Conversely, using a table of transforms (e.g. Table 7-1, p. 259-262 of Özisik, 1980), one can establish that

$$\mathcal{L}^{-1}\left[\frac{1}{\sqrt{s}}\right] = \frac{1}{\sqrt{\pi t}} \tag{7.237}$$

$$\mathcal{L}^{-1}\left[\frac{s}{s^2 + a^2}\right] = cos(at) \tag{7.238}$$

where a is a constant, and

$$\mathcal{L}^{-1}\left[\frac{exp\left(-a\sqrt{s}\right)}{s}\right] = erfc\left(\frac{a}{2\sqrt{t}}\right) \tag{7.239}$$

where *erfc*( ) is the Complementary Error function (see Chapter 6).

Note that s and t are related in such a manner that (e.g. Hershey, 1973)

$$\lim_{t\to 0} (F(t)) = \lim_{s\to\infty} (sf(s)) \tag{7.240}$$

and

$$\lim_{t\to\infty} (F(t)) = \lim_{s\to 0} (sf(s)) \tag{7.241}$$

which can be verified with the examples given above.

It would appear at first that the only function of the Laplace transform is simply to swap the variable s for t. Although this switching is part of its effect, more important is what happens to derivatives when transformed. Specifically, if we have an unknown concentration function, C(x,t), then the time-Laplace transform of this function is

$$\mathcal{L}[C(x,t)] \equiv \overline{C}(x,s) \tag{7.242}$$

The overbar is traditional to this definition, but should not be confused with Reynolds averaging. The Laplace transform of its first time-derivative is (e.g. Hildebrand, 1976, p. 59; Özisik, 1980, p. 249)

$$\mathcal{L}\left[\frac{\partial C}{\partial t}\right] = s\overline{C}(x,s) - C(x,0) \tag{7.243}$$

where C(x,0) is the initial condition. The derivative has been eliminated by changing it to a function of the unknown transformed concentration.

Equation (7.243) suggests that applying a Laplace transform to the time-dependent diffusion equation will convert this PDE to an ODE. This is exactly the case, as the Laplace transform and the diffusion equation are linear operators, i.e.

$$\mathcal{L}\left[\frac{\partial C}{\partial t}\right] = \mathcal{L}\left[D\frac{\partial^2 C}{\partial x^2}\right] - \mathcal{L}\left[v\frac{\partial C}{\partial x}\right] \tag{7.244}$$

or

$$D\frac{d^2\overline{C}}{dx^2} - v\frac{d\overline{C}}{dx} = s\overline{C}(x,s) - C(x,0) \tag{7.245}$$

The solution of the PDE is then reduced to solving an ODE and then doing an inverse transform to obtain the solution. The presence of a linear reaction term in Eq. (7.244) would not change this result.

Before giving examples of this method, I wish to describe a few additional, frequently used properties of Laplace transforms. These are:

i) *Linearity*,

$$\mathcal{L}[aF(t)] = af(s) \tag{7.246}$$

and

$$\mathcal{L}^{-1}[a\,f(s)] = a\,F(t) \tag{7.247}$$

where a is a constant (e.g. a model parameter); that is to say, a constant can be moved into and out of the transform.

ii) The *change-of-scale* property,

$$\mathcal{L}[F(at)] = \frac{1}{a}f\left(\frac{s}{a}\right) \tag{7.248}$$

iii) The *shift* property,

$$\mathcal{L}[\mathit{exp}(\pm at)\,F(t)] = f(s \mp a) \tag{7.249}$$

iv) The *translation* property, i.e. given

$$H(t-a)F(t) = \begin{cases} F(t-a) & t > a \\ 0 & t < a \end{cases} \tag{7.250}$$

which is called the *Heaviside unit function*, then

$$\mathcal{L}[H(t-a)F(t)] = \mathit{exp}(-as)\,f(s) \tag{7.251}$$

v) Finally, the *convolution* property,

$$\mathcal{L}\left[\int_0^\infty F(t-\tau)G(\tau)d\tau\right] = f(s)\cdot g(s) \tag{7.252}$$

Example 7-10. To illustrate this technique, let us reconsider bioturbation of a radioactive tracer into a semi-infinite sediment column, subject to a unit-impulse (Delta function, p. 21) source of tracer (Schink and Guinasso, 1975); we also assume that mixing is constant throughout the sediment and that advection can be ignored; thus, the conservation equation for the tracer becomes

$$\frac{\partial B}{\partial t} = D_B \frac{\partial^2 B}{\partial x^2} - kB \tag{7.253}$$

with boundary conditions,

$$-D_B \left.\frac{\partial B}{\partial x}\right|_0 = \delta(t) \tag{7.254}$$

$$\left.\frac{\partial B}{\partial x}\right|_{x\to\infty} = 0 \tag{7.255}$$

and initial condition,

$$B(x,0) = 0 \tag{7.256}$$

For this particular problem a useful, although not necessary, first step is to eliminate the decay part. This is done by defining the new variable,

$$B(x,t) = U(x,t)\, exp(-kt) \tag{7.257}$$

which generates the new problem,

$$\frac{\partial U}{\partial t} = D_B \frac{\partial^2 U}{\partial x^2} \tag{7.258}$$

with

$$-D_B \left.\frac{\partial U}{\partial x}\right|_0 = \delta(t) exp(kt) \tag{7.259}$$

$$\left.\frac{\partial U}{\partial x}\right|_{x\to\infty} = 0 \tag{7.260}$$

$$U(x,0) = 0 \tag{7.261}$$

Next take the Laplace transform of Eqs (7.258) through (7.261) with respect to the time variable:

$$D_B \frac{d^2\overline{U}}{dx^2} - s\overline{U} = 0 \tag{7.262}$$

with

$$-D_B \left.\frac{d\overline{U}}{dx}\right|_0 = 1 \tag{7.263}$$

$$\left.\frac{d\overline{U}}{dx}\right|_{x\to\infty} = 0 \tag{7.264}$$

where Eq. (7.261) is used in writing Eq. (7.262). The solution to Eq. (7.262) is (see Section 6.4.2)

$$\overline{U} = a\,exp\left(-x\sqrt{s/D_B}\right) + b\,exp\left(+x\sqrt{s/D_B}\right) \tag{7.265}$$

where a and b are integration constants. If we apply Eqs (7.263) and (7.264), Eq. (7.265) becomes

$$\overline{U} = \frac{1}{\sqrt{D_B\,s}}\,exp\left(-x\sqrt{s/D_B}\right) \tag{7.266}$$

This is the final form, i.e. f(s), that we want to invert.

In order to invert this back to U(x,t), one can use either the linearity property or the change-of-scale property. Remembering this, one can then look in the Table of Laplace transforms supplied as Table 7-1 in Özisik (1982) or Appendix A of van Genuchten and Alves (1982) to find that

$$\mathcal{L}^{-1}\left[\frac{1}{\sqrt{D_B\,s}}\,exp\left(-x\sqrt{\frac{s}{D_B}}\right)\right] = \frac{1}{\sqrt{\pi D_B t}}\,exp\left(\frac{-x^2}{\sqrt{4D_B t}}\right) \tag{7.267}$$

The final solution is then

$$B(x,t) = U(x,t)\,exp(-kt) = \frac{1}{\sqrt{\pi D_B t}}\,exp\left(\frac{-x^2}{\sqrt{4D_B t}} - kt\right) \tag{7.268}$$

which is a well-known solution (see Officer and Lynch, 1982). Thus, when the transform is in the tables, this constitutes a simple and effective method.

In a related application, Martin and Bender (1988) modelled the effects of seasonal variations in organic matter production, F(t), on labile carbon distributions in deep-sea sediments, G(x,t). Their solution to this model was of the form

$$G(x,t) = \frac{1}{\sqrt{\pi D_B}} \int_0^t G(x,\tau) F(t-\tau) d\tau \tag{7.269}$$

where

$$G(x,\tau) = \frac{1}{\sqrt{\tau}} exp\left(\frac{-x^2}{4 D_B \tau} - k\tau \right) \tag{7.270}$$

$$F(t-\tau) = f_o + f_1 \, sin\left(\frac{2\pi(t-\tau)}{T} \right) \tag{7.271}$$

where f_o, f_1, and T are various constants associated with the seasonal flux. Equation (7.269) can be recognized as the substitution of Eq. (7.268) and the boundary flux given by Eq. (7.271) into the convolution integral, Eq. (7.230). Once the unit impulse solution is obtained, many other solutions can follow, even if Eq. (7.269) must be evaluated numerically.

Example 7-11. Another application of Laplace transforms is diffusion and advection of a solute from a constant source at the sediment-water interface, i.e.

$$\frac{\partial C}{\partial t} = D' \frac{\partial^2 C}{\partial x^2} - u \frac{\partial C}{\partial x} \tag{7.272}$$

with boundary conditions,

$$C(0,t) = C_o \tag{7.273}$$

$$\left. \frac{\partial C}{\partial x} \right|_{x \to \infty} = 0 \tag{7.274}$$

and initial condition,

$$C(x,0) = C_i \tag{7.275}$$

where C_o and C_i are different constant concentrations. Middelburg and de Lange (1989) use this model to describe chlorine diffusion into a sediment that was previously deposited under freshwater conditions.

The solution is aided by introducing the new concentration variable,

$$\Theta(x,t) = C(x,t) - C_i \tag{7.276}$$

and Eq. (7.272) becomes

$$\frac{\partial \Theta}{\partial t} = D' \frac{\partial^2 \Theta}{\partial x^2} - u \frac{\partial \Theta}{\partial x} \tag{7.277}$$

with boundary conditions,

$$\Theta(0,t) = C_o - C_i \tag{7.278}$$

$$\left. \frac{\partial \Theta}{\partial x} \right|_{x \to \infty} = 0 \tag{7.279}$$

and initial condition,

$$\Theta(x,0) = 0 \tag{7.280}$$

Taking the Laplace transform of these equations, one obtains

$$D' \frac{d^2 \overline{\Theta}}{dx^2} - u \frac{d\overline{\Theta}}{dx} - s\overline{\Theta} = 0 \tag{7.281}$$

where Eq. (7.280) has been employed. The solution to Eq. (7.281) with Eq. (7.279) is

$$\overline{\Theta} = a\, exp(\alpha x) \tag{7.282}$$

where a is an integration constant and

$$\alpha = \frac{u}{2D'} - \sqrt{\left(\frac{u}{2D'}\right)^2 + \frac{s}{D'}} \tag{7.283}$$

To evaluate the constant a, apply the transformed version of the boundary condition at x = 0, Eq. (7.278),

$$a = \frac{C_o - C_i}{s} \tag{7.284}$$

so that

$$\overline{\Theta} = \frac{C_o - C_i}{s} exp\left\{ \frac{ux}{2D'} - x\sqrt{\left(\frac{u}{2D'}\right)^2 + \frac{s}{D'}} \right\} \tag{7.285}$$

or

$$\overline{\Theta} = (C_o - C_i)\, exp\left(\frac{ux}{2D'}\right)\frac{1}{s}\, exp\left\{-x\sqrt{\left(\frac{u}{2D'}\right)^2 + \frac{s}{D'}}\right\} = g(x)\cdot \bar{f}(s) \tag{7.286}$$

This form of the solution emphasizes the part that involves the s variable and that must be inverted. The g(x) part simply moves out of the inverse via the linearity property; i.e. a pure function of x looks like a parameter to the inverse.

This particular transformed function $\bar{f}(s)$ does not appear in the standard tables (e.g. Carslaw and Jaeger, 1959, Appendix V; Özisik, 1980, Table 7-1; van Genuchten and Alves, 1982, Appendix A). In fact, those $\bar{f}(s)$ that involve an exponential function are usually of the form $exp(-x\sqrt{s})$. The trick in this inversion (and nearly all such inversions) is to realize that the exponential in the $\bar{f}(s)$ of Eq. (7.286) is indeed of this form, but it has had its scale changed by 1/D´, and it has been shifted by $(u/2D')^2$; i.e. it is of the form:

$$\bar{f}(s) = \frac{1}{\left(\frac{s}{D'} + c\right) - c}\, exp\left(-x\left(\frac{s}{D'} + c\right)^{1/2}\right) \tag{7.287}$$

where $c = (u/2D')^2$ is a constant. Consequently, we are looking at a scaled and shifted version of

$$\bar{f}(s) = \frac{1}{s - c}\, exp\left(-x\sqrt{s}\right) \tag{7.288}$$

The inverse transform of Eq. (7.288) is (van Genuchten and Alves, 1982),

$$\mathcal{L}^{-1}\left[\frac{1}{s-c}\, exp\left(-x\sqrt{s}\right)\right] = \frac{exp(ct)}{2}\left\{exp\left(-x\sqrt{c}\right) erfc\left(\frac{x}{2\sqrt{t}} - c\sqrt{t}\right) + exp\left(x\sqrt{c}\right) erfc\left(\frac{x}{2\sqrt{t}} + c\sqrt{t}\right)\right\} \tag{7.289}$$

where *erfc*( ) is the complementary error function. The change of scale property means that t will be replaced by D´t in this equation, and the shift means that the inverse is multiplied by *exp*(-aD´t), so that

$$\mathcal{L}^{-1}\left[\frac{1}{\left(\frac{s}{D'} + c\right) - c}\, exp\left(-x\left(\frac{s}{D'} + c\right)^{1/2}\right)\right] = \frac{1}{2}\left\{exp\left(-x\sqrt{c}\right) erfc\left(\frac{x}{2\sqrt{D't}} - c\sqrt{D't}\right) + exp\left(x\sqrt{c}\right) erfc\left(\frac{x}{2\sqrt{D't}} + c\sqrt{D't}\right)\right\} \tag{7.290}$$

With this result, it is a mechanical task to assemble the final solution, i.e.

$$C(x,t) = C_i + \frac{(C_o - C_i)}{2}\left\{ erfc\left(\frac{x - ut}{2\sqrt{D't}}\right) + exp\left(\frac{ux}{D'}\right) erfc\left(\frac{x + ut}{2\sqrt{D't}}\right)\right\} \tag{7.291}$$

which is the same as the solution stated by Middelburg and de Lange (1989) in their study of salinity changes in Kau Bay, Indonesia. There have been some turns and twists in arriving at this result, which simply proves that adept use of Laplace transforms, like all the techniques in this chapter, requires the insight gained by sufficient practice. Other examples of 1-D diffusion-reaction problems solved by Laplace transforms can be found in van Genuchten (1981, 1985) and Toride et al. (1993). Furthermore, Laplace transforms are not restricted to 1-D Cartesian problems, but can also be applied to multi-dimensional problems in Cartesian and other coordinate systems (see Carslaw and Jaeger, 1959; Özisik, 1980; Batu and van Genuchten, 1990; Leij and Dane, 1990; Leij et al., 1991).

7.8 The Advection-Reaction Equation

The previous discussions all presumed the existence of a second-order spatial derivative to account for the effects of diffusion (e.g. molecular, biodiffusion, dispersion, etc.). Certainly, below the surficial zone of mixing, solids are transported purely by advection. In addition, some forms of mixing are nonlocal, rather than diffusive (see Chapter 3). We are then forced to consider the solution of the time-dependent advection-reaction equation, which is hyperbolic, i.e.

$$\frac{\partial B}{\partial t} + w(x)\frac{\partial B}{\partial x} = R(B, x, t) \tag{7.292}$$

where constant porosity is assumed. (Variable porosity can often be merged into B anyway.) The term on the right-hand side of Eq. (7.292) represents only reactions, as any nonlocal mixing has been removed by subtracting the equations for total solids. Added to this equation is one initial condition, say

$$B(x,0) = h(x) \tag{7.293}$$

and a single boundary condition, say

$$B(0,t) = f(t) \tag{7.294}$$

where h(x) and f(t) are prescribed functions of depth and time, respectively. (Note that the spatial domain of Eq. (7.292) seems to start from the sediment-water interface. It can start at any depth, say x_o, and the first depth can always be made to be zero by the simple translation $x = x - x_o$).

Equation (7.292) is generally solved by the method of "characteristics" (see Hildebrand, 1976, or Farlow, 1993, for more details). In this method, the differentials of time, space, and concentration are related by

$$dt = \frac{dx}{w(x)} = \frac{dB}{R(B,x,t)} \tag{7.295}$$

This relation can be interpreted as three possible ODEs, of which only two are independent, e.g.

$$dt = \frac{dx}{w(x)} \tag{7.296}$$

and

$$\frac{dx}{w(x)} = \frac{dB}{R(B,x,t)} \tag{7.297}$$

If Φ_1 and Φ_2 are solutions to Eqs (7.296) and (7.297), respectively, then the general solution to Eq. (7.292) is (Hildebrand, 1976)

$$\Phi_2 = f(\Phi_1) \tag{7.298}$$

where $f(\ )$ is an arbitrary function that must be chosen to accommodate the initial and boundary conditions. This may be difficult to visualize, so the process is illustrated with the examples given below.

Example 7-12. The simplest form of Eq. (7.292) results if w(x) is a constant and R(B,x,t) = 0, i.e.

$$\frac{\partial B}{\partial t} + w\frac{\partial B}{\partial x} = 0 \tag{7.299}$$

subject to, for example, the conditions

$$B(x,0) = 1 \tag{7.300}$$

$$B(0,t) = exp(-at) \tag{7.301}$$

For this problem, Eq. (7.296) becomes

$$dt = \frac{dx}{w} \tag{7.302}$$

or, upon integration,

$$\Phi_1 = \text{x-wt} \tag{7.303}$$

From Eq. (7.295),

$$dB = 0 \tag{7.304}$$

or

$$\Phi_2 = B \tag{7.305}$$

Thus, the general solution is

$$B(x,t) = f(x\text{-}wt) \tag{7.306}$$

We now need to find a function, $f(\ )$, that is consistent with the initial and boundary conditions. To appreciate the form that $f(\ )$ must acquire, note that Eq. (7.299) implies a total derivative (see Chapter 2)

$$\frac{dB}{dt} = 0 \tag{7.307}$$

Thus, there is no diagenesis, and there are no changes following a parcel of sediment. A sediment with any temporal variations in B is simply buried. Sediment initially (t=0) at depth x is w·t deeper at future time t. If the sediment layer being examined was not initially present, but was deposited at some time t_1 later, then it is now at a depth equal to $w \cdot (t\text{-}t_1)$. This indicates that $f(\ )$ needs to be a two-part function, one part to deal with materials initially present and one to deal with the material added subsequently.

Implementing the initial condition

$$B(x,0) = f(x) = 1 \tag{7.308}$$

and this must be true for $x\text{-}w \cdot t > 0$ at any subsequent time. For x=0,

$$B(0,t) = f(\text{-wt}) = exp(\text{-at}) \tag{7.309}$$

and a sediment now at a depth $x < w \cdot t$ must originally have had the concentration at the surface at the time t-x/w. Therefore, the complete solution is of the form,

$$B(x,t) = \begin{cases} 1 & x - wt > 0 \\ exp((x - w \cdot t)a / w) & x - wt \leq 0 \end{cases} \tag{7.310}$$

Example 7-13. The (minor) solid species in Example 7-12 was stable; let us now consider an example with first-order decay, i.e.

$$\frac{\partial B}{\partial t} + w \frac{\partial B}{\partial x} = -kB \tag{7.311}$$

subject to

$$B(x,0) = 0 \tag{7.312}$$

and

$$B(0,t) = 1 \tag{7.313}$$

Equation (7.313) represents the introduction of this tracer at a constant concentration (normalized to 1) at the sediment-water interface starting at t=0.

For this problem, the appropriate form of Eq. (7.296) is still Eq. (7.302), so that

$$\Phi_1 = x\text{-}wt \tag{7.314}$$

However, for Eq. (7.297),

$$\frac{dx}{w} = \frac{dB}{-kB} \tag{7.315}$$

or

$$\Phi_2 = B\, exp(kx/w) \tag{7.316}$$

In this case, the general solution is

$$B(x,t) = exp(-kx/w)\, f(x\text{-}wt) \tag{7.317}$$

Application of the initial and boundary conditions requires that

$$f(x\text{-}wt) = 1 \tag{7.318}$$

so that the solution in this particular case is simply

$$B(x,t) = \begin{cases} 0 & x - wt > 0 \\ exp(-kx/w) & x - wt \le 0 \end{cases} \tag{7.319}$$

More complicated problems are more easily handled numerically (see Chapter 8). In fact, the analytical treatment of PDEs is now complete, and we can move on to numerical methods.

8 NUMERICAL METHODS

"Two added to one - if that could but be done",
It said, "with one's fingers and thumbs!"
Recollecting with tears how, in earlier years,
It had taken no pains with its sums.

(Lewis Carroll, *Hunting of the Snark*, 1876)

8.1 Introduction

The fundamental purpose of a mathematical model is to calculate numbers. When model solutions are linear functions, simple polynomials, or elementary functions, a hand-held calculator may be completely sufficient to this purpose. More often, model solutions either are expressed in terms of implicit nonlinear functions or are not amenable to analytical methods at all. The modeller must then resort to the use of so-called numerical methods; these are computer-based techniques for calculating or approximating the actual solution. This chapter reviews three such techniques: root finding for nonlinear equations, such as Eqs (6.252) and (6.298), and Case 9 of Table 7.1B; numerical evaluation of integrals, such as Eqs (7.178) and (7.215); and finite-difference approximations for ODEs and PDEs. Again, only a cursory presentation is possible here, and the reader is referred to specific texts on these topics for more comprehensive treatments (e.g. Conte and deBoor, 1972; Hildebrand, 1974; Mitchell and Griffiths, 1980; Lapidus and Pinder, 1982; Davis and Rabinowitz, 1984; Kahaner et al., 1989; Press et al., 1992).

8.2 Accuracy and Stability

Prior to a discussion of numerical methods, two concepts associated with these methods must be mentioned; these are the issues of *accuracy* and *stability*. Numerical methods are approximations, and as such, there is some error in replacing exact formulas with numerical schemes, which we call the degree of accuracy. For some methods (such as finding roots), accuracy can be arbitrarily adjusted to the internal accuracy of the digital computer, termed the machine error. The accuracy of other methods is limited by the approximations made in their creation.

Stability is a completely different problem, and it reflects the fact that approximations are not the original equations and can exhibit behavior very different from that of the original equations under some conditions. Even when the accuracy is guaranteed to be high, the numerical method may fail to imitate the original equation (an example is discussed in the section on finite differences).

8.3 Finding Roots

Finding the root(s) of a single equation is one of the oldest problems of numerical mathematics. It consists of two steps: 1) *bracketing* the root and 2) *refining* the estimate. Bracketing the root is the identification of a variable range in which the root lies. This may in fact be known *a priori*, in which case this step can be omitted; nevertheless, there are a host of problems where such a range is unknown.

The safest way to *bracket* an unknown root is to conduct a simple search scheme. One starts from a lower or upper bound and increments the variable until the sign of the equation's value changes. Conceptually, if the equation is written as

$$c = F(Z) \tag{8.1}$$

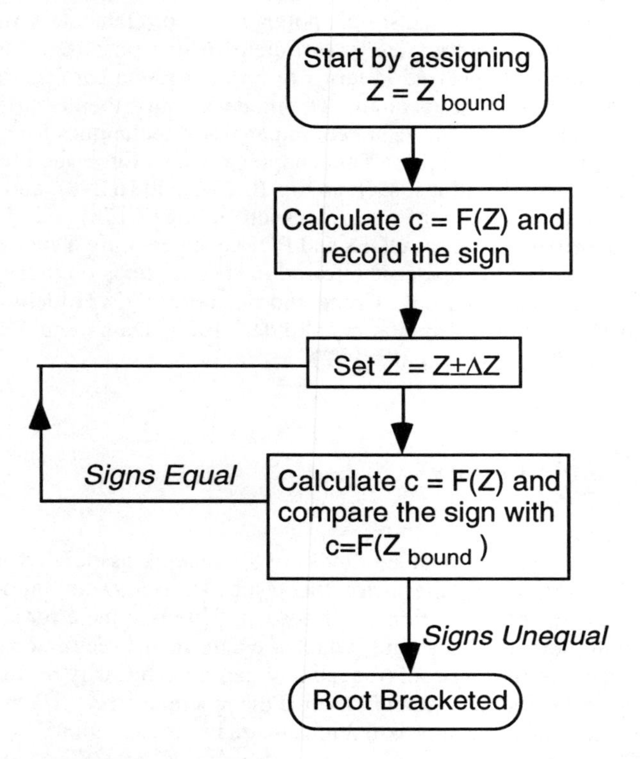

Figure 8.1. A flow-chart representation of the steps involved in bracketing an unknown root of a general algebraic equation, such as Eq. (8.1), where Z is an arbitrary independent variable.

where Z is a generic variable, usually some concentration, F( ) is the nonlinear algebraic equation in question, and c is the value of F(Z). When the value of c changes its sign as Z is systematically varied, this indicates that the root has been passed, and that it lies between the penultimate and final values assigned to the variable - see the illustrated flow chart for this process in Fig. 8.1. This scheme can be generalized to deal with coupled sets of equations.

Logic can usually supply the upper or lower bound to start the bracketing process. For example, concentrations cannot be negative, and negative eigenvalues in the SoV method (Chapter 7) don't supply any new solutions, so that zero is often a very good guess for a lower bound. In addition, concentrations of solids cannot exceed their intrinsic density, and solute concentrations cannot exceed the total salinity, so that these values may supply starting values for a downward search.

Refining a root can be done in a large number of ways (see Conte and deBoor, 1972; Hildebrand, 1974; Kahaner et al., 1989; Press et al., 1992). Two widely used methods are the *bisection* and the *Newton-Raphson* methods.

Bisection. Bisection is the most elementary method for root refinement and is simply an extension of the root-bracketing scheme described above. Given the upper and lower bounds, guess that the root is the mid-way point, and evaluate $c=F(Z_{mid})$. If the sign of c is the same as that for the upper bound, then the root is between the mid-way point and the lower bound. Set the new value of the upper bound as that mid-way value. If the sign of c is the same as that for the lower bound, then the root is closer to the upper bound, and set the new lower bound as the mid-way point. Keep repeating the process until the difference between the two bounds is reduced to some small prescribed value, usually in terms of a relative error (Fig. 8.2). Bisection is relatively slow at finding roots, but it nearly always converges on the answer.

Newton-Raphson. This refinement method is also iterative, but differs significantly from bisection in the manner each new approximation is obtained. To derive this formula, we start with a Taylor series expansion about some arbitrarily chosen point, Z, in the bracketed interval; thus, the root can be expressed as

$$0 = F(Z) + \left.\frac{\partial F}{\partial Z}\right|_Z \Delta Z + \frac{1}{2}\left.\frac{\partial^2 F}{\partial Z^2}\right|_Z \Delta Z^2 + \ldots \tag{8.2}$$

where ΔZ is the distance to the root from the chosen point. If the higher order derivatives are small compared with the first, we can approximately set

$$\Delta Z \approx -\frac{F(Z)}{\left.\frac{\partial F}{\partial Z}\right|_Z} \tag{8.3}$$

Equation (8.3) supplies an approximate correction to our first guess, Z, for the root, Z_o. Then, using this new approximation, we can again use Eq. (8.3) to obtain a second increment and a third approximation to Z_o. This process can be repeated until either

$$\frac{Z_{new} - Z_{old}}{Z_{old}} \leq \varepsilon \tag{8.4a}$$

or

$$F(Z) \leq \varepsilon \tag{8.4b}$$

where ε is a prescribed tolerance on the accuracy.

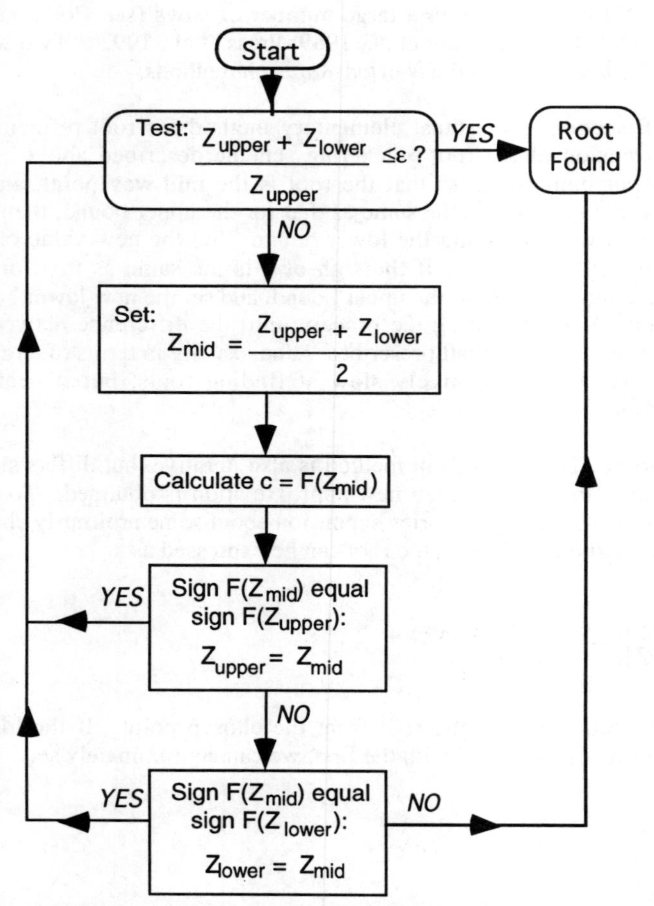

Figure 8.2. Flow chart of root refinement by the bisection method.

Newton-Raphson iteration converges rapidly on the root, but the function F must be relatively well behaved (i.e. no singular points, etc.). Also, there is a problem in calculating the derivative needed in Eq. (8.3). If F(Z) is relatively simple, an analytical (exact) derivative can be calculated; however, more commonly, F(Z) is sufficiently complicated that a numerical derivative needs to be used (see the section on finite differences below). The accuracy of both the bisection and Newton-Raphson methods is limited only by the accuracy of the machine representation of different numbers (the roundoff error), and both are completely stable.

Books on numerical analysis, such as Conte and de Boor (1972) and Press et al. (1992), provide FORTRAN codes that implement the bisection and Newton-Raphson methods, as well as other techniques. Modellers are often faced with finding the roots (solutions) of not one, but many equations simultaneously. FORTRAN computer codes are also available to solve single or sets of nonlinear algebraic equations, e.g. Moré and Cosnard, 1980; Kahaner et al., 1989; Press et al., 1992 (see also the Appendix).

8.4 Basic Numerical Integration

Some of the solutions developed in Chapter 7 require the calculation of an integral. It may be possible to evaluate such integrals analytically, and one should consult such compendia as Gradshteyn and Ryzhik (1980). If an exact solution is inconvenient because of its complexity or is simply not possible, then a numerical method can be employed. Two widely utilized numerical schemes are the *Trapezoidal* rule and *Simpson's* rule, as explained below.

To integrate a function F(Z) over an interval $Z = Z_o$ to $Z = Z_N$, the first step is to choose a finite set of N points in this interval that are equally spaced, say Z_1, Z_2, Z_3, ..., Z_{N-1}. Then evaluate the function at each of these points, including Z_o and Z_N, i.e. F_o, F_1, F_2, F_3, ..., F_{N-1}, F_N. Based on these function values, the *Trapezoidal* rule can be written as:

$$\int_{Z_o}^{Z_N} F(Z)dZ = \Delta Z\left\{\frac{F_o}{2} + F_1 + F_2 + F_3 + \ldots + F_{N-1} + \frac{F_N}{2}\right\} \tag{8.5}$$

while Simpson's rule is

$$\int_{Z_o}^{Z_N} F(Z)dZ = \frac{\Delta Z}{3}\left\{F_o + 4\left[F_1 + F_3 + \ldots + F_{N-1}\right] + 2\left[F_2 + F_4 + \ldots + F_{N-2}\right] + F_N\right\} \tag{8.6}$$

where ΔZ is the step size between points, i.e. $\Delta Z = Z_2 - Z_1 = Z_3 - Z_2$, etc.

The accuracy of the Trapezoidal rule is of the order of the square of the step size, written as $O(\Delta Z^2)$; Simpson's rule is more accurate, with errors of $O(\Delta Z^4)$. Both methods are unconditionally stable; however, great care must be exercised when dealing with rapidly oscillating functions (integrands). Enough points must be chosen in each oscillation to resolve this behavior. Note also that Simpson's rule requires an even number of points, while the Trapezoidal formula can use any number. FORTRAN computer codes for integration are available in Davis and Rabinowitz (1984), Kahaner et al. (1989), and Press et al. (1992), as well as in NETLIB. Numerical solution of integrals on semi-infinite intervals ($0 \leq Z < \infty$) has been considered by Squire (1976) and Davis and Rabinowitz (1984).

8.5 Finite Differences and Diagenetic Equations

Chapters 6 and 7 explored various ways of analytically solving diagenetic conservation equations in the form of ODEs and PDEs, respectively. Approximate solutions of such equations can also be obtained directly by numerical methods. A number of numerical techniques exist for ODEs and PDEs, including finite differences, finite elements, boundary elements, shooting methods, Runge-Kutta methods, and many others (see Lapidus and Pinder, 1982; Zwillinger, 1989; Ascher et al., 1988; Schiesser, 1994). Each approach has its strengths and weaknesses; none is superior overall, but one may be better suited to a particular problem than another. Nearly all these methods involve the replacement of the continuous derivatives in the differential equations with some form of approximation, which is valid at a chosen point or along a small finite depth-interval or over a finite surface or volume.

Finite Difference Method. This method is effective for diagenetic models, especially when implemented as the *Method of Lines* (Byrne and Hindmarsh, 1987; Schiesser, 1991). The justification for this pronouncement originates with the simple geometry needed to describe most diagenetic problems and the relatively simple implementation of this method, even in the face of complex reaction kinetics. The next sections outline the derivation, implementation, and problems associated with the finite difference method (see also Isenberg and de Vahl Davis, 1975; Smith, 1975; Nogotov, 1978; Mitchell and Griffiths, 1980; Özisik, 1980; Lapidus and Pinder, 1982; Ascher et al., 1988; Lick, 1989). This should not be construed as a disparaging statement with regard to other methods, and the reader is encouraged to explore these by consulting such texts as Lambert (1973), Lapidus and Pinder (1982), Ascher et al. (1988), Press et al. (1992), and the references given therein.

8.5.1 Lessons from Initial Value Problems

Suppose we are interested in solving the linear decay equation from some initial time that we call $t = 0$ to some other final time, designated t_f, i.e.

$$\frac{dC}{dt} = -kC \tag{8.7}$$

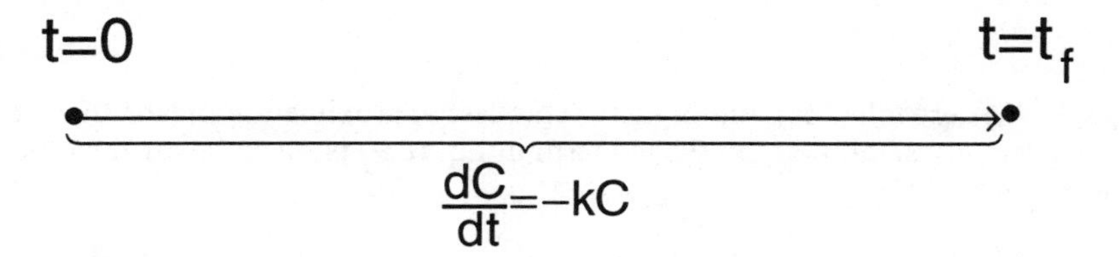

Figure 8.3. Schematic representation of the time domain over which the decay equation is to be solved.

with the initial condition,

$$C(0) = C_o \tag{8.8}$$

This problem is to be solved over the time domain shown in Fig. 8.3.

We know the analytical solution to Eq. (8.7), and it allows us to calculate continuous values of C(t) as we move along the t-axis in Fig. 8.3. Now, let us feign that we cannot obtain the analytical solution to Eq. (8.7). How might we still calculate $C(t_f)$ numerically? Instead of trying to get to t_f immediately, consider the calculation of C(t) for a time $t = t_1$ where $t_1 << t_f$ (Fig. 8.4).

How is $C(t_1)$ related to C(0)? Certainly through Eq. (8.7), but also through a Taylor series:

$$C_1 = C_o + \left.\frac{dC}{dt}\right|_{t=0} \Delta t + O(\Delta t^2) \tag{8.9}$$

where $\Delta t = t_1 - 0 = t_1$, and $O(\Delta t^2)$ incorporates all the terms in the series that are multiplied by Δt^2 and higher powers of Δt. We can rearrange Eq. (8.9) to get

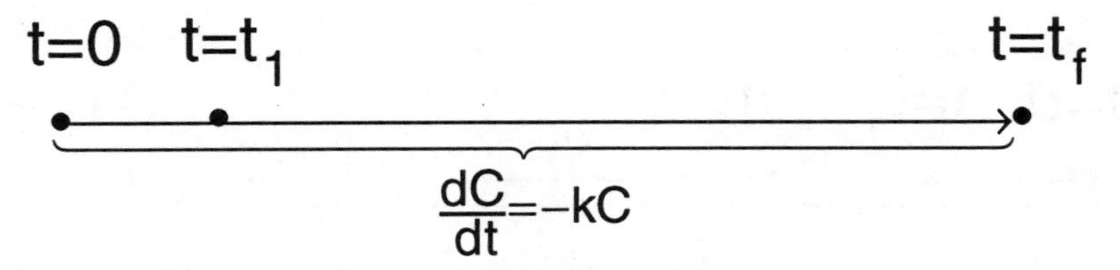

Figure 8.4. Location of a time point t_1 that is a small time interval, Δt, later than the initiation of the decay process.

$$\left.\frac{dC}{dt}\right|_{t=0} \approx \frac{C_1 - C_o}{\Delta t} \tag{8.10}$$

if we ignore the terms $O(\Delta t^2)$ and higher. This last approximation means that Δt must be sufficiently small that the second term in Eq. (8.9) is much larger than the subsequent terms.

A comparison of Eqs (8.10) and (8.7) shows that the derivatives can be eliminated between these two equations to obtain

$$\frac{C_1 - C_o}{\Delta t} \approx -k\,C_o \tag{8.11}$$

This result is called a *finite difference* formula, equation, or approximation. It is a single equation for the unknown concentration C_1 at t_1 in terms of the known concentration C_o and known parameters Δt and k. Assuming that Eq. (8.11) is an equality, then solving it for C_1 gives

$$C_1 = C_o(1 - k\Delta t) \tag{8.12}$$

Now select a second point, t_2, at the same distance Δt from t_1 (Fig. 8.5). We now calculate the concentration at this point, $C(t_2)$, in a similar fashion to that at t_1. Through a Taylor series expansion,

$$C_2 = C_1 + \left.\frac{dC}{dt}\right|_{t=t_1} \Delta t + O(\Delta t^2) \tag{8.13}$$

or, rearranged,

$$\left.\frac{dC}{dt}\right|_{t=t_1} \approx \frac{C_2 - C_1}{\Delta t} \tag{8.14}$$

if we ignore the terms of $O(\Delta t^2)$ and higher.

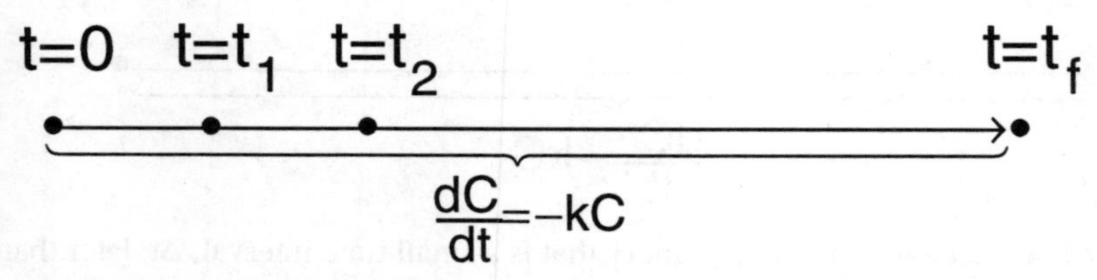

Figure 8.5. Position of a second point t_2 at distance Δt from the first, t_1.

Eliminating the derivatives between Eqs (8.7) and (8.14) produces

$$\frac{C_2 - C_1}{\Delta t} \approx -k\,C_1 \tag{8.15}$$

Like equation Eq. (8.11) before it, this is a single equation for the unknown concentration C_2 at t_2 in terms of the now known concentration C_1 and known parameters Δt and k. Solving for C_2 and assuming equality,

$$C_2 = C_1(1 - k\Delta t) \tag{8.16}$$

In general, we can keep doing this until we arrive at t_f. For the concentration C_i at the i-th time step, t_i, the finite difference formula is

$$\frac{C_i - C_{i-1}}{\Delta t} = -k\,C_{i-1} \tag{8.17}$$

or

$$C_i = C_{i-1}(1 - k\Delta t) \tag{8.18}$$

Because this formula was built with the forward Taylor series, it is called *forward differencing* or *Euler's method.* Eq. (8.18) explicitly relates C_i to the known value C_{i-1} because it can always be written in the form:

$$C_i = f(C_{i-1}, t_{i-1}) \tag{8.19}$$

Forward differences are thus easily used, even if Eq. (8.19) is nonlinear, e.g.

$$\frac{dC}{dt} = -kC^n \tag{8.20}$$

It can still be written as

$$\frac{C_i - C_{i-1}}{\Delta t} = -k\,C_{i-1}^n \tag{8.21}$$

or

$$C_i = C_{i-1}\left(1 - kC_{i-1}^{n-1}\Delta t\right) \tag{8.22}$$

First-order ODEs in space, rather than time, can be solved in precisely the same way. For example, for the familiar advection-decay equation,

$$w\frac{dC}{dx} = -kC \tag{8.23}$$

with

$$C(x=0) = C_o \tag{8.24}$$

Then we have that

$$\frac{C_i - C_{i-1}}{\Delta x} = -\frac{k}{w} C_{i-1} \tag{8.25}$$

or

$$C_i = C_{i-1}\left(1 - \frac{k}{w}\Delta x\right) \tag{8.26}$$

where C_i is the concentration at the spatial point i in the sediment.

As convenient as the forward difference formula might appear to be, it has a major drawback. Notice that, for the example given by Eq. (8.26), if

$$\frac{k}{w}\Delta x > 1 \tag{8.27}$$

then the concentration C_i will be negative, which is physically impossible. On the next step, C_{i-1} is negative so that Eq. (8.26) has a chance of predicting a positive C_i. Consequently, once Eq. (8.27) is violated, the solution may, thereafter, oscillate. This phenomenon is called *stiffness* (Shampine and Gear, 1979). So, as k/w becomes larger, Δx must be set to a smaller and smaller value in order to avoid this problem. However, there is a limit to this refinement process, because there is a limit to the smallest number that a computer can represent, and because a user needs an answer in a reasonable real time interval.

A *stable scheme* is required. Numerical analysts have discovered that if you use a backward Taylor series instead of a forward one, this goal can be achieved; e.g. let

$$C_1 = C_2 - \left.\frac{dC}{dx}\right|_{x_2} \Delta x + O(\Delta x^2) \tag{8.28}$$

or

$$\left.\frac{dC}{dx}\right|_{x_2} = \frac{C_2 - C_1}{\Delta x} \tag{8.29}$$

Now with Eq. (8.7), we obtain

$$\frac{C_2 - C_1}{\Delta x} = -\frac{k}{w} C_2 \tag{8.30}$$

or

$$C_2 = \frac{C_1}{\left(1 + \frac{k}{w}\Delta x\right)} \tag{8.31}$$

Notice that C_2 is never negative for any value of k and Δx. Equation (8.31) is called a *backward difference* formula (BDF).

Although the stability problem has been fixed, there is a downside to BDFs. Consider the case in which the right-hand side is nonlinear, as in Eq. (8.20), so that

$$\frac{C_i - C_{i-1}}{\Delta x} = -\frac{k}{w} C_i^n \tag{8.32}$$

In general, Eq. (8.32) cannot be written in an explicit form for the unknown C_i as a function of the known C_{i-1}, as Eq. (8.22), i.e.

$$C_i - \frac{k}{w} C_i^n \Delta x = C_{i-1} \tag{8.33}$$

BDFs are thus said to be *implicit* for nonlinear equations. Fortunately, there are various strategies to get around this problem, and this topic is discussed later.

The next question is, how accurate are the forward and backward difference formulas given above? To answer, start by re-examining the forward Taylor series evaluated at x_i, i.e.

$$C_i + 1 = C_i + \left.\frac{dC}{dx}\right|_{x_i} \Delta x + \frac{1}{2}\left.\frac{d^2C}{dx^2}\right|_{x_i} (\Delta x)^2 + O(\Delta x^3) \tag{8.34}$$

where $O(\Delta x^3)$ means all the other terms in the series that are multiplied by Δx^3 and progressively higher powers of Δx. We can rearrange Eq. (8.34) to get

$$\left(\left.\frac{dC}{dx}\right|_{x_i}\right)_{\text{forward}} = \frac{C_{i+1} - C_i}{\Delta x} - \frac{1}{2}\left.\frac{d^2C}{dx^2}\right|_{x_i} \Delta x - O(\Delta x^2) \tag{8.35}$$

Notice that the second derivative term, which we ignored in the forward difference formula, is multiplied by Δx. Equation (8.35) is accurate only to order Δx. This means that the accuracy obtained with an approximation that ignores the second derivative and higher terms is of the size of Δx. Therefore, Δx must be kept small in order to obtain good accuracy, but this may generate round-off errors if Δx becomes too small.

Conversely, from the backward difference formula evaluated at x_i,

$$C_{i-1} = C_i - \left.\frac{dC}{dx}\right|_{x_i} \Delta x + \frac{1}{2}\left.\frac{d^2C}{dx^2}\right|_{x_i} (\Delta x)^2 - O(\Delta x^3) \tag{8.36}$$

we obtain:

$$\left(\left.\frac{dC}{dx}\right|_{x_i}\right)_{\text{backward}} = \frac{C_i - C_{i-1}}{\Delta x} + \frac{1}{2}\left.\frac{d^2C}{dx^2}\right|_{x_i} \Delta x - O(\Delta x^2) \tag{8.37}$$

Notice again that the first term being ignored, i.e. second derivative term, is multiplied by Δx; therefore, Eq. (8.37) is also accurate only to order Δx.

We can do better. There is something very interesting about Eqs (8.35) and (8.37). Specifically, if we take the mean of these equations, we get a second-order accurate difference scheme, i.e.

$$\overline{\left.\frac{dC}{dx}\right|_{x_i}} = \frac{1}{2}\left[\left(\left.\frac{dC}{dx}\right|_{x_i}\right)_{\text{forward}} + \left(\left.\frac{dC}{dx}\right|_{x_i}\right)_{\text{backward}}\right] \tag{8.38}$$

or

$$\overline{\left.\frac{dC}{dx}\right|_{x_i}} = \frac{1}{2}\left[\frac{C_{i+1} - C_i}{\Delta x} + \frac{C_i - C_{i-1}}{\Delta x} + O(\Delta x^2)\right] \tag{8.39}$$

because the second derivative terms cancel. Simplifying this last result,

$$\overline{\left.\frac{dC}{dx}\right|_{x_i}} = \frac{1}{2}\left[\frac{C_{i+1} - C_{i-1}}{\Delta x} + O(\Delta x^2)\right] \tag{8.40}$$

Normally we drop the overbar on the left-hand side and assume equality to define the *central difference* formula:

$$\left.\frac{dC}{dx}\right|_{x_i} = \frac{C_{i+1} - C_{i-1}}{2\Delta x} \tag{8.41}$$

which is second-order accurate in Δx (or Δt if it's a time problem).

How stable is this new central difference? Consider again the steady-state advection-decay equation

$$\frac{dC}{dx} = -\frac{k}{w}C \tag{8.42}$$

Substitution of central differences produces

$$\frac{1}{2}\left[\frac{C_{i+1}-C_{i-1}}{\Delta x}\right]=-\frac{k}{w}C_i \qquad (8.43a)$$

or

$$C_{i+1}=C_{i-1}-\frac{2k}{w}\Delta x C_i \qquad (8.43b)$$

First, note that this formula is *explicit*, as both C_{i-1} and C_i are known. Secondly, if

$$C_{i-1}<\frac{2k}{w}\Delta x C_i \qquad (8.44)$$

then the calculated concentration C_{i+1} will again be negative. Therefore, the central difference scheme is not universally stable. Large values of k/w will cause problems. In fact, all explicit difference formulas are potentially unstable to large reaction terms.

There is a second-order backward difference formula that is always stable. For a general initial value problem of the form

$$\frac{dC}{dx}=f(C,x) \qquad (8.45)$$

where f(C,x) is an arbitrary function of C and x (or t), then the second-order BDF can be constructed by taking the difference between the forward Taylor series expansion of C_{i+1} in terms of C_i, i.e. Eq. (8.34), and the backward Taylor series expansion for C_i in terms of C_{i+1}, i.e.

$$C_i=C_{i+1}-\left.\frac{dC}{dx}\right|_{x_{i+1}}\Delta x+\frac{1}{2}\left.\frac{d^2C}{dx^2}\right|_{x_{i+1}}\Delta x^2-\ldots \qquad (8.46)$$

By taking this difference, one obtains

$$2(C_{i+1}-C_i)=\left.\frac{dC}{dx}\right|_{x_i}\Delta x+\left.\frac{dC}{dx}\right|_{x_{i+1}}\Delta x+\frac{1}{2}\left.\frac{d^2C}{dx^2}\right|_{x_i}\Delta x^2-\frac{1}{2}\left.\frac{d^2C}{dx^2}\right|_{x_{i+1}}\Delta x^2+\ldots \qquad (8.47)$$

Rearranging Eq. (8.47), dropping terms multiplied by Δx^2 and higher powers, and assuming equality produce

$$C_{i+1}=C_i+\frac{\Delta x}{2}\left[\left.\frac{dC}{dx}\right|_{x_i}+\left.\frac{dC}{dx}\right|_{x_{i+1}}\right] \qquad (8.48)$$

With Eq. (8.45), Eq. (8.48) becomes

$$C_{i+1} = C_i + \frac{\Delta x}{2}\left[f(C_{i+1}, x_{i+1}) + f(C_i, x_i)\right] \tag{8.49}$$

which uses the mean value of the right-hand side of Eq. (8.45) at the current position (or time, if x is replaced by t) and at the next position (or time), which is unknown.

Equation (8.49) is implicit if f(C,x) is nonlinear. Nevertheless, this equation and higher order variants are the basis for advanced numerical programs for the solution of stiff ODEs. In fact, perhaps the best way to deal with complicated initial value problems is through the use of standard programs (integrators). Byrne and Hindmarsh (1987) review the progress that has been made in this direction and some of the codes that are available. An outstanding example is the FORTRAN code known by the acronym VODE (Brown et al., 1989); it can be obtained from NETLIB (also see the Appendix).

8.5.2 Dealing with Second-Order Derivatives

Diffusion processes introduce second-order derivatives into diagenetic equations. How are these terms approximated by finite differences? As a first illustration of this type of approximation, let us consider the following simple steady-state diagenetic equation with constant coefficients on a finite depth interval $0 \le x \le x_f$, i.e.

$$D\frac{d^2C}{dx^2} - v\frac{dC}{dx} + R(C, x) = 0 \tag{8.50}$$

where D, v, and R(C,x) are generic.

To approximate Eq. (8.50) with finite differences, we need to replace the space domain on which Eq. (8.50) is valid with a finite number of points, and then represent the derivatives with difference formulas. Let us assume that we are at some arbitrary depth x_i that corresponds to the i-th point of the grid and where the concentration is C_i (Fig. 8.6).

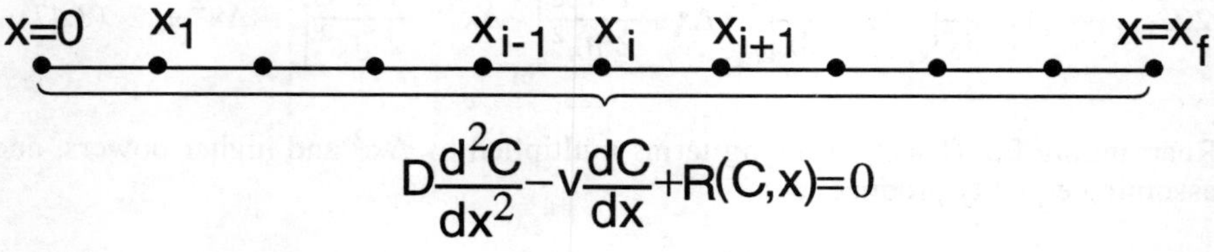

$x=0$ x_1 x_{i-1} x_i x_{i+1} $x=x_f$

$$D\frac{d^2C}{dx^2} - v\frac{dC}{dx} + R(C,x) = 0$$

Figure 8.6. Numerical grid that approximates the continuous depth interval on which Eq. (8.50) is valid.

Now consider two points adjacent to x_i that are each separated by a distance Δx from x_i. The concentrations C_{i-1} and C_{i+1} at x_{i-1} and x_{i+1}, respectively, can be related to the concentration at x_i by two Taylor series:

$$C_{i-1} = C_i - \left.\frac{dC}{dx}\right|_{x_i} \Delta x + \frac{1}{2}\left.\frac{d^2C}{dx^2}\right|_{x_i} \Delta x^2 - \frac{1}{6}\left.\frac{d^3C}{dx^3}\right|_{x_i} \Delta x^3 + O(\Delta x^4) \tag{8.51}$$

and

$$C_{i+1} = C_i + \left.\frac{dC}{dx}\right|_{x_i} \Delta x + \frac{1}{2}\left.\frac{d^2C}{dx^2}\right|_{x_i} \Delta x^2 + \frac{1}{6}\left.\frac{d^3C}{dx^3}\right|_{x_i} \Delta x^3 + O(\Delta x^4) \tag{8.52}$$

If Eqs (8.51) and (8.52) are added together, we find that

$$C_{i+1} + C_{i-1} = 2C_i + \left.\frac{d^2C}{dx^2}\right|_{x_i} \Delta x^2 + O(\Delta x^4) \tag{8.53}$$

or, after some reorganization and dropping of higher order derivatives,

$$\left.\frac{d^2C}{dx^2}\right|_{x_i} = \frac{C_{i+1} - 2C_i + C_{i-1}}{\Delta x^2} \tag{8.54}$$

Equation (8.54) is the justly famous *central difference* formula for the second derivative. The central difference is second-order accurate; i.e. the terms dropped off are of order Δx^2 and higher. (Difference formulas of higher order accuracy can be constructed, e.g. Lau (1981) or Kubicek and Hlavacek (1983), yet these are difficult to use because of problems near boundaries.)

If we introduce the central difference formulas for the first and second derivatives, Eqs (8.41) and (8.54), into Eq. (8.50), we obtain

$$D\frac{C_{i+1} - 2C_i + C_{i-1}}{\Delta x^2} - v\frac{C_{i+1} - C_{i-1}}{2\Delta x} + R(C_i, x_i) = 0 \tag{8.55}$$

which is now an algebraic equation, valid at $x = x_i$. If, for example, the boundary conditions on Eq. (8.50) are

$$C(0) = C_o \tag{8.56}$$

$$C(x_f) = C_L \tag{8.57}$$

then we have created a set of n algebraic equations at each of the n points: x_1, x_2, x_3, x_4, ...x_i, ..., x_{n-1}, x_n , where $x_1 = \Delta x$, ..., $x_n = n\Delta x = x_f$, as shown in Fig. 8.7. Notice that each equation that is not at a boundary depends on the value of C at that point and the two adjacent points. There is no way to solve an isolated equation like Eq. (8.55) by itself. You must solve all the equations at all the n points simultaneously. I will return to methods of solution later.

Let us now examine the *stability* of the difference approximation given by Eq. (8.55), starting with the special case in which $R(C,x) = 0$, i.e.

$$D\frac{d^2C}{dx^2} - v\frac{dC}{dx} = 0 \tag{8.58}$$

over a small interval $0 \le x \le 2\Delta x$ with arbitrary boundary conditions:

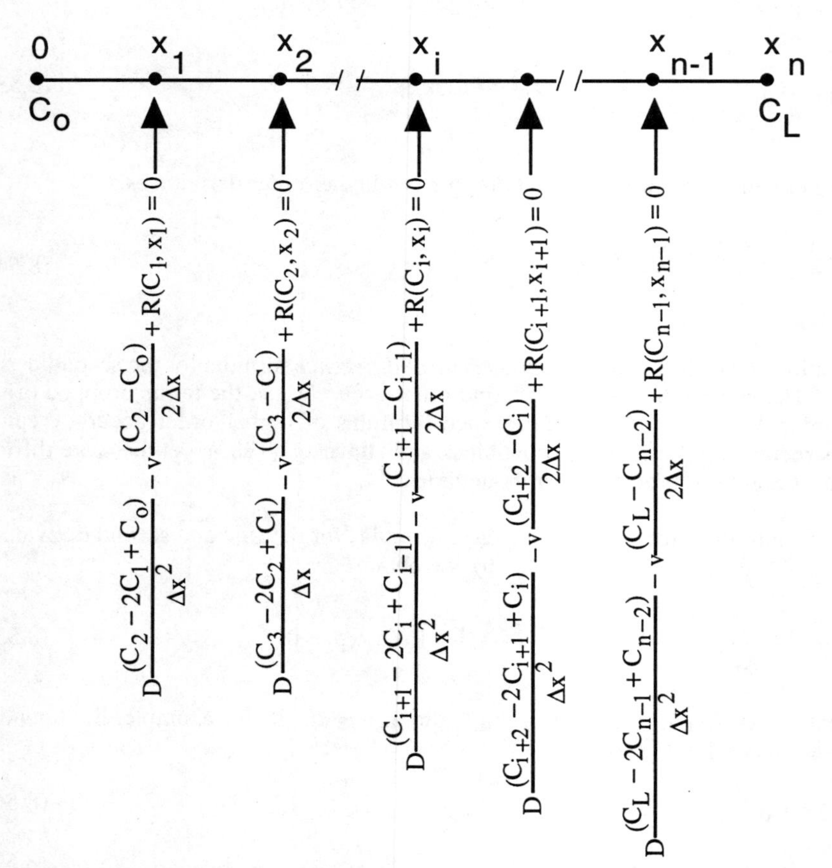

Figure 8.7. Finite difference equations that approximate Eq. (8.50) on an even grid, subject to the boundary conditions given by Eqs (8.56) and (8.57). (Note the breaks along this x-axis.)

$$C(0) = C_o \tag{8.59}$$

$$C(2\Delta x) = C_o + \Delta C \tag{8.60}$$

The exact solution to Eq. (8.58) is

$$C(x) = C_o + \Delta C\left(\frac{exp(vx/D)-1}{exp(2v\Delta x/D)-1}\right) \tag{8.61}$$

For later reference, let us determine the form of Eq. (8.61) as $2v\Delta x/D$ becomes small. Expanding the exponentials in this equation using Taylor series produces

$$C(x) = C_o + \frac{x\Delta C}{2\Delta x}\left(\frac{1+\frac{1}{2}\frac{vx}{D}+...}{1+\frac{v\Delta x}{D}+...}\right) \tag{8.62}$$

A second approximation is reached with a Binomial (Chapter 2) expansion of the denominator of Eq. (8.62); if we retain only first-order terms, then at $x = \Delta x$,

$$C(\Delta x) = C_o + \frac{\Delta C}{2}\left(1-\frac{v\Delta x}{2D}\right) \tag{8.63}$$

The numerical approximation of Eq. (8.58) with central differences is

$$D\frac{C_{i+1}-2C_i+C_{i-1}}{\Delta x^2} - v\frac{C_{i+1}-C_{i-1}}{2\Delta x} = 0 \tag{8.64}$$

Let us consider the case in which there are only three points in the grid (see Fig. 8.8):

$$\text{grid}\begin{cases} i=0 & x=0 & C=C_o \text{ (known)} \\ i=1 & x=\Delta x & C=C_1 \text{ (unknown)} \\ i=2 & x=2\Delta x & C=C_o+\Delta C \text{ (known)} \end{cases} \tag{8.65}$$

In this simple case with only one unknown (C_1), Eq. (8.64) becomes

$$2C_o + \Delta C - 2C_1 - \frac{v\Delta x}{2D}\Delta C = 0 \tag{8.66}$$

or

$$C_1 = C_o + \frac{\Delta C}{2}\left(1-\frac{v\Delta x}{2D}\right) \tag{8.67}$$

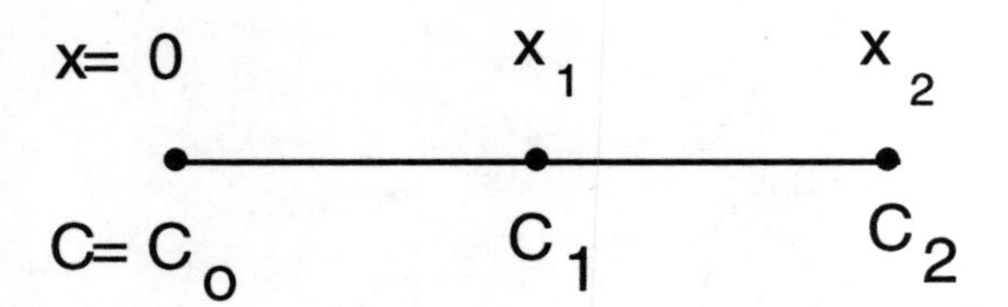

Figure 8.8. A simple grid with only one unknown concentration, C_1, used to establish the stability of the difference scheme given by Eq. (8.64).

This is identical to the second level approximation to the analytical solution, i.e. Eq. (8.63); however, from the exact solution given by Eq. (8.61),

$$C \rightarrow \begin{cases} \text{increases if } \Delta C > 0 \\ \text{decreases if } \Delta C < 0 \end{cases} \tag{8.68}$$

Yet, this will only be true of the central difference solution, Eq. (8.67), if

$$\frac{v\Delta x}{2D} < 1 \tag{8.69}$$

This result has also been proven for an arbitrary point on a larger grid by several other investigators (e.g. Fiadeiro and Veronis, 1977; Hemker, 1977).

In contrast, if we use the backward difference formula for the advection term, then the approximation of Eq. (8.58) is

$$D\left(\frac{C_{i+1} - 2C_i + C_{i-1}}{\Delta x^2}\right) - v\left(\frac{C_i - C_{i-1}}{\Delta x}\right) = 0 \tag{8.70}$$

and the solution for our simple grid problem is

$$C_1 = C_o + \frac{\Delta C}{2\left(1 + \frac{v\Delta x}{2D}\right)} \tag{8.71}$$

This latest result is quite similar to the first level approximation to Eq. (8.62), if $v\Delta x/D << 1$. However, unlike Eq. (8.67), Eq. (8.71) never generates negative concentrations for any value of $v\Delta x/(2D)$; that is to say, it is completely stable. The behavior seen in the approximation of first-order ODEs continues with second-order ODEs. Unfortunately, approximation Eq. (8.70) is only Δx accurate, while the central difference approximation, Eq. (8.64), is Δx^2 accurate.

Is there a way that the differencing scheme can be both second-order accurate and stable? A number of investigators have derived such formulas (Shyy, 1985; Patel et al., 1985), but that of Fiadeiro and Veronis (1977) is one of the simplest to use. They derive the blended approximation,

$$D\left(\frac{C_{i+1}-2C_i+C_{i-1}}{\Delta x^2}\right)-v\left(\frac{(1-\sigma)C_{i+1}+2\sigma C_i-(1+\sigma)C_{i-1}}{2\Delta x}\right)=0 \tag{8.72}$$

which is a blend of backward (upstream) and central differences. The amount of blending is dictated by the value of the parameter σ, defined as

$$\sigma=\coth\left(\frac{v\Delta x}{2D}\right)-\frac{2D}{v\Delta x} \tag{8.73}$$

where the parameter grouping, $Pe_{\Delta x}$ = $v\Delta x/(2D)$, is called the grid Peclet number. The parameter σ has the property that

$$\sigma=\begin{cases}0 & \dfrac{v\Delta x}{2D}\rightarrow 0\\ 1 & \dfrac{v\Delta x}{2D}\rightarrow \infty\end{cases} \tag{8.74}$$

Thus, if $\sigma = 0$ (diffusion dominated), then pure central differencing is obtained, and if $\sigma = 1$ (advection dominated), then pure backward differencing results. Interestingly enough, this blended or weighted scheme is second-order accurate even as it switches to backward differencing (Fiadeiro and Veronis, 1977).

The existence of a formulation like Eq. (8.72) is very important when dealing with problems in which D and/or v are functions of depth. For example, biodiffusion coefficients often decrease to zero with depth. A central difference approximation will be unstable and/or inaccurate in this case. While Eq. (8.72) was derived for constant D, it will deal with a spatially dependent D(x) when applied locally (e.g. Boudreau, 1986a). My past experience indicates that this weighing appears to apply equally well to time-dependent problems.

The second special case of Eq. (8.50) occurs when $v = 0$. To simplify this presentation, let us also assume that $R(C,x) = -kC$, i.e. a simple decay, so that

$$D\frac{d^2C}{dx^2}-kC=0 \tag{8.75}$$

with simplified boundary conditions:

$$C(0)=1 \qquad \text{and} \qquad C(x_f)=1 \tag{8.76a,b}$$

The analytical solution to this problem can be written in the form (Hemker, 1977)

$$C(x) = \frac{exp\left[\frac{\Delta x\sqrt{k}}{\sqrt{D}}\frac{(-2x+x_f)}{2\Delta x}\right] + exp\left[\frac{\Delta x\sqrt{k}}{\sqrt{D}}\frac{(2x-x_f)}{2\Delta x}\right]}{exp\left[\frac{\Delta x\sqrt{k}}{\sqrt{D}}\frac{x_f}{2\Delta x}\right] + exp\left[\frac{\Delta x\sqrt{k}}{\sqrt{D}}\frac{(-x_f)}{2\Delta x}\right]} \tag{8.77}$$

What happens to a central difference approximation of Eq. (8.75) as k/D gets large? Hemker (1977) studied the stability of this problem. Using an even grid of N points (i=0,1,2,3,...,N), the central difference approximation to Eq. (8.75) is

$$D\left[\frac{C_{i+1} - 2C_i + C_{i-1}}{\Delta x^2}\right] - kC_i = 0 \tag{8.78}$$

From this, Hemker (1977) obtained a solution for C_i of the form

$$C_i = \frac{\Lambda^{-i+N/2} + \Lambda^{i-N/2}}{\Lambda^{N/2} + \Lambda^{-N/2}} \tag{8.79}$$

where

$$\Lambda = 1 + \frac{k\Delta x^2}{2D} + \sqrt{\left(1 + \frac{k\Delta x^2}{2D}\right)^2 - 1} \tag{8.80}$$

and where it is assumed that $N >> 1$. The question of the stability of Eq. (8.79) is simple to answer, as no positive value of the parameter grouping $k\Delta x^2/(2D)$ can lead to a negative value of C_i; therefore, the scheme is always stable, but this does not establish that it is a good approximation to Eq. (8.77).

Equation (8.79) becomes a good approximation to Eq. (8.77) in the limits as

$$\frac{k\Delta x^2}{D} << 1 \tag{8.81}$$

In this limit, a Taylor series expansion gives that

$$exp\left(\Delta x\sqrt{k/D}\right) \approx 1 + \Delta x\sqrt{\frac{k}{D}} \tag{8.82}$$

and use of the Binomial Theorem gives that

$$\Lambda \approx 1 + \Delta x\sqrt{\frac{k}{D}} \tag{8.83}$$

In addition, by definition, at the grid points,

$$\frac{-2x + x_f}{2\Delta x} \equiv -i + \frac{N}{2} \tag{8.84}$$

and

$$\frac{2x - x_f}{2\Delta x} \equiv i - \frac{N}{2} \tag{8.85}$$

If all these results are introduced into both Eqs (8.77) and (8.79), these two equations become identical. The limiting condition, Eq. (8.81), requires that the time scale for generic diffusion over the distance of the grid spacing, i.e. $\Delta x^2/D$, be very much shorter than the time scale for reaction, i.e. $1/k$. As the reaction becomes fast, i.e. k becomes large, a smaller grid spacing, Δx, must be used to ensure a satisfactory approximation. This condition remains in force even if the problem is time-dependent and advection is present. This is a further expression of *stiffness* in the problem. (One way to deal effectively with stiffness is to take advantage of standard computer codes, as explained below in the section on the *Method of Lines*.)

Finally, if the equation in question is time-dependent, but contains no diffusion, (sharp) step inputs should be preserved as sharp fronts as they are advected through the sediment. To preserve this behavior it may be necessary to use backward difference schemes or special higher order variants (see Section 8.9.2).

8.5.3 Finite Differences in 3-D Coordinate Systems

The Cartesian System. The sediment volume in 3-D or the surface in 2-D being modelled is replaced with a 3-D/2-D grid of points, to which we assign the indices i for x-direction, j for the y-direction and k for the z-direction. (Don't confuse the index k with a rate constant.) Assume that the point of current interest on the grid has indices i, j, and k, so that the unknown concentration is $C_{i,j,k}$. The three possible finite difference representations of first-order derivatives are the forward difference, e.g.

$$\left.\frac{\partial C}{\partial x}\right|_{x=x_i} \approx \frac{C_{i+1,j,k} - C_{i,j,k}}{\Delta x} \tag{8.86}$$

the backward difference, e.g.

$$\left.\frac{\partial C}{\partial x}\right|_{x=x_i} \approx \frac{C_{i,j,k} - C_{i-1,j,k}}{\Delta x} \tag{8.87}$$

and the central difference, e.g.

$$\left.\frac{\partial C}{\partial x}\right|_{x=x_i} \approx \frac{C_{i+1,j,k} - C_{i-1,j,k}}{2\Delta x} \tag{8.88}$$

where the first two formulas are accurate to order Δx and the last is accurate to Δx^2. For second-order derivatives, we have the central difference, e.g.

$$\left.\frac{\partial^2 C}{\partial x^2}\right|_{x=x_i} \approx \frac{C_{i+1,j,k} - 2C_{i,j,k} + C_{i-1,j,k}}{\Delta x^2} \tag{8.89}$$

which is also accurate to order Δx^2. The backward and central differences can be blended as in Eq. (8.72). The extension to y and z-coordinate derivatives involves simply replacing the x and Δx in these formulas with y or z and Δy or Δz, respectively, and incrementing the appropriate index for that chosen variable. Remember that the accuracy of the scheme will depend on the largest value of Δx, Δy and Δz.

The Cylindrical System. The need to model such problems as irrigation by tube-dwelling organisms requires us to have finite-difference approximations for derivatives in the cylindrical coordinate system; see Eq. (3.105). The depth (x-direction) derivatives are dealt with in the same way as in the Cartesian case. To deal with the second-order radial diffusion operator, expand this term to obtain

$$\frac{1}{r}\frac{\partial}{\partial r}\left(r\frac{\partial C}{\partial r}\right) = \frac{\partial^2 C}{\partial r^2} + \frac{1}{r}\frac{\partial C}{\partial r} \tag{8.90}$$

Some practitioners advise against this step, stating that direct representation of the left-hand side of Eq. (8.90) is somewhat more accurate (Nogotov, 1978). My own practical experience (e.g. Boudreau and Marinelli, 1994) argues that this effect is quite small and is outweighed by the ease in dealing with the expanded form.

The diffusion process represented by the second derivative is a symmetric process; i.e. it depends on the gradient on both sides of the point of interest. Therefore, with the central difference approximations,

$$\left.\frac{\partial^2 C}{\partial r^2}\right|_{r=r_j} \approx \frac{C_{i,j+1,k} - 2C_{i,j,k} + C_{i,j-1,k}}{\Delta r^2} \tag{8.91}$$

and

$$\left.\frac{\partial C}{\partial r}\right|_{r=r_j} \approx \frac{C_{i,j+1,k} - C_{i,j-1,k}}{2\Delta r} \tag{8.92}$$

we obtain,

$$\left[\frac{\partial^2 C}{\partial r^2} + \frac{1}{r}\frac{\partial C}{\partial r}\right]_{r=r_j} \approx \frac{C_{i,j+1,k} - 2C_{i,j,k} + C_{i,j-1,k}}{\Delta r^2} + \frac{1}{r_j}\frac{C_{i,j+1,j} - C_{i,j-1,k}}{2\Delta r} \tag{8.93}$$

which cannot hold at $r_j = 0$. At this point, two possible courses of action present themselves (Smith, 1975; Özisik, 1980). If the problem is not radially symmetric, then

$$\left[\frac{\partial^2 C}{\partial r^2}+\frac{1}{r}\frac{\partial C}{\partial r}\right]_{r=0} \approx 4\left(\frac{C_m - C_{i,0}}{\Delta r^2}\right) \tag{8.94}$$

where C_m is the arithmetic average of $C_{i,1,k}$ as the index k goes from 1 to its greatest values (i.e. the number of radial lines), and $C_{i,0}$ is the concentration at r=0 (i.e. the index k has no meaning and is dropped).

Alternatively, if the problem is radially symmetric, then

$$\underset{r\to 0}{\text{limit}}\left\{\frac{1}{r}\frac{\partial C}{\partial r}\right\} = \frac{\partial^2 C}{\partial r^2} \tag{8.95a}$$

so that

$$\left[\frac{\partial^2 C}{\partial r^2}+\frac{1}{r}\frac{\partial C}{\partial r}\right]_{r=0} \approx 4\left(\frac{C_{i,1} - C_{i,0}}{\Delta r^2}\right) \tag{8.95b}$$

where $C_{i,0}$ is the concentration at r = 0, and where it has been assumed that $C_{i,1,k} = C_{i,-1,k}$.

All diagenetic uses of the cylindrical coordinate system have so far assumed radial symmetry; that is to say, the problem has no ϑ-direction dependence. This lack of ϑ-direction dependence will not always be the case; consequently, it is worth discussing the treatment of diffusion in the direction of the radial angle Finite-difference representation of the second derivative in the ϑ-direction at all points except r = 0 requires no new treatment, i.e.

$$\frac{1}{r^2}\frac{\partial^2 C}{\partial \vartheta^2} \approx \frac{1}{r_j^2}\left(\frac{C_{i,j,k+1} - 2C_{i,j,k} + C_{i,j,k-1}}{\Delta\vartheta^2}\right) \tag{8.96}$$

If the line r=0 is part of the model, this term cannot be represented by this type of finite difference. In fact, there is no simple or even entirely satisfactory treatment, which just proves that finite differences aren't a panacea. (There is no similar problem when dealing with analytical solutions, as in Chapters 6 and 7. At r = 0, there will be two spatial solutions, only one of which will be bounded along this line, and that's the one we adopt for concentration.)

The suggested treatments for the r = 0 problem include (Thibault et al., 1987): a) simply leaving out the radial-angle term at r = 0; b) excluding the line r=0 by surrounding it with a cylinder of small radius, $\delta r < \Delta r$, and assigning a zero-flux condition at $r = \delta r$ (of course, you cannot do this if this line is a source material as in the hydrological "well problem"); and c) using Cartesian coordinates locally near $r = \delta r$, and then trying to match this mesh with the radial mesh elsewhere. While necessary, none of these actions is very palatable.

The Spherical System. Some diagenetic problems (e.g. concretion growth) may necessitate the use of finite differences in the spherical coordinate system; see Eq. (3.111). Again the diffusion terms are of primary interest here (see Thibault et al., 1987). For the radial component,

$$\left[\frac{1}{r^2}\frac{\partial}{\partial r}\left(r^2\frac{\partial C}{\partial r}\right)\right]_{r=r_i} \approx \frac{C_{i+1,j,k} - 2C_{i,j,k} + C_{i-1,j,k}}{\Delta r^2} + \frac{C_{i+1,j,k} - C_{i-1,j,k}}{r_i \Delta r} \tag{8.97}$$

At $r = 0$, we again encounter problems; nevertheless, an approximation analogous to that of Eq. (8.95) is available,

$$\underset{r\to 0}{\text{limit}}\left[\frac{1}{r^2}\frac{\partial}{\partial r}\left(r^2\frac{\partial C}{\partial r}\right)\right] = 3\frac{\partial^2 C}{\partial r^2} \approx 6\left(\frac{C_{1,j,k} - C_{0,j,k}}{\Delta r^2}\right) \tag{8.98}$$

where i, j, and k are the r, ϑ (latitudinal), and ϕ (longitudinal) indices, respectively.

For many diagenetic problems, an assumption of radial symmetry is quite acceptable, and the latitudinal and longitudinal diffusion terms can be ignored. When this cannot be done, finite difference approximations must also be introduced for the ϑ and ϕ components. First, we have that

$$\frac{1}{r^2 \sin\vartheta}\frac{\partial}{\partial\vartheta}\left(\sin\vartheta\frac{\partial C}{\partial\vartheta}\right) \approx \frac{C_{i,j+1,k} - 2C_{i,j,k} + C_{i,j-1,k}}{r_i^2 \Delta\vartheta^2} + \frac{\cos\vartheta_j\left(C_{i,j+1,k} - C_{i,j-1,k}\right)}{2r_i^2 \sin\vartheta_j \Delta\vartheta} \tag{8.99}$$

except at $r = 0$, $\vartheta = 0$, and $\vartheta = \pi$. The singularities at $\vartheta = 0$ and $\vartheta = \pi$ can be eliminated (Thibault et al., 1987) because

$$\underset{\vartheta\to 0,\pi}{\text{limit}}\left[\frac{1}{r^2 \sin\vartheta}\frac{\partial}{\partial\vartheta}\left(\sin\vartheta\frac{\partial C}{\partial\vartheta}\right)\right] = \frac{2}{r^2}\frac{\partial^2 C}{\partial\vartheta^2} \tag{8.100}$$

and, consequently, Eq. (8.96) can be employed. However, the singularity at $r = 0$ is essential in nature, and can be dealt with only by the *ad hoc* methods mentioned for the spherical case.

The longitudinal term is treated similarly,

$$\frac{1}{r^2 \sin^2\vartheta}\frac{\partial^2 C}{\partial\phi^2} \approx \frac{C_{i,j,k+1} - 2C_{i,j,k} + C_{i,j,k+1}}{r_i^2 \sin^2\vartheta_j \Delta\phi^2} \tag{8.101}$$

All three singularities for this term, i.e. $r = 0$ and $\vartheta = 0,\pi$ are essential and need *ad hoc* treatment.

8.5.4 Depth-Dependent Diffusivity

It is often the case that the diffusion coefficients in diagenetic equations are functions of depth. This can occur because of depth-dependent porosity and tortuosity for molecular/ionic coefficients, or decreasing intensity of biodiffusion with depth (see Chapters 3 and 4). As an example, consider the generic case

$$\frac{d}{dx}\left(D(x)\frac{dC}{dx}\right) - v\frac{dC}{dx} + R(C,x) = 0 \tag{8.102}$$

This example is based on a steady-state ODE, but the method is similar when time-derivatives are present.

Equation (8.102) can be treated in two ways. The first is to leave the equation in this form and use differences of the form (Nogotov, 1978)

$$\frac{1}{\Delta x}\left[D_{i+1/2}\left(\frac{C_{i+1}-C_i}{\Delta x}\right) - D_{i-1/2}\left(\frac{C_i - C_{i-1}}{\Delta x}\right)\right] - v\left(\frac{(1-\sigma_i)C_{i+1} + 2\sigma_i C_i - (1+\sigma_i)C_{i-1}}{2\Delta x}\right) - R(C_i, x_i) = 0 \tag{8.103}$$

where $D_{i+1/2}$ and $D_{i-1/2}$ are the values of $D(x)$ at $x = (x_{i+1}+x_i)/2$ and $x = (x_i+x_{i-1})/2$, respectively, and σ_i is the blending parameter defined in Eq. (8.73) applied to the point x_i. The second method is to expand the second derivative in order to get

$$D(x)\frac{d^2C}{dx^2} + \left(\frac{dD}{dx} - v\right)\frac{dC}{dx} + R(C,x) = 0 \tag{8.104}$$

and then use standard finite differences to obtain

$$D_i \frac{C_{i+1} - 2C_i + C_{i-1}}{\Delta x^2} + \left(\left.\frac{dD}{dx}\right|_{x_i} - v\right)\left(\frac{(1-\sigma_i)C_{i+1} + 2\sigma_i C_i - (1+\sigma_i)C_{i-1}}{2\Delta x}\right) - R(C_i, x_i) = 0 \tag{8.105}$$

Spatial variations in the generic velocity, v, can be handled in much the same way. The treatment of purely advective transport of solids is outlined in Section 8.9.2 below.

8.6 Boundary Conditions

Not much has been said yet about implementing boundary conditions in a finite difference scheme. This section illustrates how the various types of boundary conditions (see Chapter 5) fit into the difference formulas derived above.

8.6.1 Dirichlet (Concentration) Conditions

When the concentration is known at some boundary, the known value there just slips into the difference equation at the grid point next to the boundary. For example, a general 1-D diagenetic equation can be written in terms of spatial finite differences at an arbitrary grid point as

$$\frac{dC_i}{dt} = D\frac{C_{i+1} - 2C_i + C_{i-1}}{\Delta x^2} - v\frac{C_{i+1} - C_{i-1}}{2\Delta x} + R(C_i, x_i) \tag{8.106}$$

The time derivative can be retained without affecting the following results.

If $C(0) = C_o$ and $C(L) = C_L$, then at $x = x_1$ (i=1), Eq. (8.106) becomes

$$\frac{dC_1}{dt} = D\frac{C_2 - 2C_1 + C_o}{\Delta x^2} - v\frac{C_2 - C_o}{2\Delta x} + R(C_1, x_1) \tag{8.107}$$

and, at $x = x_{n-1}$ (i.e. L-Δx), it reads

$$\frac{dC_{n-1}}{dt} = D\frac{C_L - 2C_{n-1} + C_{n-2}}{\Delta x^2} - v\frac{C_L - C_{n-2}}{2\Delta x} + R(C_{n-1}, x_{n-1}) \tag{8.108}$$

8.6.2 Neumann (No Flux) Boundary Conditions

In this case, either

$$\left.\frac{\partial C}{\partial x}\right|_0 = 0 \tag{8.109}$$

or

$$\left.\frac{\partial C}{\partial x}\right|_L = 0 \tag{8.110}$$

In either case, the concentrations at x = 0 or x = L, i.e. C_o (C_0 of the grid) and C_L (C_n of the grid), are not known and must be calculated as part of the solution to the problem.

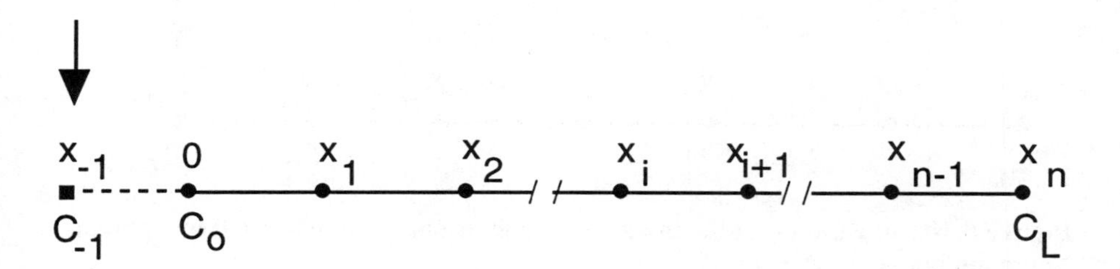

Figure 8.9. Illustration of the imaginary/nonexistent point generated by applying a Neumann boundary at x = 0. (Note that there are breaks in this 1-D grid.)

Let us begin with Eq. (8.109). This condition refers to a diffusive flux, not an advective flux; thus, it is reasonable to take the central difference approximation:

$$\frac{C_1 - C_{-1}}{2\Delta x} = 0 \tag{8.111}$$

or

$$C_1 = C_{-1} \tag{8.112}$$

where C_{-1} is the concentration at an imaginary point at x = $-\Delta x$ (Fig. 8.9); such a point does not really exist, as it is outside the domain of validity of the model. We need to rid the equations of it.

Fortunately, the diagenetic equation itself applies at x = 0; after substituting finite differences, we secure another equation containing C_{-1}, i.e.

$$\frac{dC_0}{dt} = D\frac{C_1 - 2C_0 + C_{-1}}{\Delta x^2} - v\frac{C_1 - C_{-1}}{2\Delta x} + R(C_0, x_0) \tag{8.113}$$

If we substitute the boundary condition, Eq. (8.112), into Eq. (8.113), we get an equation for C_0 that no longer contains C_{-1}:

$$\frac{dC_0}{dt} = 2D\frac{C_1 - C_0}{\Delta x^2} + R(C_0, x_0) \tag{8.114}$$

This new equation for the boundary point contains no values at the imaginary point.

Next, at x = L with Eq. (8.110), we can follow the same procedure. A central difference approximation of Eq. (8.110) gives that

$$C_{n+1} = C_{n-1} \tag{8.115}$$

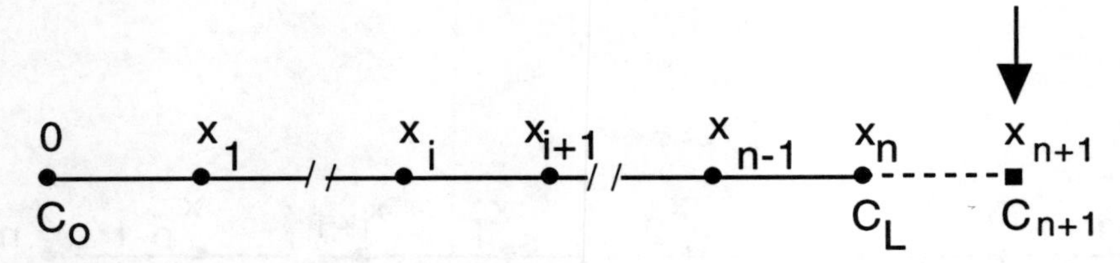

Figure 8.10. Illustration of the imaginary/nonexistent point generated by applying a Neumann boundary at x = L.

where C_{n+1} is the concentration at an imaginary point (depth), i.e. $x = L + \Delta x$, which lies outside the domain (Fig. 8.10).

Next we apply the diagenetic equation at x = L and "discretize" (substitute finite differences); this produces another equation containing the concentration at the imaginary point, C_{n+1}, i.e.

$$\frac{dC_n}{dt} = D\frac{C_{n+1} - 2C_n + C_{n-1}}{\Delta x^2} - v\frac{C_{n+1} - C_{n-1}}{2\Delta x} + R(C_n, x_n) \tag{8.116}$$

If we substitute the boundary condition Eq. (8.115) into Eq. (8.116), we now obtain an equation for C_n that no longer contains C_{n+1}:

$$\frac{dC_n}{dt} = 2D\frac{C_{n-1} - C_n}{\Delta x^2} + R(C_n, x_n) \tag{8.117}$$

8.6.3 Robin's (Prescribed Flux) Boundary Conditions

Regardless of their exact form, flux boundary conditions are handled in the very same way as Neumann boundary conditions (see immediately above). For example,

$$vC(0) - D\left.\frac{\partial C}{\partial x}\right|_0 = F_o \tag{8.118}$$

is the condition at the sediment-water interface for a prescribed flux, F_o. Inserting a central difference for the diffusion term leads to

$$vC_o - D\frac{C_1 - C_{-1}}{2\Delta x} = F_o \tag{8.119}$$

or

$$C_{-1} = C_1 + 2\Delta x\left(\frac{F_o - vC_o}{D}\right) \tag{8.120}$$

which, like its predecessors, expresses the imaginary concentration C_{-1} as a function of concentrations at real points in the grid.

Next, apply the relevant diagenetic equation at x = 0 and insert chosen finite differences, i.e. Eq. (8.106). As before, substitute Eq. (8.120) into this equation,

$$\frac{dC_0}{dt} = 2D\frac{C_1 - C_0 + \Delta x\left(\frac{F_o - vC_0}{D}\right)}{\Delta x^2} - v\left(\frac{F_o - vC_0}{D}\right) + R(C_0, x_0) \tag{8.121}$$

which is the desired result.

8.6.4 Unknown Points

This type of problem is called a *free boundary problem* in the literature on numerical methods. As stated in Chapter 5, customarily we set both the concentration and its gradient at some unknown point in 1-D, along some curve in 2-D and across some surface in 3-D. Implementation of these conditions is often done in what appears to be a very primitive way (Berger et al., 1975; Evans and Gourlay, 1977; Rogers, 1977; Rogers et al., 1979; Crank, 1984). Specifically, one chooses a fixed domain (i.e. depth range, surface, or volume of sediment) that is guaranteed to contain the unknown point (or curve or surface) during the time period of interest. After each time step, the concentration of the species of interest will be negative below the unknown point (i.e. beyond the unknown curve or volume). One simply takes the last grid point with non-negative concentration as the position of the unknown point at that time. The negative concentrations are then all set to zero, and another time step is taken. This method works best with a fully-discrete differencing method (see below). Boudreau and Taylor (1989) solve the problem of O_2 diagenesis beneath surficial objects in this way.

8.7 Uneven Grids

For many reasons, a modeller may want to place more points in a given area; e.g. gradients are often sharpest near the sediment-water interface and more points there will help resolve these gradients. Adding grid points thus reduces the effects of stiffness. Past conventional wisdom was that finite differences on unevenly spaced grids were not as accurate as with evenly spaced grids. While this may be true of random spacings, recent work has shown that regular uneven spacings are as accurate as even grids (e.g. effectively second order, Manteuffel and White, 1986).

The procedure to generate a smooth uneven grid is to imagine an even grid in a hypothetical space, ξ, and deduce what mathematical formula will translate that to the uneven grid that suits your purposes in real space. For example, say you want fine spacing near x = 0 and coarse as x becomes large. Then an even grid in (arbitrary) ξ-space, i.e. $\xi_1, \xi_2, \xi_3, \xi_4, ..., \xi_i, ..., \xi_{n-1}, \xi_n$, separated by $\Delta\xi$, has the desired spacing in (real) x-space if, for example,

$$x_i = \omega(\xi_i)^2 \tag{8.122}$$

where ω is a scaling factor between the two grids. There are innumerable transformations like Eq. (8.122), and the only requirement is that they suit your purpose.

The corresponding second-order finite difference formulas for an uneven grid were deduced by Pearson (1968):

i) The first derivative via backward differences:

$$\left.\frac{\partial C}{\partial x}\right|_{x_i} = \frac{C_i - C_{i-1}}{\Delta x_i} \tag{8.123}$$

where

$$\Delta x_i = x_i - x_{i-1} \tag{8.124}$$

ii) The first derivative via central differences:

$$\left.\frac{\partial C}{\partial x}\right|_{x_i} = -\frac{\Delta x_{i+1}}{\Delta x_i(\Delta x_i + \Delta x_{i+1})}C_{i-1} + \frac{\Delta x_{i+1} - \Delta x_i}{\Delta x_i \Delta x_{i+1}}C_i + \frac{\Delta x_i}{\Delta x_{i+1}(\Delta x_i + \Delta x_{i+1})}C_{i+1} \tag{8.125}$$

where

$$\Delta x_{i+1} = x_{i+1} - x_i \tag{8.126}$$

iii) Finally, the second derivative via central differences:

$$\left.\frac{\partial^2 C}{\partial x^2}\right|_{x_i} = \frac{2}{\Delta x_i(\Delta x_i + \Delta x_{i+1})}C_{i-1} - \frac{2}{\Delta x_i \Delta x_{i+1}}C_i + \frac{2}{\Delta x_{i+1}(\Delta x_i + \Delta x_{i+1})}C_{i+1} \tag{8.127}$$

Other formulas for uneven grids can be found in Lau (1981), Ascher et al. (1988), and Fornberg (1988).

The implementation of boundary conditions is quite similar to that for even grids. In fact, for Dirichlet conditions it is exactly as before. For Neumann and Robin's conditions, the technique can be illustrated with the example:

$$\left.\frac{\partial C}{\partial x}\right|_L = 0 \tag{8.128}$$

The problem lies at the point x = L (i.e. x_n), because the approximation for the second derivative of a diffusion term contains the imaginary concentration C_{i+1} and the imaginary spacing Δx_{n+1}, i.e.

$$\left.\frac{\partial^2 C}{\partial x^2}\right|_{x_n} = \frac{2}{\Delta x_n(\Delta x_n + \Delta x_{n+1})} C_{n-1} - \frac{2}{\Delta x_n \Delta x_{n+1}} C_n + \frac{2}{\Delta x_{n+1}(\Delta x_n + \Delta x_{n+1})} C_{n+1} \tag{8.129}$$

Fortunately, the central difference approximation of the boundary condition also contains these imaginary values, i.e.

$$-\frac{\Delta x_{n+1}}{\Delta x_n(\Delta x_n + \Delta x_{n+1})} C_{n-1} + \frac{\Delta x_{n+1} - \Delta x_n}{\Delta x_n \Delta x_{n+1}} C_n + \frac{\Delta x_n}{\Delta x_{n+1}(\Delta x_n + \Delta x_{n+1})} C_{n+1} = 0 \tag{8.130}$$

Multiplying Eq. (8.129) by Δx_i and Eq. (8.130) by 2 and subtracting produces

$$\left.\frac{\partial^2 C}{\partial x^2}\right|_{x_n} = \frac{2}{(\Delta x_n)^2}(C_{n-1} - C_n) \tag{8.131}$$

This replaces Eq. (8.127) as the appropriate finite difference at x = L, and it leads to an equation for C_n without involving imaginary values and points.

As another example, say that

$$\left[D\frac{\partial C}{\partial x}\right]_0 = F_o \tag{8.132}$$

Following the same procedure as above generates

$$\left.\frac{\partial^2 C}{\partial x^2}\right|_0 = \frac{2}{(\Delta x_1)^2}(C_1 - C_0 - \Delta x_1 F_o) \tag{8.133}$$

8.8 Solution by the Method of Lines (MoL)

8.8.1 Diffusion Present

Given that we can replace the spatial derivatives of diagenetic equations expressed as both ODEs and PDEs by finite differences, we now need to examine how the resulting equations can be solved. One of the most powerful solution methods is variously called the numerical Method of Lines (Ramos, 1986a; Byrne and Hindmarsh, 1987; Schiesser, 1991), semi or partial discretization (Huntley, 1986), and sometimes the method of false transient (Kubicek and Hlavacek, 1983). This method naturally solves time-dependent PDEs, but the solution of ODEs is attained simply as the long-term (steady-state) solution of an appropriate PDE. Consequently, it can solve all types of diagenetic equations.

A single time and space-dependent PDE becomes a set of time-dependent ODEs, valid at each grid point, when spatial finite differences are substituted for spatial derivatives in the original equation. To illustrate, consider the following examples.

Example 8-1. Consider the time-dependent diagenetic equation

$$\frac{\partial C}{\partial t} = D\frac{\partial^2 C}{\partial x^2} - v\frac{\partial C}{\partial x} + R(C,x) \qquad (8.134)$$

subject to the conditions $C(0,t) = C_o$, $C(L,t) = C_L$, and $C(x,0) = f(x)$. Replacing the spatial derivatives with central differences gives a set of ODEs of the form

$$\frac{dC_i}{dt} = D\frac{C_{i+1} - 2C_i + C_{i-1}}{\Delta x^2} - v\frac{C_{i+1} - C_{i-1}}{2\Delta x} + R(C_i, x_i) \qquad (8.135)$$

with one equation governing the concentration C_i at each of the n-1 interior points of the grid (Fig. 8.11). Each equation is dependent on the concentrations at the two adjacent points, C_{i-1} and C_{i+1}, and so the set of equations is coupled and must be solved together. Such systems of ODEs, i.e. Eq. (8.135) with i=1,2,3,...,n-1, can now be solved without recourse to substituting finite differences for the time derivative. As recommended in the section on "Lessons from Initial Value Problems" (see p. 302), the answer is to use a standard integrator for stiff ODEs to solve the resulting system of ODEs (Byrne and Hindmarsh, 1987).

There are significant advantages in solving diagenetic equations by the MoL: i) stiff problems caused by fast reactions are handled readily; ii) highly nonlinear problems can be solved with no fuss; iii) the initial guess is not all that important when solving for the steady state; iv) this procedure is usually very stable and convergent; v) the number of space dimensions is irrelevant; and vi) both transient and steady-state problems can be solved.

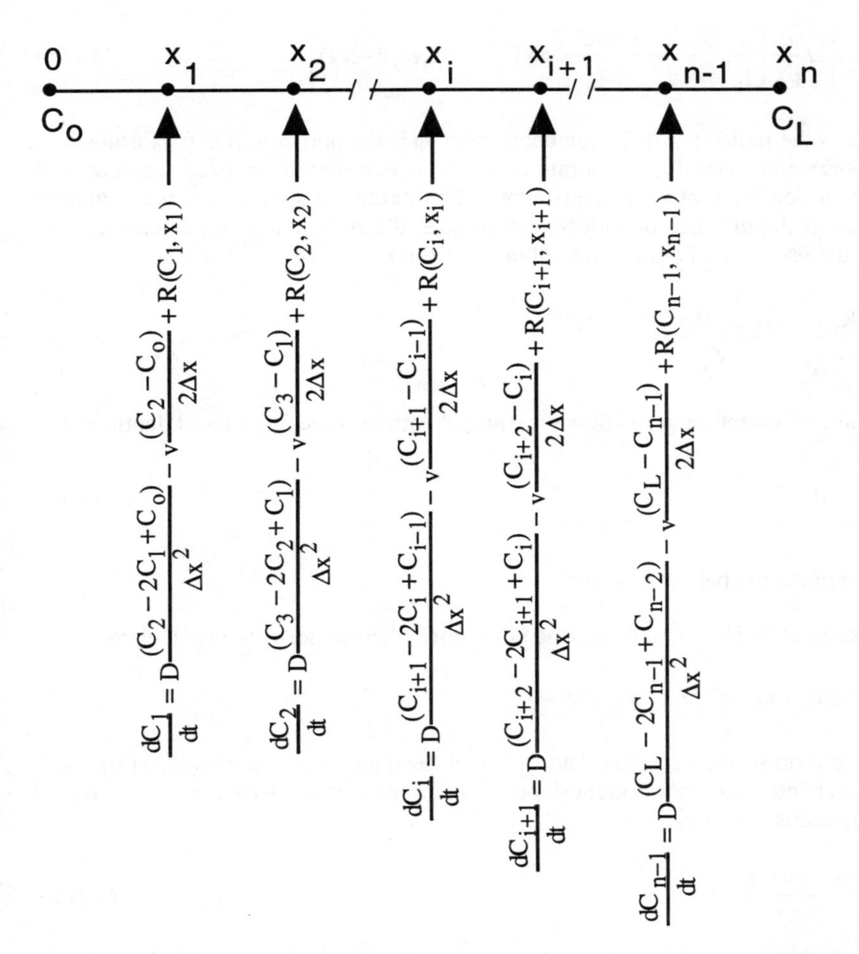

Figure 8.11. Illustration of the ODEs that result from the partial (spatial) discretization of the PDE Eq. (8.134), which can then be solved by the Method of Lines (MoL).

As a further illustration of this technique, a series of typical diagenetic problems is detailed below, along with their conversion to MoL form. Example FORTRAN codes for the solution of these problems using the integrator VODE (Byrne et al., 1989), can then be obtained, as outlined in the Appendix.

Example 8-2. Consider now a single solute species that is subject to consumption by an m-order homogeneous reaction, production by an inhomogeneous reaction in the form of an exponential, and transport by molecular diffusion and porewater advection, all in a variable porosity sediment. The governing diagenetic equation is

$$\frac{\partial C}{\partial t} = \frac{1}{\varphi}\left\{D\frac{\partial}{\partial x}\left(\frac{\varphi}{\theta^2}\frac{\partial C}{\partial x}\right) - \frac{\partial \varphi u C}{\partial x} - \varphi k C^m + \varphi R\, exp(-\alpha x)\right\} \tag{8.136}$$

where D is the molecular diffusion coefficient, φ is the porosity, θ is the tortuosity, u is the porewater velocity, k is a rate constant, R is a maximum production at x=0, and α is a depth-attenuation coefficient. The parameters φ, θ, and u are general functions of depth. The boundary condition at the sediment-water interface, x=0, reflects the presence of a diffusive sublayer of thickness δ_D,

$$-\left[\frac{\varphi}{\theta^2}\frac{\partial C}{\partial x}\right]_0 = \frac{(C_w - C(0))}{\delta_D} \tag{8.137}$$

while the concentration gradient is set equal to zero at some large fixed depth, x=L,

$$\left[\frac{\partial C}{\partial x}\right]_L = 0 \tag{8.138}$$

to approximate the behavior $x \to \infty$.

The parameters D, k, and R are true constants, whereas porosity is calculated as

$$\varphi(x) = [\varphi(0) - \varphi(\infty)]\; exp(-\gamma x) + \varphi(\infty) \tag{8.139}$$

where γ is a prescribed constant and $\varphi(0)$ and $\varphi(\infty)$ are known porosities at the sediment-water interface and at great depth. The porewater velocity can be calculated from this porosity function as

$$u(x) = \frac{\varphi(\infty)u(\infty)}{\varphi(x)} \tag{8.140}$$

and we assume for tradition's sake that tortuosity is well represented by an empirical Archie-type law (Chapter 4),

$$\theta^2 = \varphi^{-p} \tag{8.141}$$

Introducing spatial finite differences at each of the interior points of the grid (i=1,2,3,...,n-1), the PDE Eq. (8.136) generates a set of ODEs of the form

$$\begin{aligned}\frac{dC_i}{dt} = \frac{D}{\varphi_i}&\left\{\varphi_i^{p+1}\frac{C_{i+1} - 2C_i + C_{i-1}}{\Delta x^2} + \left((p+1)\varphi_i^p \left.\frac{d\varphi}{dx}\right|_{x_i}\right)\frac{C_{i+1} - C_{i-1}}{2\Delta x}\right\} \\ &- \left\{u_i\frac{C_{i+1} - C_{i-1}}{2\Delta x} + C_i\left(\frac{u_i}{\varphi_i}\left.\frac{d\varphi}{dx}\right|_{x_i} + \left.\frac{du}{dx}\right|_{x_i}\right)\right\} - kC_i^m + R\, exp(-\alpha x_i)\end{aligned} \tag{8.142}$$

where the subscript i indicates the value at $x = x_i$, the i-th node of the grid. The values of porosity and velocity, as well as for their derivatives, can be calculated from Eqs (8.139) and (8.140). The advective term is approximated with a central difference because $(u\Delta x)/(2D) < 2$ for most solutes.

At the sediment-water interface, i=0, Eq. (8.137) gives that

$$C_{-1} = C_1 + \frac{2\Delta x}{\varphi_0^{p+1}\delta_D}(C_w - C_0) \tag{8.143}$$

which can be substituted into Eq. (8.142) at this point,

$$\frac{dC_0}{dt} = D\left\{2\varphi_0^p \frac{C_1 - C_0 + \dfrac{\Delta x}{\varphi_0^{p+1}\delta_D}(C_w - C_0)}{\Delta x^2} - \left((p+1)\frac{d\varphi}{dx}\bigg|_0\right)\frac{(C_w - C_0)}{\varphi_0^2\delta_D}\right\}$$
$$-\left\{u_0\frac{(C_0 - C_w)}{\varphi_0^{p+2}\delta_D} + C_0\left(\frac{u_0}{\varphi_0}\frac{d\varphi}{dx}\bigg|_0 + \frac{du}{dx}\bigg|_0\right)\right\} - kC_0^m + R\,exp(-\alpha x_i) \tag{8.144}$$

At the lower boundary, Eq. (8.138) gives that

$$C_{n+1} = C_{n-1} \tag{8.145}$$

and Eq. (8.142) applied at i = n becomes

$$\frac{dC_n}{dt} = \frac{D}{\varphi_n}\left\{2\varphi_n^{p+1}\frac{C_{n-1} - C_n}{\Delta x^2}\right\} - \left\{C_n\left(\frac{u_n}{\varphi_n}\frac{d\varphi}{dx}\bigg|_L + \frac{du}{dx}\bigg|_L\right)\right\} - kC_n^m + R\,exp(-\alpha L) \tag{8.146}$$

which completes the specification of the numerical equations for this example.

Example 8-3. Let us assume that the species in Example 8-2 is instead a trace component of the solids, subject to biodiffusion, burial, m-th order decay, and exponential production with depth-variable porosity,

$$\frac{\partial B}{\partial t} = \frac{1}{\varphi_s}\left\{\frac{\partial}{\partial x}\left(\varphi_s D_B(x)\frac{\partial B}{\partial x}\right) - \frac{\partial \varphi_s w B}{\partial x} - \varphi_s k B^m + \varphi_s R\,exp(-\alpha x)\right\} \tag{8.147}$$

with boundary conditions

$$-\varphi_s(0) D_B(0)\frac{\partial B}{\partial x}\bigg|_0 + \varphi_s(0)\,w(0)\,B(0) = F_B(t) \tag{8.148}$$

and

$$\left.\frac{\partial B}{\partial x}\right|_L = 0 \tag{8.149}$$

Here φ_s is the solid volume fraction, 1-φ, where φ is the porosity as given by Eq. (8.139), and w(x) is the burial velocity

$$w(x) = \frac{\varphi_s(\infty)w(\infty)}{\varphi_s(x)} \tag{8.150}$$

The key to this example is the appropriate treatment of the biodiffusion term. Because $D_B(x)$ is allowed to disappear with depth, it is necessary to make use of the blended difference scheme (central to backward differencing) given by Eq. (8.72). Expanding the diffusion term and substituting these finite differences produce at the interior grid points:

$$\begin{aligned}\frac{dB_i}{dt} &= D_B(x_i)\frac{B_{i+1} - 2B_i + B_{i-1}}{\Delta x^2} + \left(\frac{D_B(x_i)}{(\varphi_s)_i}\left.\frac{d\varphi_s}{dx}\right|_{x_i} + \left.\frac{dD_B}{dx}\right|_{x_i}\right)\frac{B_{i+1} - B_{i-1}}{2\Delta x} \\ &- w_i\left(\frac{(1-\sigma_i)B_{i+1} + 2\sigma_i B_i - (1+\sigma_i)B_{i-1}}{2\Delta x}\right) - \left(\frac{w_i}{(\varphi_s)_i}\left.\frac{d\varphi_s}{dx}\right|_{x_i} + \left.\frac{dw}{dx}\right|_{x_i}\right)B_i \\ &- kB_i^m + R\,exp(-\alpha x_i)\end{aligned} \tag{8.151}$$

where

$$\sigma_i = coth\left(\frac{w_i\Delta x}{2D_B(x_i)}\right) - \frac{2D_B(x_i)}{w_i\Delta x} \tag{8.152}$$

Notice that σ_i is calculated for each grid point, and not for the grid as a whole. The derivatives of the solid volume fraction and biodiffusion coefficient can be calculated from their prescribed functions, and the values substituted into Eq. (8.151).

At the sediment-water interface, i=0, Eq. (8.148) gives that

$$B_{-1} = B_1 + \frac{2\Delta x\left[F_B(t) - \varphi_s(0)w(0)B_0\right]}{\varphi_s(0)D_B(0)} \tag{8.153}$$

This result can be substituted in Eq. (8.151) when it is applied at i=0 to remove the imaginary concentration B_{-1}, as was done in Eq. (8.144).

At the base of the model, i=n, Eq. (8.149) again gives that

$$B_{n+1} = B_{n-1} \tag{8.154}$$

so that we have

$$\frac{dB_n}{dt} = 2D_B(L)\left(\frac{B_{n-1}-B_n}{\Delta x^2}\right) - w_n\sigma\left(\frac{B_n - B_{n-1}}{\Delta x}\right)$$
$$-\left(\frac{w_n}{(\varphi_s)_n}\frac{d\varphi_s}{dx}\bigg|_L + \frac{dw}{dx}\bigg|_L\right)B_n - kB_n^m + R\,exp(-\alpha L) \tag{8.155}$$

Example 8-4. Let us reconsider Example 8-2, but solved on an uneven grid. This grid concentrates points near the sediment-water interface, while not increasing the total number of points relative to an even grid. One possible choice to achieve this uneven grid is the mapping

$$x_i = \frac{L\left\{\left[\xi_i^2 + \xi_c^2\right]^{1/2} - \xi_c\right\}}{\left\{\left[L^2 + \xi_c^2\right]^{1/2} - \xi_c\right\}} \tag{8.156}$$

where x_i is the location of the i-th grid point in the (real) sediment we are modelling, L is the thickness of the sediment zone being modelled with the uneven grid, ξ_i is a point on a hypothetical even grid on the same depth interval, and ξ_c is a user-chosen depth relative to which the x_i are quadratically distributed for $\xi_i << \xi_c$ and linearly distributed for $\xi_i >> \xi_c$. The actual spacing size between the points x_{i-1} and x_i, i.e. Δx_i, is simply their difference.

We can now use Pearson's formulas to approximate the spatial derivatives in the conservation equation to obtain,

$$\frac{dC_i}{dt} = D\left\{\varphi_i^p\left[\frac{2C_{i-1}}{\Delta x_i(\Delta x_i + \Delta x_{i+1})} - \frac{2C_i}{\Delta x_i \Delta x_{i+1}} + \frac{2C_{i+1}}{\Delta x_{i+1}(\Delta x_i + \Delta x_{i+1})}\right]\right.$$
$$+\frac{1}{\varphi_i}\left((p+1)\varphi_i^p\frac{d\varphi}{dx}\bigg|_{x_i} - \frac{\varphi_i u_i}{D}\right)\left[-\frac{\Delta x_{i+1}C_{i-1}}{\Delta x_i(\Delta x_i + \Delta x_{i+1})} + \frac{\Delta x_{i+1} - \Delta x_i}{\Delta x_i \Delta x_{i+1}}C_i\right.$$
$$\left.\left.+\frac{\Delta x_i C_{i+1}}{\Delta x_{i+1}(\Delta x_i + \Delta x_{i+1})}\right]\right\} - C_i\left(\frac{u_i}{\varphi_i}\frac{d\varphi}{dx}\bigg|_{x_i} + \frac{du}{dx}\bigg|_{x_i}\right) - kC_i^m + R\,exp(-\alpha x_i) \tag{8.157}$$

Boundary conditions are dealt with as indicated in the section on uneven grids.

Example 8-5. So far I have discussed only a single species, governed by a single PDE. Coupling between species is commonplace in diagenetic reactions, so that the modeller must often consider the simultaneous solution of two or more diagenetic equations. As an example of the way such problems can be handled by the MoL, we consider the case of a two-species reaction. These two species, A and B, can be either solutes or solids. This example further assumes that both species are subject to constant diffusion (whether molecular or mixing), but advection and porosity changes are ignored; consequently, the conservation equations for these species are

$$\frac{\partial C_A}{\partial t} = D_A \frac{\partial^2 C_A}{\partial x^2} - kC_A C_B \tag{8.158}$$

and

$$\frac{\partial C_B}{\partial t} = D_B \frac{\partial^2 C_B}{\partial x^2} - \gamma k C_A C_B \tag{8.159}$$

where γ is a unit-conversion and stoichiometry constant, k is a suitable rate constant, and D_A and D_B are appropriate diffusion coefficients. For boundary conditions, assume that

$$C_A(x=0) = C_o \tag{8.160}$$

$$-D_B \frac{\partial C_B}{\partial x}\bigg|_0 = F_B \tag{8.161}$$

$$\frac{\partial C_A}{\partial x}\bigg|_L = \frac{\partial C_B}{\partial x}\bigg|_L = 0 \tag{8.162}$$

where L is an arbitrary large value of x, chosen to approximate $x \to \infty$. This model is one representation of the dissolution of soluble opal (see Schink and Guinasso, 1977b; Boudreau, 1990a).

The trick here is to recognize that when Eqs (8.158) and (8.159) are replaced with finite differences on a chosen grid with n+1 points, there will be 2n+1 unknown concentrations at these n+1 points and 2n+1 corresponding conservation equations. Each of the two species contributes an equation at each point; plus species B contributes a single equation at x=0 because of Eq. (8.161).

It is best to use two different indices, one for the grid points, say i, and another for the unknown concentrations at each grid point, say j. Then at any grid point i, $1<i<n$, there are two ODEs. The first is for the species A,

$$\frac{dC_j}{dt} = D_A\left(\frac{C_{j-2} - 2C_j + C_{j+2}}{\Delta x^2}\right) - kC_j C_{j+1} \tag{8.163}$$

and the second for species B,

$$\frac{dC_{j+1}}{dt} = D_B\left(\frac{C_{j-1} - 2C_{j+1} + C_{j+3}}{\Delta x^2}\right) - \gamma k C_j C_{j+1} \tag{8.164}$$

We begin counting the equation index with B at x = 0 rather than A because only B needs to be considered at this first grid point, i=1 and j=1 (see Eq. (8.161)),

$$\frac{dC_1}{dt} = 2D_B\left(\frac{\frac{\Delta x F_B}{D_B} - C_1 + C_3}{\Delta x^2}\right) - \gamma k C_o C_1 \tag{8.165}$$

Note that for i=2 and j=2, then $C_{j-2} = C_0 = C_o$ from Eq. (8.160).

The same methodology is followed in the event that there are more that two dependent variables, as illustrated in the program CANDI by Boudreau (1996), which can be obtained as described in that paper or the Appendix.

8.8.2 Pure Advective Transport

While molecular diffusion is always around to mix solutes (and place a second derivative term in their diagenetic equations), the presence of biodiffusion is not inevitable. Below the surface mixed-zone in "normal" sediments and for sediments deposited in anoxic waters, advection caused by burial and compaction is the only mode of solid transport. Even in the bio-active surface layer, bioturbation is not necessarily diffusive, but may be essentially advective (e.g. conveyor-belt mixing, as described in Chapter 3). We need to be able to solve equations of the form

$$\frac{\partial B}{\partial t} = -\frac{\partial wB}{\partial x} \pm R(B, x) \tag{8.166}$$

where the advective velocity, w, is always assumed to be positive and may be a function of x, and where R(B,x) is an arbitrary reaction term.

The discussion surrounding the blended difference scheme, Eq. (8.72), indicated that a central difference approximation for advection terms is unstable when advection is dominant. Logically, this remains true when diffusion is totally absent. As a consequence, the simplest partial discretization of Eq. (8.166) uses the backward difference formula,

$$\frac{dB_i}{dt} = -\frac{(w_i B_i - w_{i-1} B_{i-1})}{\Delta x} \pm R(B_i, x_i) \tag{8.167}$$

if the derivative is not expanded, and

$$\frac{dB_i}{dt} = -w_i \frac{(B_i - B_{i-1})}{\Delta x} - B_i \frac{dw}{dx}\bigg|_{x_i} \pm R(B_i, x_i) \tag{8.168}$$

with expansion. The derivative of the advective velocity can be calculated either analytically or numerically from the prescribed w(x).

Equations (8.167) and (8.168) are both formally accurate only to $O(\Delta x)$, although in practice they seem to deliver acceptable results. There exist second-order accurate backwards approximations (Özisik, 1980),

$$\frac{\partial wB}{\partial x} \approx \frac{(wB)_{i-2} - 4(wB)_{i-1} + 3(wB)_i}{2\Delta x} \tag{8.169}$$

and

$$\frac{\partial B}{\partial x} \approx \frac{B_{i-2} - 4B_{i-1} + 3B_i}{2\Delta x} \tag{8.170}$$

that can be substituted for the first-order approximations in Eqs (8.167) and (8.168). These have one drawback however; at the first spatial point, say i=1, the nonexistent value B_{-1} arises. Khaliq and Twizell (1984) suggest using Eq. (8.167) and (8.168) at the first point, but this may lower the formal accuracy of the scheme.

Schiesser (1994) offers a differencing scheme that is formally $O(\Delta x^3)$ accurate and that he also claims to be upwind weighted (i.e. stable); specifically, he proposes that for an n-point grid,

$$\frac{\partial (wB)}{\partial x}\bigg|_{x_i} = \frac{1}{3!\,\Delta x}\left[(wB)_{i-2} - 6(wB)_{i-1} + 3(wB)_i + 2(wB)_{i+1}\right] \tag{8.171}$$

for i = 3, 4, ..., n-2, n-1. Meanwhile at the border points, he advances

$$\frac{\partial (wB)}{\partial x}\bigg|_{x_1} = \frac{1}{3!\,\Delta x}\left[-11(wB)_1 + 18(wB)_2 - 9(wB)_3 + 2(wB)_4\right] \tag{8.172}$$

for i=1,

$$\frac{\partial (wB)}{\partial x}\bigg|_{x_2} = \frac{1}{3!\,\Delta x}\left[-2(wB)_1 - 3(wB)_2 + 6(wB)_3 - (wB)_4\right] \tag{8.173}$$

for i=2, and

$$\frac{\partial (wB)}{\partial x}\bigg|_{x_n} = \frac{1}{3!\,\Delta x}\left[-2(wB)_{n-3} + 9(wB)_{n-2} - 18(wB)_{n-1} + 11(wB)_n\right] \tag{8.174}$$

at the end-point of the grid, i=n.

The equivalent formulas for B alone are obtained simply by removing w from within the parentheses in these expressions. Furthermore, while variable porosity is not explicitly included in these equations, the variable porosity versions are recovered simply by replacing B everywhere with φB. This may result in the addition of a number of φ_s to the reaction term(s), but that changes nothing from a numerical point of view.

A word of warning: the above schemes may or may not accurately conserve sharp concentration fronts (gradients) as they move through the sediment. The user should test this for himself/herself. To this aim, Osher and Chakravarthy (1984) offer a different second-order scheme that claims to track such fronts when there is no reaction, i.e. R(B,x) = 0,

$$\frac{dB_i}{dt} = -\frac{1}{\Delta x}\left[F_i - F_{i-1} + \frac{1}{2}\left\{\Phi(r_i^+)(F_{i+1} - F_i) - \Phi(r_{i-1}^+)(F_i - F_{i-1})\right\}\right] \tag{8.175}$$

where F_i is the flux $(wB)_i$,

$$r_i^+ = \frac{F_i - F_{i-1}}{F_{i+1} - F_i} \tag{8.176}$$

$$\Phi(r_i^+) = \begin{cases} 0 & \text{if } r_i^+ < 0 \\ r_i^+ & \text{if } r_i^+ \le 1 \\ 1 & \text{if } r_i^+ > 1 \end{cases} \tag{8.177}$$

Equation (8.176) also solves the case with linear decay kinetics because the equation

$$\frac{\partial B}{\partial t} = -\frac{\partial wB}{\partial x} - kB \tag{8.178}$$

becomes the new equation,

$$\frac{\partial \Theta}{\partial t} = -\frac{\partial w\Theta}{\partial x} \tag{8.179}$$

under the dependent-variable transformation

$$B(x,t) = \Theta(x,t)\mathit{exp}(-kt) \tag{8.180}$$

Remember to also apply this last transformation to the top boundary condition. Chalabi (1992) offers an extension to Eq. (8.175) that includes an arbitrary nonlinear reaction term.

8.9 Full Differencing

8.9.1 Diffusive Transport Present

There are a number of reasons that partial discretization and the MoL may not be the optimum method for the solution of a particular problem. For example, one may simply not have access to a good ODE integrator, or in the case of multi-dimensional problems, it is sometimes faster to solve them by *fully explicit methods.* The primary characteristic of these alternative techniques is that they replace both spatial and temporal derivatives with finite differences.

By far the most common way of dealing with the time derivative is to replace it with a forward difference,

$$\left.\frac{\partial C}{\partial t}\right|_{t=t_k} = \frac{C_i^{k+1} - C_i^k}{\Delta t} \tag{8.181}$$

where t_k is time at the k-th time step, Δt is a time step, and C_i^k is the concentration at the grid point $x = x_i$ at time t_k. (Don't confuse the time step k with a rate constant.) Less commonly use is also made of the temporal central difference,

$$\left.\frac{\partial C}{\partial t}\right|_{t=t_k} = \frac{C_i^{k+1} - C_i^{k-1}}{2\Delta t} \tag{8.182}$$

With the introduction of user control over time, there results some choice in the concentration that appears in the spatial derivatives. Specifically, do we use the concentrations at the time step we have just completed, e.g. C_i^k, which is a known quantity, or do we use the values at the time we are about to calculate, e.g. C_i^{k+1}, which is an unknown? Likewise for the reaction term, should it be given as $R(C_i^{k+1},x_i)$ or as $R(C_i^k,x_i)$?

Actually, these two alternatives are but end-members of a continuum of choices we can make. In general, any spatial derivative can be represented by a weighted average of these two extreme cases, i.e. with forward differences,

$$\frac{\partial C}{\partial x} = (1-\varepsilon)\left(\frac{C_i^k - C_{i-1}^k}{\Delta x}\right) + \varepsilon\left(\frac{C_i^{k+1} - C_{i-1}^{k+1}}{\Delta x}\right) \tag{8.183}$$

or, with central differences,

$$\frac{\partial C}{\partial x} = (1-\varepsilon)\left(\frac{C_{i+1}^k - C_{i-1}^k}{2\Delta x}\right) + \varepsilon\left(\frac{C_{i+1}^{k+1} - C_{i-1}^{k+1}}{2\Delta x}\right) \tag{8.184}$$

and

$$\frac{\partial^2 C}{\partial x^2} = (1-\varepsilon)\left(\frac{C_{i+1}^k - 2C_i^k + C_{i-1}^k}{\Delta x^2}\right) + \varepsilon\left(\frac{C_{i+1}^{k+1} - 2C_i^{k+1} + C_{i-1}^{k+1}}{\Delta x^2}\right) \tag{8.185}$$

where ε is a new user-set *weighting parameter* that determines how much the difference approximation depends on past (known) values and how much on future (unknown) values of the concentration ($0 \le \varepsilon < 1$).

To illustrate the application of this weighted scheme, we start with a generic diagenetic equation of the form

$$\frac{\partial C}{\partial t} = D\frac{\partial^2 C}{\partial x^2} - v\frac{\partial C}{\partial x} + R(C, x) \tag{8.186}$$

Substitution of the weighted central differences for spatial derivatives and forward differencing in time arrives at

$$\frac{C_i^{k+1} - C_i^k}{\Delta t} = D\left[(1-\varepsilon)\left(\frac{C_{i+1}^k - 2C_i^k + C_{i-1}^k}{\Delta x^2}\right) + \varepsilon\left(\frac{C_{i+1}^{k+1} - 2C_i^{k+1} + C_{i-1}^{k+1}}{\Delta x^2}\right)\right]$$
$$- v\left[(1-\varepsilon)\left(\frac{C_{i+1}^k - C_{i-1}^k}{2\Delta x}\right) + \varepsilon\left(\frac{C_{i+1}^{k+1} - C_{i-1}^{k+1}}{2\Delta x}\right)\right] + (1-\varepsilon)R(C_i^k, x_i) + \varepsilon R(C_i^{k+1}, x_i) \tag{8.187}$$

There are three special cases of this approximation for three limiting values of ε.

Explicit Differencing ($\varepsilon = 0$). In this limit, Eq. (8.187) reduces to

$$\frac{C_i^{k+1} - C_i^k}{\Delta t} = D\left(\frac{C_{i+1}^k - 2C_i^k + C_{i-1}^k}{\Delta x^2}\right) - v\left(\frac{C_{i+1}^k - C_{i-1}^k}{2\Delta x}\right) + R(C_i^k, x_i) \tag{8.188}$$

Notice that the unknown C_i^{k+1} is present only in the linear difference term on the left-hand side of this equation. Therefore, we can write an *explicit* equation for C_i^{k+1}, i.e.

$$C_i^{k+1} = C_i^k + \Delta t\left[D\left(\frac{C_{i+1}^k - 2C_i^k + C_{i-1}^k}{\Delta x^2}\right) - v\left(\frac{C_{i+1}^k - C_{i-1}^k}{2\Delta x}\right) + R(C_i^k, x_i)\right] \tag{8.189}$$

even if the reaction term $R(C_i^k, x_i)$ is nonlinear.

This equation is Δt accurate in time and Δx^2 accurate in space. To get the full solution, simply solve Eq. (8.189) for all the n points in the grid for each of the k time steps needed to reach the time of interest, as illustrated in Fig. 8.12.

Stability analysis (e.g. Nogotov, 1978; Mitchell and Griffiths, 1980; Lapidus and Pinder, 1982) shows that explicit formulas such as Eq. (8.189) are stable only if the time and space steps satisfy certain inequalities. If $R(C,x) = 0$, then the first such inequality is

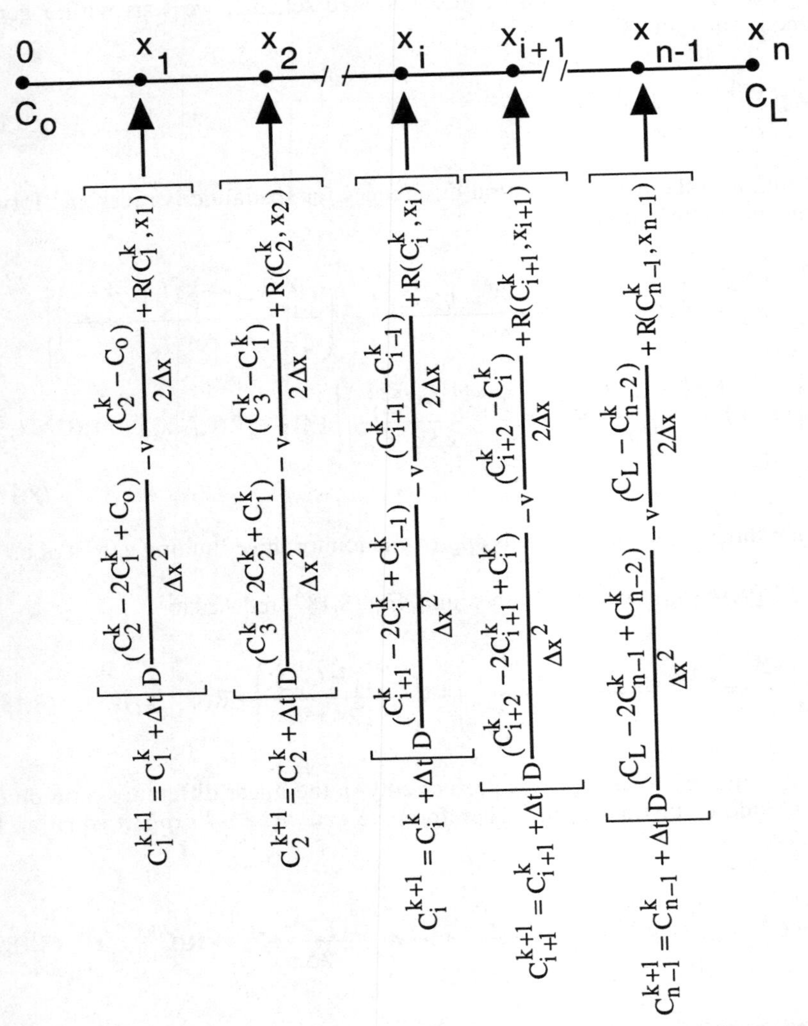

Figure 8.12. Illustration of equations that result from substituting explicit finite differences, Eq. (8.189), into the diagenetic equation given by Eq. (8.186).

$$\Delta t \le \frac{\Delta x^2}{2D} \tag{8.190a}$$

If, on the other hand, the reaction term is linear, i.e. $R(C,x) = -\kappa\, C$, then

$$\Delta t\left(\kappa + \frac{2D}{\Delta x^2}\right) \le 1 \tag{8.190b}$$

where κ in this equation is a rate constant. These requirements restrict the allowed time steps to very small values when the grid is fine; i.e. Δx is small, or the reaction fast, i.e. large κ. Therefore, while the explicit method is easy to use, potentially, it can take a great deal of computer time.

Fully-Implicit Differencingε = 1). With $\varepsilon = 1$, Eq. (8.187) reduces to

$$\frac{C_i^{k+1} - C_i^k}{\Delta t} = D\left(\frac{C_{i+1}^{k+1} - 2C_i^{k+1} + C_{i-1}^{k+1}}{\Delta x^2}\right) - v\left(\frac{C_{i+1}^{k+1} - C_{i-1}^{k+1}}{2\Delta x}\right) + R(C_i^{k+1}, x_i) \tag{8.191}$$

where we wish to calculate the unknown C_i^{k+1}. However, this equation involves three unknown concentrations at the grid points i-1, i, and i+1. There is no way to write a single equation for only one of them. Consequently, this is an *implicit* equation for C_i^{k+1}.

The set of equations generated by applying Eq. (8.191) at each point on an even grid must be solved simultaneously to get all the unknown C_i^{k+1}, i=1,2,3,4,...,n-1. If and only if $R(C_i^{k+1},x,t)$ is linear in concentration, then the set of such equations can be written as a matrix that is said to be *tri-diagonal.* In a tri-diagonal matrix, there are only three non-zero diagonal columns, the center and the two adjacent columns in each row (Fig. 8.13).

Well-known algorithms exist for the solutions of such matrices (Conte and de Boor, 1972; Nogotov, 1978; Mitchell and Griffiths, 1980; Lapidus and Pinder, 1982; Kahaner et al., 1989), and these have been implemented in a variety of computer codes (e.g. Conte and de Boor, 1972; Press et al., 1992). The best known of these methods is *Gaussian elimination.* To outline this method, we can rewrite the system of equations illustrated in Fig. 8.13 as

$$\left.\begin{array}{l} b_1C_1 - c_1C_2 = d_1 \\ -a_2C_1 + b_2C_2 - c_2C_3 = d_2 \\ -a_3C_2 + b_3C_3 - c_3C_4 = d_3 \\ \vdots \\ -a_{n-1}C_{n-2} + b_{n-1}C_{n-1} - c_{n-1}C_n = d_{n-1} \\ -a_nC_{n-1} + b_nC_n = d_n \end{array}\right\} \tag{8.192}$$

$$
\begin{bmatrix}
\left(\frac{2D\Delta t}{\Delta x^2}+\lambda+1\right) & -\left(\frac{D\Delta t}{\Delta x^2}-\frac{v\Delta t}{2\Delta x}\right) & 0 & 0 & \cdot & \cdot \\
-\left(\frac{D\Delta t}{\Delta x^2}+\frac{v\Delta t}{2\Delta x}\right) & \left(\frac{2D\Delta t}{\Delta x^2}+\lambda+1\right) & -\left(\frac{D\Delta t}{\Delta x^2}-\frac{v\Delta t}{2\Delta x}\right) & 0 & \cdot & \cdot \\
\cdot & \cdot & \cdot & \cdot & \cdot & \cdot \\
\cdot & \cdot & \cdot & \cdot & \cdot & \cdot \\
\cdot & \cdot & 0 & -\left(\frac{D\Delta t}{\Delta x^2}+\frac{v\Delta t}{2\Delta x}\right) & \left(\frac{2D\Delta t}{\Delta x^2}+\lambda+1\right) & -\left(\frac{D\Delta t}{\Delta x^2}-\frac{v\Delta t}{2\Delta x}\right) \\
\cdot & \cdot & 0 & 0 & -\left(\frac{D\Delta t}{\Delta x^2}+\frac{v\Delta t}{2\Delta x}\right) & \left(\frac{2D\Delta t}{\Delta x^2}+\lambda+1\right)
\end{bmatrix}
\begin{bmatrix}
C_1^{k+1} \\ C_2^{k+1} \\ \cdot \\ \cdot \\ C_{n-2}^{k+1} \\ C_{n-1}^{k+1}
\end{bmatrix}
=
\begin{bmatrix}
C_1^k + \left(\frac{D\Delta t}{\Delta x^2}+\frac{v\Delta t}{2\Delta x}\right) C_o^k \\ C_2^k \\ \cdot \\ \cdot \\ C_{n-2}^k \\ C_{n-1}^k + \left(\frac{D\Delta t}{\Delta x^2}-\frac{v\Delta t}{2\Delta x}\right) C_L^k
\end{bmatrix}
$$

Figure 8.13. Typical tridiagonal matrix generated by implicit finite differencing of Eq. (8.191) when the reaction term is linear, i.e. $R(C.x) = -\lambda C$, where λ is a first-order rate constant. The concentrations Co and C_L at the boundaries, x=0 and x=L, are treated as known functions of time (which includes constants).

where $b_1 = 2D\Delta t/(\Delta x^2)+k+1$, etc., as in Fig. 8.13, and these are known constants. The first equation in this set can be used to eliminate C_1 from the second, then the resulting second equation can be used to eliminate C_2 from the third, and so on. After the last elimination, the last equation will involve only C_n, and we can solve for this unknown concentration. Once C_n is known, we can substitute back into the previous equation to get C_{n-1}, and so on down to the first equation. The explicit formulas for this solution technique are (Smith, 1975)

$$C_n = \frac{s_n}{e_n} \tag{8.193}$$

and

$$C_i = \frac{(s_i + c_i C_{i+1})}{e_i} \qquad i = n-1, n-2, \ldots, 2, 1 \tag{8.194}$$

where

$$e_1 = b_1 \tag{8.195}$$

$$e_i = b_i - \frac{a_i c_{i-1}}{e_{i-1}} \tag{8.196}$$

$$s_1 = d_1 \tag{8.197}$$

$$s_i = d_i + \frac{a_i s_{i-1}}{e_{i-1}} \tag{8.198}$$

In many problems, such as the one above, the e_i and a_i/e_{i-1} are independent of time and need be calculated only once and used, thereafter, in each time step.

Fully implicit formulas are stable for all values of Δt and Δx, and they are accurate to order Δt and Δx^2 in time and space, respectively. It remains that you still cannot take too large a time step or the resulting solutions become inaccurate, even though they are stable. Moreover, it is difficult to deal with nonlinear reactions using this method because this results in a nonlinear set of equations.

Crank-Nicolson Finite Differencing (ε = 1/2). Equation (8.186) becomes

$$\begin{aligned} \frac{C_i^{k+1} - C_i^k}{\Delta t} &= \frac{D}{2}\left[\left(\frac{C_{i+1}^k - 2C_i^k + C_{i-1}^k}{\Delta x^2}\right) + \left(\frac{C_{i+1}^{k+1} - 2C_i^{k+1} + C_{i-1}^{k+1}}{\Delta x^2}\right)\right] \\ &\quad - \frac{v}{2}\left[\left(\frac{C_{i+1}^k - C_{i-1}^k}{2\Delta x}\right) + \left(\frac{C_{i+1}^{k+1} - C_{i-1}^{k+1}}{2\Delta x}\right)\right] + \frac{1}{2}\left[R(C_i^k) + R(C_i^{k+1})\right] \end{aligned} \tag{8.199}$$

Again, this equation involves three unknown concentrations; consequently, this is an implicit equation that also generates a tri-diagonal system of equations. So what is the advantage? It is accurate to both Δt^2 and Δx^2; that is to say, it is potentially far more accurate than either the fully implicit or the explicit schemes.

Equation (8.192) exemplifies the justly famous Crank-Nicolson finite differencing. This particular scheme is widely used, but we still face the difficulties resulting from a nonlinear reaction term. Some sort of linearization is necessary, if we cannot use a nonlinear equation solver; e.g. it may be both possible and permissible to replace the nonlinear reaction term with a linear Taylor series expansion (e.g. Ramos, 1987):

$$R(C_i^{k+1}, x_i) \approx R(C_i^k, x_i) + \left.\frac{\partial R(C, x)}{\partial C}\right|_{C_i^k} \left(C_i^{k+1} - C_i^k\right) \tag{8.200}$$

8.9.2 Diffusive Transport Absent

The absence of diffusion in a problem can substantially complicate a numerical solution. This may at first seem puzzling, but diffusion adds a certain stability to a problem. Without diffusion, some finite difference schemes are unstable or cannot accurately reproduce sharp concentration fronts or gradients (Rood, 1987). Keeping this in mind, let us start by reviewing some simple or classic schemes for the solution of the simple advection equation,

$$\frac{\partial B}{\partial t} = -w \frac{\partial B}{\partial x} \tag{8.201}$$

or, more conveniently,

$$\frac{\partial B}{\partial t} = -\frac{\partial F}{\partial x} \tag{8.202}$$

where F is the advective flux, wB. (Note that B may include variable porosity; i.e. B = φB) The term *conservation form* is often employed for Eq. (8.202).

Arguably the simplest explicit approximation to Eq. (8.202) is forward in time and backward in space,

$$\frac{B_i^{k+1} - B_i^k}{\Delta t} = -\frac{F_i^k - F_{i-1}^k}{\Delta x} \tag{8.203}$$

This formula is only accurate to $O(\Delta t)$ and $O(\Delta x)$ and stable as long as

$$\frac{w \Delta t}{\Delta x} \leq 1 \tag{8.204}$$

This condition is widely known as the *Courant number* or the *Courant-Fredrichs-Lewy parameter.* In addition, sharp gradients are not well preserved by this scheme. To see this, substitute wB for F in Eq. (8.203) and rearrange to obtain,

$$B_i^{k+1} = B_i^k \underbrace{- \frac{w\Delta t}{2\Delta x}\left(B_{i+1}^k - B_{i-1}^k\right)}_{\text{advection by central difference}} + \underbrace{\frac{w\Delta t}{2\Delta x}\left(B_{i+1}^k - 2B_i^k + B_{i-1}^k\right)}_{\text{correction}} \tag{8.205}$$

The third term on the right-hand side of Eq. (8.205), designated "correction", looks like the finite difference equivalent of a diffusion term; i.e. compare with Eq. (8.189). This "diffusion", known as *numerical dispersion*, is fictional; it is added to the solution to enhance the stability of the method, but at the price of some inaccuracy, as gradients will eventually diffuse away.

To increase the accuracy in the spatial direction, one is first tempted to substitute central differencing in x,

$$\frac{B_i^{k+1} - B_i^k}{\Delta t} = -\frac{F_{i+1}^k - F_{i-1}^k}{2\Delta x} \tag{8.206}$$

Unfortunately, this formulation is always unstable (Lapidus and Pinder, 1982; Rood, 1987), as argued previously, and so we must seek a different approach.

The next device, known as the *Lax-Wendroff scheme* (for details, see Lapidus and Pinder, 1982; Rood, 1987; Press et al., 1992), attempts to retain the advantages of Eq. (8.203) and increase the accuracy, but reduce the necessary numerical diffusion. When applied to Eq. (8.201), the Lax-Wendroff formula asserts that

$$B_i^{k+1} = B_i^k - \frac{w\Delta t}{2\Delta x}\left(B_{i+1}^k - B_{i-1}^k\right) + \frac{1}{2}\left(\frac{w\Delta t}{\Delta x}\right)^2\left(B_{i+1}^k - 2B_i^k + B_{i-1}^k\right) \tag{8.207}$$

which is second-order accurate in both Δt and Δx. This form cannot be derived by replacing derivatives with finite differences, but Rood (1987) furnishes a rather simple derivation based on Taylor series expansions. Rood's derivation further proves that the scheme is second order in time and space.

When applied to the conservation form, Eq. (8.202), the Lax-Wendroff scheme does not generate a single equation. Instead, it leads to a so-called two-step method (Press et al., 1992). Specifically, one first calculates a concentration at a hypothetical half-time and half-space step using the formula

$$B_{i+1/2}^{k+1/2} = \frac{1}{2}\left(B_{i+1}^k + B_i^k\right) - \frac{\Delta t}{2\Delta x}\left(F_{i+1}^k - F_i^k\right) \tag{8.208}$$

Then one calculates the flux at this hypothetical time for all the hypothetical half-steps, i.e.

$$F_{i+1/2}^{k+1/2} = w_{i+1/2}\, B_{i+1/2}^{k+1/2} \tag{8.209}$$

with i=0,1,2,3,...,n-1. With these intermediate flux values, the final concentration values are calculated with a centered difference scheme,

$$B_{i+1}^{k+1} = B_i^k - \frac{\Delta t}{\Delta x}\left(F_{i+1/2}^{k+1/2} - F_{i-1/2}^{k+1/2}\right) \tag{8.210}$$

The Lax-Wendroff scheme has found considerable use in various applications; yet it is still somewhat diffusive and does not preserve sharp gradients as they are advected (see Rood, 1987, for an example). Another problem is what to do at the base of the grid (i=n). In an advective problem there is no boundary condition here; so if we apply the Lax-Wendroff at this point, it will contain the nonexistent concentration B_{i+1} that cannot be eliminated with a boundary condition. One can apply Eq. (8.203) at this point and hope it does not perturb the scheme.

The final classic scheme consists of substituting centered difference approximations for both time and space derivatives in Eqs (8.201) and (8.202), i.e.

$$\frac{B_i^{k+1} - B_i^{k-1}}{\Delta t} = -w\frac{B_{i+1}^k - B_{i-1}^k}{\Delta x} \tag{8.211}$$

and

$$\frac{B_i^{k+1} - B_i^{k-1}}{\Delta t} = -\frac{F_{i+1}^k - F_{i-1}^k}{\Delta x} \tag{8.212}$$

These formulas are known as *Leapfrog schemes* because their time-derivative approximations skip (or jump) over concentration values at the current time, $t=k\Delta t$.

The Leapfrog formulas are second order in both time and space, and they have no numerical dispersion. Nevertheless, they are not perfect because they generate spurious oscillations when they encounter sharp gradients, termed *computational modes*. These can grow and come to dominate the solution. Furthermore, the Leapfrog scheme inapplicable at the first time step; so a different start-up scheme needs to be used to generate the first set of data. This will introduce some small amount (we hope) of the bias in the result. In addition, as with the Lax-Wendroff, leapfrogging has a problem at the final point of the grid. Accepting that no classic scheme is perfect, Press et al. (1992) still prefer the Leapfrog scheme, at least when sharp gradients (shocks) aren't involved.

Preserving and resolving sharp gradients (capturing shocks) is sometimes important in the solution to Eq. (8.202). For example, we may want to model the abrupt introduction of a tracer or contaminant (e.g. Boudreau, 1986a,b). In this situation, it is possible to use so-called flux-limiting or flux-corrected schemes (van Leer, 1974; Sweby, 1984; Rood, 1987; Swanson and Turkel, 1992). These corrections adjust the local amount of diffusion so that sharp gradients are not smoothed out. The Sweby (1984) scheme is typical of this genre, i.e.

$$B_i^{k+1} = B_i^k - \frac{\Delta t}{\Delta x}\left[F_i^k - F_{i-1}^k\right] + \frac{\Delta t}{2\Delta x}\left[\Phi(r_i)\nu_i(F_{i+1}^k - F_i^k) - \Phi(r_{i-1})\nu_{i-1}(F_i^k - F_{i-1}^k)\right] \quad (8.213)$$

where

$$\nu_i = 1 - \frac{\Delta t}{\Delta x}\left[\frac{F_{i+1}^k - F_i^k}{B_{i+1}^k - B_i^k}\right] \quad (8.214)$$

unless the denominator is equal to zero, in which case use

$$\nu_i = 1 - \frac{\Delta t}{\Delta x}\left[w_{i+1} - w_i\right] \quad (8.215)$$

which will be equal to one if the velocities are also equal. Continuing,

$$r_i = \frac{\nu_{i-1}\left[F_i^k - F_{i-1}^k\right]}{\nu_i\left[F_{i+1}^k - F_i^k\right]} \quad (8.216)$$

and, finally,

$$\Phi(r_i) = \begin{cases} 0 & \text{if } r_i < 0 \\ r_i & \text{if } r_i \le 1 \\ 1 & \text{if } r_i > 1 \end{cases} \quad (8.217)$$

which has the same behavior as $\Phi(r_i^+)$ in Eq. (8.178).

Equation (8.213) will also solve the advection equation with linear decay, i.e.

$$\frac{\partial B}{\partial t} = -\frac{\partial wB}{\partial x} - kB \quad (8.218)$$

because the change of variable B(x,t) = U(x,t)*exp*(-kt) will transform this equation into

$$\frac{\partial U}{\partial t} = -\frac{\partial wU}{\partial x} \quad (8.219)$$

with F now defined as wU. Again, the transform must also be applied to the boundary condition at x=0.

The solution of advection equations with kinetic expressions of general form has only recently constituted an explicit topic of mathematical research, and the type and number of suitable schemes remain small. Nevertheless, single-step second-order differencing schemes, which capture shocks, have been introduced by Chalabi (1992) and Glaister (1993). For the conservation equation

$$\frac{\partial B}{\partial t} = -\frac{\partial F}{\partial x} - R(B, x) \tag{8.220}$$

Glaister's formula is

$$\begin{aligned} B_i^{k+1} = B_i^k &+ \frac{\Delta t}{2}\left[R_{i+1/2} - \frac{F_{i+1}^k - F_i^k}{\Delta x}\right] + \frac{\Delta t}{2}\left[R_{i-1/2} - \frac{F_i^k - F_{i-1}^k}{\Delta x}\right] \\ &+ \frac{\Delta t^2}{4}\left[\left.\frac{\partial R}{\partial B}\right|_i \left(R_{i+1/2} - \frac{F_{i+1}^k - F_i^k}{\Delta x}\right) + \left.\frac{\partial R}{\partial B}\right|_i \left(R_{i-1/2} - \frac{F_i^k - F_{i-1}^k}{\Delta x}\right)\right] \\ &- \frac{\Delta t^2}{2\Delta x}\left[F'_{i+1/2}\left(R_{i+1/2} - \frac{F_{i+1}^k - F_i^k}{\Delta x}\right) - F'_{i-1/2}\left(R_{i-1/2} - \frac{F_i^k - F_{i-1}^k}{\Delta x}\right)\right] \end{aligned} \tag{8.221}$$

where

$$F'_{i+1/2} = \begin{cases} \dfrac{F_{i+1}^k - F_i^k}{B_{i+1}^k - B_i^k} & B_{i+1}^k \neq B_i^k \\ w_{i+1} - w_i & B_{i+1}^k = B_i^k \end{cases} \tag{8.222}$$

$$F'_{i-1/2} = \begin{cases} \dfrac{F_i^k - F_{i-1}^k}{B_i^k - B_{i-1}^k} & B_i^k \neq B_{i-1}^k \\ w_i - w_{i-1} & B_i^k = B_{i-1}^k \end{cases} \tag{8.223}$$

$$R_{i\pm 1/2} = R(B_i^k, x_i \pm \frac{\Delta x}{2}) \tag{8.224}$$

and the derivatives of the reaction term R with respect to the concentration B, i.e. $\partial R/\partial B$, are both evaluated with B_i and x_i. Because this is an explicit scheme, keep in mind that the Courant number must be satisfied, i.e. Eq. (8.204). Alternatively, multi-step methods, that contend with stiff homogeneous kinetics, have been advanced by LeVeque and Yee (1990) and Loureiro and Rodrigues (1991).

8.10 Diagenesis in 2 and 3 Dimensions

This section deals with the finite-difference solution of PDEs in two and three dimensions and with time dependence, e.g.

$$\frac{\partial C}{\partial t} = D_x \frac{\partial^2 C}{\partial x^2} + D_y \frac{\partial^2 C}{\partial y^2} - v_x \frac{\partial C}{\partial x} - v_y \frac{\partial C}{\partial y} + R(C, x, y) \tag{8.225}$$

and

$$\frac{\partial C}{\partial t} = D' \frac{\partial^2 C}{\partial x^2} + \frac{D'}{r} \frac{\partial}{\partial r}\left(r \frac{\partial C}{\partial r}\right) + R(C, x, r) \tag{8.226}$$

Application of (semi) implicit methods (e.g. Crank-Nicolson) to these equations can potentially generate a vast set of equations, whose solution would be demanding of both computer time and memory. Fortunately, numerical methodologies exist that are relatively easy to implement and are much less demanding on a computer; three such schemes are: 1) Explicit Methods, 2) Operator Splitting Methods (OS), and 3) Alternating-Direction-Implicit Methods.

8.10.1 Explicit Method

This approach is a straightforward extension of the 1-D method. For example, Eq. (8.225) with explicit central differences reads

$$\begin{aligned} \frac{C_{i,j}^{k+1} - C_{i,j}^{k}}{\Delta t} &= D_x \frac{C_{i+1,j}^{k} - 2C_{i,j}^{k} + C_{i-1,j}^{k}}{\Delta x^2} + D_y \frac{C_{i,j+1}^{k} - 2C_{i,j}^{k} + C_{i,j-1}^{k}}{\Delta y^2} \\ &- v_x \frac{C_{i+1,j}^{k} - C_{i-1,j}^{k}}{2\Delta x} - v_y \frac{C_{i,j+1}^{k} - C_{i,j-1}^{k}}{2\Delta y} + R(C_{i,j}^{k}, x_i, y_j) \end{aligned} \tag{8.227}$$

or, upon rearrangement,

$$\begin{aligned} C_{i,j}^{k+1} = C_{i,j}^{k} + \Delta t \Bigg[& D_x \frac{C_{i+1,j}^{k} - 2C_{i,j}^{k} + C_{i-1,j}^{k}}{\Delta x^2} + D_y \frac{C_{i,j+1}^{k} - 2C_{i,j}^{k} + C_{i,j-1}^{k}}{\Delta y^2} \\ & - v_x \frac{C_{i+1,j}^{k} - C_{i-1,j}^{k}}{2\Delta x} - v_y \frac{C_{i,j+1}^{k} - C_{i,j-1}^{k}}{2\Delta y} + R(C_{i,j}^{k}, x_i, y_j) \Bigg] \end{aligned} \tag{8.228}$$

Equation (8.228) allows direct calculation of the unknown concentration at each of the grid points in the 2-D grid that approximates the sediment area being modelled. The stability criterion in this case is

$$\Delta t\left[\frac{D_x}{\Delta x^2}+\frac{D_y}{\Delta y^2}\right]\leq\frac{1}{2} \tag{8.229}$$

which can severely limit the size of the time step for a fine grid.

8.10.2 Operator Splitting - Locally 1-D Method

This is an apparently crazy idea that works rather well in practice. It greatly simplifies the numerical solution of a multi-dimensional diagenetic equation by splitting it into separate 1-D parts that can be solved easily using explicit or implicit formulas or even the MoL. The particular variant explained here is known as the locally 1-D method or LOD (Nogotov, 1978; Mitchell and Griffiths, 1980; Lapidus and Pinder, 1982; Ramos, 1986b, 1987; Marchuk, 1990).

Example 8-6. To illustrate the LOD method, we begin with the diagenetic equation for Aller's (1980) tube model as an example,

$$\frac{\partial C}{\partial t}=D'\frac{\partial^2 C}{\partial x^2}+\frac{D'}{r}\frac{\partial}{\partial r}\left(r\frac{\partial C}{\partial r}\right)+R(C,x,r) \tag{8.230}$$

Then we assume that this can be written as the sum of three separate parts:

$$\frac{1}{3}\frac{\partial C}{\partial t}=D'\frac{\partial^2 C}{\partial x^2} \tag{8.231}$$

$$\frac{1}{3}\frac{\partial C}{\partial t}=\frac{D'}{r}\frac{\partial}{\partial r}\left(r\frac{\partial C}{\partial r}\right) \tag{8.232}$$

$$\frac{1}{3}\frac{\partial C}{\partial t}=R(C,x,r) \tag{8.233}$$

Each part is now solved numerically at every time step. The advantage of this approach is that, at most, we are solving a 1-D diffusion problem.

Step #1. The choice of the numerical method is optional; however, if we choose to employ the Crank-Nicolson formula for Eq. (8.231), we obtain

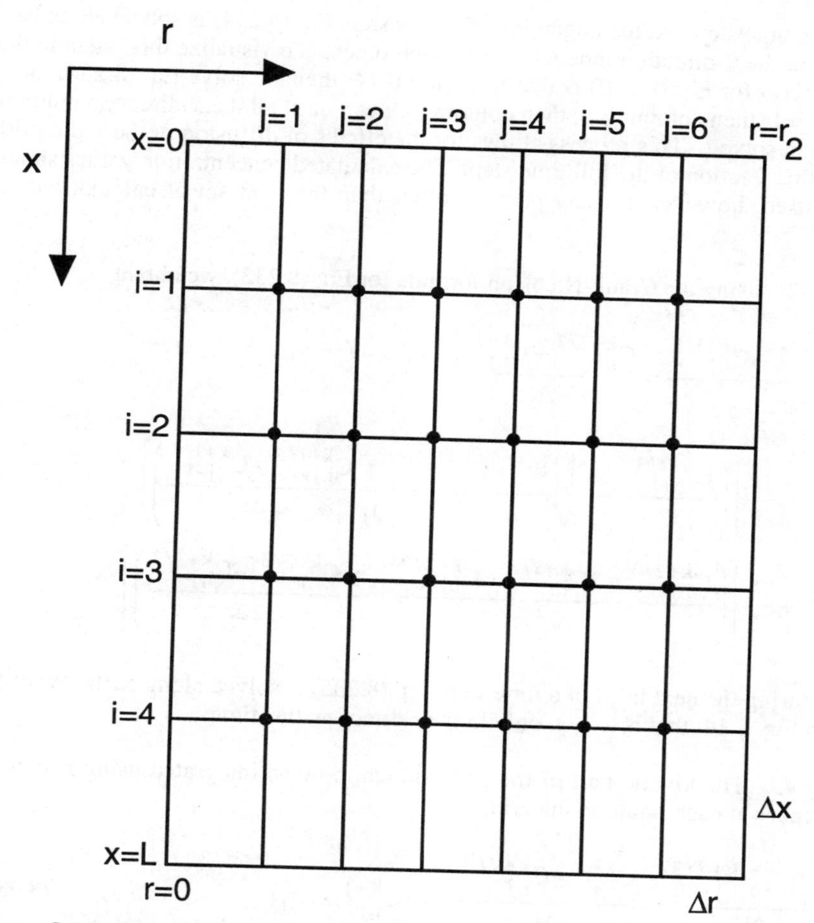

Figure 8.14. Simplified 2-D grid used in solving Eq. (8.234). Notice that the grid spacings Δx and Δx need not be equal, but the accuracy of the scheme is determined by the larger of the two.

$$\frac{1}{3}\frac{C_{i,j}^{k+1/3}-C_{i,j}^{k}}{\frac{\Delta t}{3}}=\frac{C_{i,j}^{k+1/3}-C_{i,j}^{k}}{\Delta t}$$

$$=\frac{D'}{2}\left[\left(\frac{C_{i+1,j}^{k}-2C_{i,j}^{k}+C_{i-1,j}^{k}}{\Delta x^2}\right)+\left(\frac{C_{i+1,j}^{k+1/3}-2C_{i,j}^{k+1/3}+C_{i-1,j}^{k+1/3}}{\Delta x^2}\right)\right] \quad (8.234)$$

The time index k+1/3 indicates that solution of Eq. (8.234) constitutes only one-third of the calculations that must be made to advance the solution of this problem one complete time step. At the beginning of a time step, Eq. (8.234) is solved along each column in the x-direction independent of each other! To visualize this, assume that the 2-D grid for Eq. (8.230) is that in Figure 8.14; then we solve Eq. (8.234) along column j=1, then column j=2, then column j=3, etc. until all the x-direction columns have been solved. This process allows for the effects of diffusion in the x-direction during this fraction of the full time step. The calculated concentration values should not be used, however, for any purpose other than the next set of calculations, as outlined below.

Step #2. Using the Crank-Nicolson formula for Eq. (8.232), we obtain,

$$\frac{1}{3}\frac{C_{i,j}^{k+2/3}-C_{i,j}^{k+1/3}}{\frac{\Delta t}{3}}=\frac{C_{i,j}^{k+2/3}-C_{i,j}^{k+1/3}}{\Delta t}$$

$$=\frac{D'}{2}\left[\left(\frac{C_{i,j+1}^{k+1/3}-2C_{i,j}^{k+1/3}+C_{i,j-1}^{k+1/3}}{\Delta r^2}+\frac{1}{r_j}\frac{C_{i,j+1}^{k+1/3}-C_{i,j-1}^{k+1/3}}{2\Delta r}\right)\right] \tag{8.235}$$

$$+\frac{D'}{2}\left[\left(\frac{C_{i,j+1}^{k+2/3}-2C_{i,j}^{k+2/3}+C_{i,j-1}^{k+2/3}}{\Delta r^2}+\frac{1}{r_j}\frac{C_{i,j+1}^{k+2/3}-C_{i,j-1}^{k+2/3}}{2\Delta r}\right)\right]$$

Then during the next third of a time step, Eq. (8.235) is solved along each row of the grid in Fig. 8.14; that is to say, only in the r-direction this time.

Step #3. The kinetic part of the problem can now be integrated using backward differences at each point on the grid, i.e.

$$\frac{1}{3}\frac{C_{i,j}^{k+1}-C_{i,j}^{k+2/3}}{\frac{\Delta t}{3}}=\frac{C_{i,j}^{k+1}-C_{i,j}^{k+2/3}}{\Delta t}=R(C_{i,j}^{k+1},x_i,r_j) \tag{8.236}$$

This allows handling of any stiffness in Eq. (8.233), independent of the method used to deal with the diffusion steps. Instead of supplying an integration scheme for Eq. (8.236), one could as well use a standard stiff ODE integration code (e.g. Verwer, 1994; Verwer and van Loon, 1994).

At the end of Step #3, we have the solution at the end of one full time step. To get to the final solution, we keep repeating until the sum of the time steps equals the desired end time. Valocchi and Malmstead (1992), Kaluarachchi and Morshed (1995), and Morshed and Kaluarachchi (1995) note that mass conservation and accuracy for the LOD method requires that the order of Eqs (8.234), (8.235), and (8.236) be reversed each full time step.

8.10.3 Alternating-Direction-Implicit Method

The Alternating-Direction-Implicit (ADI) is the third and last technique for multi-dimensional diagenetic equations that will be considered in this text. Like the LOD method, the ADI is a fractional multi-step scheme for advancing in time that isolates the calculations in the various spatial directions. The ADI approach to a problem is to divide a full time step into two fractional steps (three for a fully 3-D problem). At the first fractional time step, an implicit finite difference approximation is assigned to the diffusion-advection operator in one direction only, while an explicit form is assigned to the other(s). The resulting set of equations is then solved. (At worst, this will be a collection of tri-diagonal systems, and if these are linear, the scheme given by Eqs (8.192) through (8.198) can be utilized.) At the next fractional time step, the implicit and explicit approximations are switched, and the solution process is repeated. This completes one full time step. The difference with the LOD method is that the operators are never separated, only treated differently at any given fractional time step (see Mitchell and Griffiths, 1980). This concept is once again best illustrated with an example application.

Example 8-7. To allow full comparison with the LOD methods, we revisit Aller's (1980) tube model as given in Example 8-6, i.e.

$$\frac{\partial C}{\partial t} = D'\frac{\partial^2 C}{\partial x^2} + \frac{D'}{r}\frac{\partial}{\partial r}\left(r\frac{\partial C}{\partial r}\right) + R(C, x, r) \tag{8.237}$$

At the first fractional time step, i.e. k+1/2, Eq. (8.237) is approximated by

$$\begin{aligned}\frac{C_{i,j}^{k+1/2} - C_{i,j}^{k}}{\Delta t} &= D'\left(\frac{C_{i+1,j}^{k+1/2} - 2C_{i,j}^{k+1/2} + C_{i-1,j}^{k+1/2}}{\Delta x^2}\right)\\ &+ D'\left[\left(\frac{C_{i,j+1}^{k} - 2C_{i,j}^{k} + C_{i,j-1}^{k}}{\Delta r^2} + \frac{1}{r_j}\frac{C_{i,j+1}^{k} - C_{i,j-1}^{k}}{2\Delta r}\right)\right] + R(C_{i,j}^{k}, x_i, r_j)\end{aligned} \tag{8.238}$$

The unknowns are those concentrations at k+1/2, and these are coupled only in the x-direction; therefore, we can solve the resulting tri-diagonal equations along each row of the grid (Fig. 8.14).

To complete the calculations, we make the r-direction calculations implicit and the x-direction explicit using the now known values at k+1/2, i.e.

$$\begin{aligned}\frac{C_{i,j}^{k+1} - C_{i,j}^{k+1/2}}{\Delta t} &= D'\left(\frac{C_{i+1,j}^{k+1/2} - 2C_{i,j}^{k+1/2} + C_{i-1,j}^{k+1/2}}{\Delta x^2}\right)\\ &+ D'\left[\left(\frac{C_{i,j+1}^{k+1} - 2C_{i,j}^{k+1} + C_{i,j-1}^{k+1}}{\Delta r^2} + \frac{1}{r_j}\frac{C_{i,j+1}^{k+1} - C_{i,j-1}^{k+1}}{2\Delta r}\right)\right] + R(C_{i,j}^{k+1/2}, x_i, r_j)\end{aligned} \tag{8.239}$$

The solution to Eq. (8.239) follows that of Eq. (8.238), but this time the implicit coupling is in the r-direction. Fortunately, once more we have a set of tri-diagonal equations along each column of the grid (Fig. 8.14), which can be solved as before. Another example of the application of the ADI method can be found in Boudreau and Taylor (1989).

One major disadvantage of the ADI scheme as presented above is the use of known values of C in the reaction term. Even if the kinetics are linear, this approach will not be stable/accurate for stiff problems. Using unknown concentrations helps to fix this problem, but may create other problems if the kinetics are nonlinear.

8.10.4 Boundary Conditions

When the boundaries of the sediment region being modelled can be described as regular lines or surfaces in one of the standard coordinate systems, then boundary conditions can be dealt with in the same way as in 1-D modelling. If, however, the boundary is irregular in shape, this procedure must be altered, as described in Hunt (1978), Özisik (1980), and Noye and Arnold (1990). For complex and irregular geometries it may be far better to switch to the finite-element method (see Lapidus and Pinder, 1982).

8.11 Canned Codes for ODEs and PDEs

Practitioners of statistical analysis have long employed standard computer codes (mostly of a commercial origin) to solve their problems. Even though they are less commonly used, both public-domain and commercial software exist also exists for solving ODEs and PDEs. Such "canned" codes can save considerable time and effort if they are suited to the problem being considered. Like the statistical software, they should not be utilized in a "black-box" manner; i.e. users should be well versed in the specific solution technique utilized, along with its particular strengths and weaknesses.

There are a large number of public domain codes, and most of them are available at NETLIB (see also the Appendix). Among the ODE codes, COLSYS by Ascher et al. (1981; ACM Algorithm 569) is particularly useful in solving steady-state diagenetic problems (e.g. Boudreau, 1991). This code is based on a method called collocation (as explained in Ascher et al., 1988). It can solve mixed advection-reaction and ADR equations, and it can handle fairly stiff problems by a technique called numeric continuation. (This involves an iterative process in which the result of a less stiff problem is used as the initial starting guess for a somewhat stiffer problem, and this continued until the actual problem in question is solved. Specifically, start with a small value for the rate constant, and keep making it bigger until it reaches the actual value.)

For time-dependent 1-D ADR problems, wide use has been made of the code PDECOL (Madsen and Sincovec, 1979; ACM Algorithm 540) and now its new variant EPDCOL (Keast and Muir, 1991; ACM Algorithm 688). These codes implement finite-element collocation with the MoL for time stepping. EPDCOL is easy to set up, and examples are provided with that code; however, it cannot deal with advection-reaction equations, which limits its utility. Very recently, Blom and Zegeling (1994) have introduced a set of codes (ACM Algorithm 731) which is said to deal with systems of 1-D PDEs with steep temporal and spatial gradients. The equations need not include diffusion, but abrupt fronts may not be captured (J.M. Blom, pers. commun.). This code is too new at this time for a full appreciation of its strengths and weaknesses. Two codes, ACM Algorithm 565 (Melgaard and Sincovec, 1981) and ACM Algorithm 621 (Sommeijer and van der Houwen, 1984), have been developed to solve 2-D ADR problems, both using the MoL for time integration. All these codes are available at NETLIB.

Finally, a number of codes have been developed for modelling diagenetic and hydrological systems. Codes for diagenetic modelling have been offered by Gardner and Lerche (1987, 1990), Rabouille and Gaillard (1991a,b), Van Cappellen et al. (1993), Blackburn et al. (1994), Dhakar and Burdige (1996), van Cappellen and Wang (1995, 1996), Soetaert et al. (1996), and Boudreau (1996). Recent examples of hydrological programs include Miller and Benson (1983), Cederberg et al. (1985), Crittenden et al. (1986); Brusseau et al. (1989), Liu and Narasimhan (1989a,b), Yeh and Tripathi (1991) and Engesgaard and Kipp (1992). Some of these hydrological codes may be of use in diagenetic situations; however, hydrologists don't consider common diagenetic processes such as bioturbation, and they often assume that redox reactions are at equilibrium, which is not the case in diagenesis.

8.12 Differential-Algebraic Equations

The differential-algebraic equations (DAEs) discussed in Chapter 3 are not differential equations (Petzold, 1982a), so that some modification of the preceding schemes is necessary to obtain their numerical solution. From the point of view of a user, however, this change in procedure is not serious.

A generic problem of this type starts with a system of M differential diagenetic (transport-reaction) equations and N algebraic (equilibrium) equations, for a total of M+N species. That is to say there are M equations of the form

$$\sum_{p=1}^{M+N} \omega_p \left(\frac{\partial(C_p)}{\partial t} - \frac{\partial}{\partial x}\left[D_p \frac{\partial(C_p)}{\partial x} - C_p \right] \right) = \pm R\left(C_1, C_2, \ldots, C_{M+N}, x, t\right) \tag{8.240}$$

and N equations of the form

$$f_q\left(C_1, C_2, \ldots, C_{N+M}\right) = 0 \qquad q = N, N+1, \quad , N+M \tag{8.241}$$

where C_p is the concentration of the p-th species, D, v, and R( ) are generic diffusivity, advective velocity, and reaction kinetics, respectively, and $f_q(\,)$ is the q-th conservation equation in the form of an algebraic equation that involves any or all of the dependent variables. Here ω_p is a weighting function related to the stoichiometry of the equilibrium reactions (see the example below).

The simplest way to tackle these equations numerically is by a modified MoL. Specifically, the diagenetic PDEs are first converted to time-dependent ODEs by replacing the spatial derivatives with appropriate finite differences, i.e.

$$\sum_{p=1}^{M+N} \omega_p \left(\frac{d(C_p)_i}{dt} \right) = \sum_{p=1}^{M+N} \omega_p f\left[(C_p)_{i-1}, (C_p)_i, (C_p)_{i+1}, \Delta x, D_p, v\right] \pm R\left((C_1)_i, (C_2)_i, \ldots, (C_{N+M})_i, x_i, t\right) \tag{8.242}$$

and again

$$f_q\left(C_1, C_2, C_3, \ldots, C_{N+M}\right) = 0 \qquad q = N, N+1, \ldots, N+M \tag{8.243}$$

where i is the index of the node point in the finite-difference grid and $f[\,]$ is an appropriate finite-difference operator (i.e. formula) for the diffusion and advection terms. (While the indexing with p and q is convenient for discussing these equations, an actual finite-difference solution requires a somewhat altered indexing scheme, as illustrated below.)

In general, standard ODE integrators, such as VODE (Brown et al., 1989), cannot work with this set of equations§. Fortunately, there exist integrators designed for these equations, of which the most frequently employed is the code DASSL by Petzold (1982b). The construction of a complete solution program is nearly identical to a MoL solution for a PDE. Further ideas about the solution of transport-reaction DAEs can be found in Yeh and Tripathi (1989).

Example 8-8. To place this discussion in concrete terms, we can consider the simplest possible model for diagenesis of the dissolved carbonate system with CO_2 production (e.g. by oxic organic-matter decay). Each dissolved species diffuses by molecular diffusion, but advection is ignored and porosity is treated as constant. In addition, the CO_2 production is modelled as an exponential function of depth, and no other reactions, except species interconversion as given by Eq. (6.287), affect the system. The resulting equations for CO_2, HCO_3^-, and $CO_3^=$ are, respectively,

$$\frac{\partial [CO_2]}{\partial t} = D'_{CO2} \frac{\partial^2 [CO_2]}{\partial x^2} - R_c + R_o\, exp(-\alpha x) \tag{8.244}$$

---

§ Only if the equations are completely decoupled, which is a trivial case.

$$\frac{\partial[HCO_3]}{\partial t} = D'_{HCO3}\frac{\partial^2[HCO_3]}{\partial x^2} + 2R_c \tag{8.245}$$

$$\frac{\partial[CO_3]}{\partial t} = D'_{CO3}\frac{\partial^2[CO_3]}{\partial x^2} - R_c \tag{8.246}$$

where R_c is the net rate of species interconversion, which is assumed to be unknown. Eliminating this unknown rate from these equations, we obtain the following two equations:

$$\frac{\partial[CO_2]}{\partial t} + \frac{\partial[HCO_3]}{\partial t} + \frac{\partial[CO_3]}{\partial t}$$
$$= D'_{CO2}\frac{\partial^2[CO_2]}{\partial x^2} + D'_{HCO3}\frac{\partial^2[HCO_3]}{\partial x^2} + D'_{CO3}\frac{\partial^2[CO_3]}{\partial x^2} + R_o\, exp(-\alpha x) \tag{8.247}$$

and

$$\frac{\partial[HCO_3]}{\partial t} + 2\frac{\partial[CO_3]}{\partial t} = D'_{HCO3}\frac{\partial^2[HCO_3]}{\partial x^2} + 2D'_{CO3}\frac{\partial^2[CO_3]}{\partial x^2} \tag{8.248}$$

The first of these equations expresses conservation of total dissolved CO_2, and the second is a statement of charge (alkalinity) conservation.

There are now only two equations for three dependent variables. To remove this indeterminacy, we also assume that thermodynamic equilibrium is locally valid; i.e. the interconversion reaction is fast and reversible on the time scale of diffusion and reaction in the modelled system; thus,

$$K_{eq}[CO_2][CO_3] = [HCO_3]^2 \tag{8.249}$$

Equations (8.247), (8.248), and (8.249) constitute the example DAE system which is equivalent to generic Eqs (8.240) and (8.241). The fact that the time-derivative terms on the left-hand sides are sums, rather than single derivatives, is another feature of diagenetic DAE systems that shows that they possess a peculiar coupling which is not traditional to PDEs. (This example has an analytical solution (Boudreau, 1987), which can be used to verify the numerical solution. In general, such solutions do not exist.)

To be solved numerically, Eqs (8.247) and (8.249) need to be placed in the form of Eq. (8.241), via some choice of differencing scheme. With standard central differences on an even grid, there results

$$\left(\frac{dC_j}{dt}+\frac{dC_{j+1}}{dt}+\frac{dC_{j+2}}{dt}\right)_i$$

$$=D'_{CO2}\left(\frac{C_{j+3}-2C_j+C_{j-3}}{\Delta x^2}\right)_i+D'_{HCO3}\left(\frac{C_{j+4}-2C_{j+1}+C_{j-2}}{\Delta x^2}\right)_i \tag{8.250}$$

$$+D'_{CO3}\left(\frac{C_{j+5}-2C_{j+2}+C_{j-1}}{\Delta x^2}\right)_i+R_o\, exp(-\alpha x_i)$$

$$\left(\frac{dC_{j+1}}{dt}+2\frac{dC_{j+2}}{dt}\right)_i=$$

$$D'_{HCO3}\left(\frac{C_{j+4}-2C_{j+1}+C_{j-2}}{\Delta x^2}\right)_i+2D'_{CO3}\left(\frac{C_{j+5}-2C_{j+2}+C_{j-1}}{\Delta x^2}\right)_i \tag{8.251}$$

and

$$K_{eq}(C_j)(C_{j+2})-(C_{j+1})^2=0 \tag{8.252}$$

Here the index i indicates the node point, and the index j labels the equations. If there are i=N nodes where concentrations are unknown, then there is a total of j=M=3N equations, i.e. three species at each node. (This is the change in indexing alluded to above.)

The solution of this example using the code DASSL by Petzold (1982b) can be obtained as described in the Appendix. The user simply supplies a subroutine containing Eqs (8.250) through (8.252), at each grid point. The only complication arises from the necessity to supply initial estimates of the time derivatives of each concentration at each grid point. Formulas for these initial time derivatives can be calculated from Eqs (8.250) through (8.252) and the assumed initial distributions of CO_2, HCO_3^-, and $CO_3^=$. For example, if the initial concentration distributions are simply constants, i.e.

$$CO_2(x,0)=[CO_2]_o \tag{8.253}$$

$$HCO_3(x,0)=[HCO_3]_o \tag{8.254}$$

$$CO_3(x,0)=[CO_3]_o \tag{8.255}$$

then Eqs (8.250) and (8.251) give that

$$\left(\frac{dC_j}{dt}+\frac{dC_{j+1}}{dt}+\frac{dC_{j+2}}{dt}\right)_i^0=R_o\, exp(-\alpha x_i) \tag{8.256}$$

$$\left(\frac{dC_{j+1}}{dt}+2\frac{dC_{j+2}}{dt}\right)_i^0=0 \tag{8.257}$$

In addition, differentiation of Eq. (8.252) generates

$$\left(K_{eq}\left\{[CO_3]_o\frac{dC_j}{dt}+[CO_2]_o\frac{dC_{j+2}}{dt}\right\}-2[HCO_3]_o\frac{dC_{j+1}}{dt}\right)_i^0=0 \tag{8.258}$$

where the o superscript indicates initial values. Equations (8.256) through (8.258) are linear in the initial derivatives, and they can be solved easily to obtain,

$$\left(\frac{dC_{j+1}}{dt}\right)_i^0=-\frac{K_{eq}[CO_3]_o R_o e^{-\alpha x_i}}{\left\{\frac{K_{eq}}{2}([CO_3]_o-[CO_2]_o)+2[HCO_3]_o\right\}} \tag{8.259}$$

$$\left(\frac{dC_j}{dt}\right)_i^0=\frac{1}{2}\left(\frac{dC_{j+1}}{dt}\right)_i^0+R_o\,exp(-\alpha x_i) \tag{8.260}$$

and

$$\left(\frac{dC_{j+2}}{dt}\right)_i^0=-\frac{1}{2}\left(\frac{dC_{j+1}}{dt}\right)_i^0 \tag{8.261}$$

If calculation of the initial time derivatives appears to be complicated, DASSL will try to do this numerically, if desired. (The program also requests the so-called "bandwidth" of the system of equations if you wish to reduce the storage requirements. The bandwidth is the difference between the largest and smallest values of the index j that apply at a given i value, plus one. In the current example, the smallest j value is j-3 and the largest is j+5, which means the total bandwidth is 9. The upper part of the band is 5 and the lower is 3. For more details, see the introductory comments in DASSL and the example code from the Appendix.)

8.13 Calculating Diffusive Fluxes

An integral part of diagenetic modelling is the calculation of diffusive fluxes at interfaces, either from model results or from observational data. The following deals only with the first case.

Some numerical technique must be employed to calculate the value of a gradient at a chosen interface. (Note that the formulas that follow should not be used indiscriminately in a finite-difference solution, as discussed previously. As stated here, they are intended for one time flux calculations.) Given a set of equally spaced model-generated values, the gradient in a direction parallel to one of the main coordinates can be obtained through finite-difference formulas. For such a purpose, one should use at least a second-order accurate approximation, e.g. at the sediment-water interface (x=0),

$$\left.\frac{\partial C}{\partial x}\right|_0 = \frac{-3C_0 + 4C_1 - C_2}{2\Delta x} \tag{8.262}$$

where C_0, C_1, and C_2 are the concentration in the x-direction at x=0, the first interior grid point, and the second interior grid point, respectively. At a bottom boundary, x=L, this formula becomes,

$$\left.\frac{\partial C}{\partial x}\right|_L = \frac{-3C_N + 4C_{N-1} - C_{N-2}}{2\Delta x} \tag{8.263}$$

where N is the grid index at x=L.

Higher order approximations use even more interior points and provide, in principle, better estimates. The third and fourth-order formulas at x=0 are, respectively,

$$\left.\frac{\partial C}{\partial x}\right|_0 = \frac{-11C_0 + 18C_1 - 9C_2 + 2C_3}{6\Delta x} \tag{8.264}$$

and

$$\left.\frac{\partial C}{\partial x}\right|_0 = \frac{-25C_0 + 48C_1 - 36C_2 + 16C_3 - 3C_4}{12\Delta x} \tag{8.265}$$

Kubicek and Hlavacek (1983, p. 45) tabulate even higher order formulas. The equivalent equations for a bottom boundary can be obtained easily by analogy.

If lateral gradients are strong and/or there is a discontinuity in the boundary condition at an interface, the simple formulas suggested above may not provide an accurate estimate of the flux. It may then be necessary first to fit the concentration field (local distribution) with a multi-dimensional interpolating function, followed by differentiation to obtain the gradient (e.g. see Evans and Gourlay, 1977).

9 EPILOGUE

No book can exhaustively cover any scientific field. There are various topics related to diagenetic modelling that the present book has failed to address, if for no other reason than the author's accumulated weariness.

A prime example of such an omission is the area of so-called "inverse" models and methods; this branch of modelling includes everything from data fitting to calculating the unknown time-dependent flux of a species at the sediment-water interface based on the preserved sedimentary record after diagenesis. In this respect, I can only refer the reader to the excellent papers by Officer and Lynch (1983), Lynch and Officer (1984), Banks (1987), Schiffelbein (1985), Christensen and Goetz (1987), Christensen and Osuma (1989), and Fukumori et al. (1992) for examples of such work in a diagenetic context. (One might also wish to consult Albarède, 1995.) Furthermore, new mathematical methods are continuously being developed in applied mathematics and other branches of science, and these may very well have important applications in diagenesis, e.g. stochastic methods, such as random-walk and cellular automata simulations, or the concept of fractals (see for example Nicolis, 1995). Regrettably, these must await another monograph or a future edition of the present one.

Finally, I wish to leave you with an admonition; specifically, do not to take the results of your models **too** seriously. This may be a splash of cold water after 360 pages of models and mathematics, but models are only abstractions of nature and not reality itself. Models are a reflection of our often flawed or imperfect understanding. Discretion needs to be exercised.

> "As far as the laws of Mathematics refer to reality;
> they are not certain,
> and as far as they are certain,
> they do not refer to reality."
>
> (Attributed to Albert Einstein)

10 APPENDIX: HOW TO OBTAIN FORTRAN CODES

This appendix describes the availability of software in the form of FORTRAN computer programs that 1) implement the examples in Chapter 8; 2) can be used to calculate the special functions mentioned in Chapters 6 and 7; and 3) provide parameter values using the relations given in Chapter 4. This software resides at an electronic node (i.e. site) on the Internet. The Internet is a world-wide communication system that links computers. Access is gained by either direct connection via an Ethernet card or indirectly by being connected to a computer that is on the net. For details about gaining access to the Internet, see your computing services department in universities and industries, or ask at your local computer clubs and computer stores. Once on the Internet, you can use FTP to connect to the site. All programs at this site that are written by the author are considered to be in the public domain. As best as this author can determine, all the other programs at this site are also in the public domain.

The Internet site for the software that accompanies this text is the computer at the address mudchem.ocean.dal.ca. Log-in as "anonymous" with your own Internet address (i.e. your form of user@your.machine.name) as the password. The file structure at that software site is illustrated in Fig. A.1.

Upon entering the site, you will be in a root-like directory that presently (i.e. 1996) has four subdirectories or file folders of interest: Chapter 8, Integrators, Functions, and Utilities. The directory named Chapter 8 (i.e. /Chapter 8 in FTP) contains FORTRAN implementations of Examples 8-1 through Examples 8-5 and also Example 8-7. This directory further contains a subdirectory CANDI, where a copy of the MoL general diagenesis code CANDI.f (Boudreau, 1996) can be found. The directory Integrators contains the four main ODE, PDE, and DAE integrator codes alluded to in the text, i.e. COLSYS.f (Ascher et al., 1981a,b), DASSL.f (Petzold, 1982), EPDCOL (Keast and Muir, 1991), and VODE.f (Brown et al., 1989). The directory Functions is divided into three subdirectories, Bessel that has FORTRAN codes for various Bessel functions, Error that has codes for the Error function, and Hypergeometric for codes that calculate the confluent hypergeometric function. Finally, there is a directory Utilities that contains various FORTRAN codes to calculate the density of seawater (DENSITY.f), diffusion coefficients and kinematic viscosity (DIFCOEF.f), and so-called stoichiometric thermodynamic constants. Occasionally, new pieces of software (and perhaps directories) may be added.

Additional software can be obtained at the Internet site netlib.att.com. This is a vast repository of well-tested FORTRAN software for numerical calculations. Log-in again as "anonymous" with your own address as the password.

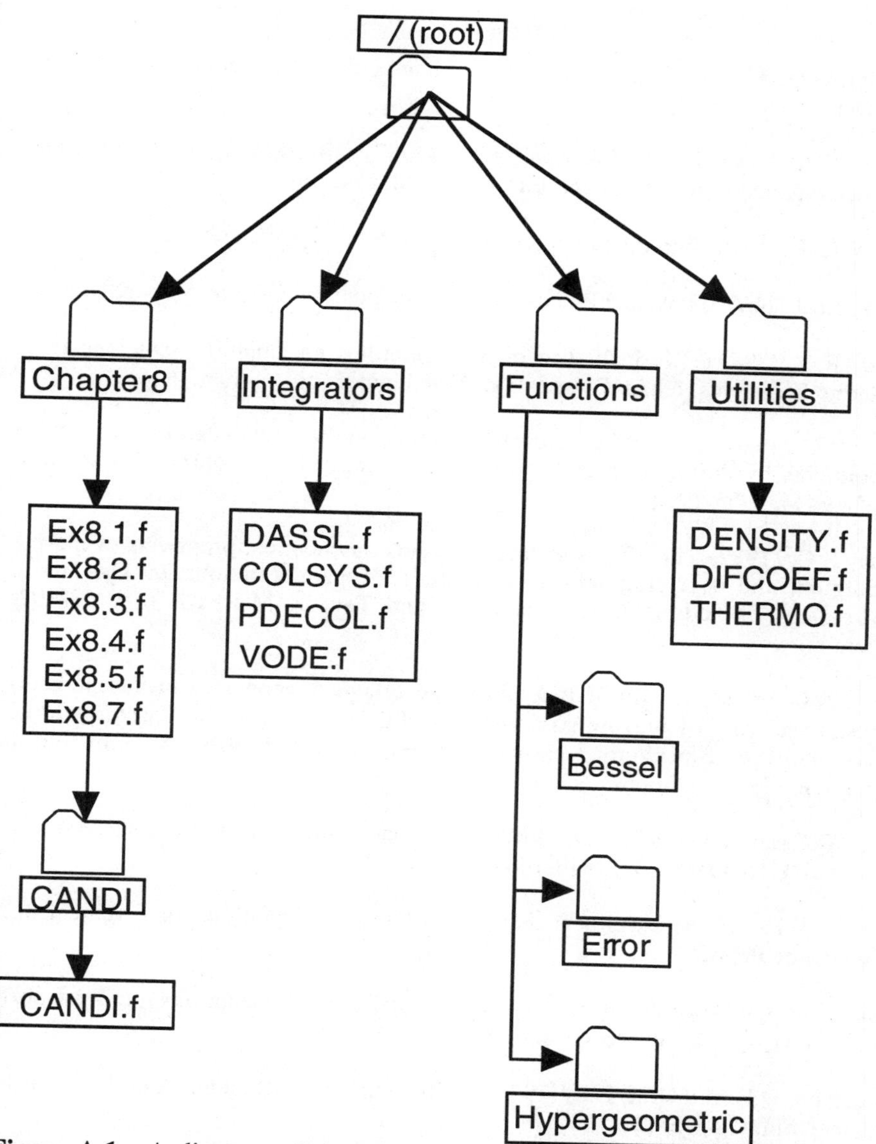

Figure A.1. A diagram of the organization of the directories and files of the Internet site mudchem.ocean.dal.ca that contains the software associated with this book.

11 REFERENCES

Abramowitz, M. and Stegun, I.A. (1972) Handbook of Mathematical Functions. Dover.

Akani, K.A., Evans J.W. and Abramson, I.S. (1987) Effective transport coefficients in heterogeneous media. Chem. Eng. Sci. v. 42, 1945-54.

Albarède, F. (1995) Introduction to Geochemical Modeling. Cambridge.

Allen, J.R.L. (1970) Physical Processes of Sedimentation. George Allen & Unwin.

Aller, R.C. (1977) The Influence of Macrobenthos on Chemical Diagenesis of Marine Sediments. Ph.D. Dissertation, Yale University.

Aller, R.C. (1980) Quantifying solute distributions in the bioturbated zone of marine sediments by defining an average microenvironment. Geochim. Cosmochim. Acta v. 44, 1955-1965.

Aller, R.C. (1982) The effects of macrobenthos on chemical properties of marine sediment and overlying water; *in* Animal-Sediment Relations, *in* The Biotic Alteration of Sediments (McCall, P.L. and Tevesz, M.J., eds.), p. 53-102. Plenum.

Aller, R.C. (1988) Benthic fauna and biogeochemical processes in marine sediments: the role of burrow structures, *in* Nitrogen Cycling in Coastal Marine Environments (Blackburn, T.H. and Sørensen, J., eds), p. 301-338. John Wiley and Sons.

Aller, R.C. and Aller, J.Y. (1992) Meiofauna and solute transport in marine muds. Limnol. Oceanogr. v. 37, 1018-1033.

Ames, W.F. (1965) Nonlinear Partial Differential Equations in Engineering. Academic Press.

Ames, W.F. (1968) Nonlinear Ordinary Differential Equations in Transport Processes. Academic Press.

Anderson, D.E. and Graf, D.L. (1976) Multicomponent diffusion. Ann. Rev. Earth Planet. Sci. v. 4, 95-121.

Applin, K.R. (1987) The diffusion of dissolved silica in dilute aqueous solution. Geochim. Cosmochim. Acta v. 51, 2147-2151.

Applin, K.R. and Lasaga, A.C. (1984) The determination of $SO_4^{=}$, $NaSO_4^-$, and $MgSO_4^0$ tracer diffusion coefficients and their application to diagenetic flux calculations. Geochim. Cosmochim. Acta v. 48, 2151-2162.

Archer, D. and Devol, A. (1992) Benthic oxygen fluxes on the Washington shelf and slope: A comparison of in situ microelectrode and chamber flux measurements. Limnol. Oceanogr. v. 37, 614-629.

Archer, D., Emerson, S. and Reimers, C. (1989) Dissolution of calcite in deep-sea sediments: pH and O_2 microelectrode results. Geochim. Cosmochim. Acta v. 53, 2831-2845.

Archie, G.E. (1942) The electrical resistivity log as an aid in determining some reservoir characteristics. Petrol. Tech. v. 1, 55-62.

Aris, R. (1968) Prolegomena to the rational analysis of systems of chemical reactions, II. Some addenda. Arch. Rational Mech. Anal. v. 27, 356-364.

Aris, R. (1989) Reactions in continuous mixtures. Amer. Inst. Chem. Eng. Jour. v. 35, 539-548.

Aris, R. and Gavalas, G.R. (1966) On the theory of reactions in continuous mixtures. Royal Soc. Phil. Trans. v. A260, 351-393.

Ascher, U., Christensen, J. and Russell, R.D. (1981) Collocation software for boundary-value ODEs. ACM Trans. Math. Software v. 7, 209-222.

Ascher, U. M. , Mattheij, R.M.M. and Russell, R.D. (1988) Numerical Solution of Boundary Value Problems for Ordinary Differential Equations. Prentice-Hall.

Atkin, R.J. and Craine, R.E. (1976) Continuum theories of mixtures: basic theory and historical development. Quart. Jour. Mech. Appl. Math. v. 29, 209-244.

Bahr, J.M. (1990) Kinetically influenced terms for solute transport affected by heterogeneous and homogeneous classical reactions. Water Resour. Res. v. 26, 21-34.

Bahr, J.M. and Rubin, J. (1987) Direct comparison of kinetic and local equilibrium formulations for solute transport affected by surface reactions. Water Resour. Res. v. 23, 438-452.

Banks, H.T. and Rose, I.G. (1987) Numerical schemes for the estimation of functional parameters in distributed models for mixing mechanisms in lake and sea sediment cores. Inverse Problems v. 3, 1-23.

Basha, H.A. and El-Habel, F.S. (1993) Analytical solution of the one-dimensional time-dependent transport equation. Water Resour. Res. v. 29, 3209-3214.

Batu, V. and van Genuchten, M. Th. (1990) First- and third-type boundary conditions in two-dimensional solute transport modeling. Water Resour. Res. v. 26, 339-350.

Bear, J. (1972) Dynamics of Fluids in Porous Media. American Elsevier.

Bear, J. and Bachmat, Y. (1992) Deletion of nondominant effects in modeling transport in porous media. Transport Porous Media v. 7, 15-38.

Beavers, G.S. and Joseph, D.D. (1967) Boundary conditions at a naturally permeable wall. Jour. Fluid Mech. v. 30, 197-207.

Bedford, A. (1978) On the balance equations for chemically reacting mixtures. Acta Mechanica v. 30, 275-282.

Bedford, A. and Drumheller, D.S. (1983) Theories of immiscible and structured mixtures. Internat. Jour. Eng. Sci. v. 21, 863-960.

Beekman, J.W. (1990) Mathematical description of heterogeneous materials. Chem. Eng. Sci. v. 45, 2603-2610.

Bender, C. M. and Orszag, S.A. (1978) Advanced Mathematical Methods for Scientists and Engineers. McGraw-Hill.

Bennett, R.H., Fischer, K.M., Lavois, D.L., Bryant, W.R. and Rezak, R. (1989) Porometry and fabric of marine clay and carbonate sediments: determinants of porosity. Mar. Geol. v. 89, 127-152.

Berger, A.E., Ciment, M. and Rogers, C.W. (1975) Numerical solution of a diffusion consumption problem with a free boundary. SIAM Jour. Numer. Anal. v. 12, 646-672.

Berner, R.A. (1968) Rate of concretion growth. Geochim. Cosmochim. Acta v. 32, 477-483.

Berner, R.A. (1971) Principles of Chemical Sedimentology. McGraw-Hill.

Berner, R.A. (1972) Chemical kinetic models of early diagenesis. Jour. Geol. Edu. v. 20, 267-272.

Berner, R.A. (1974) Kinetic models for the early diagenesis of nitrogen, sulfur, phosphorus, and silicon in anoxic marine sediments, *in* The Sea, v. 5 (Goldberg, E.D., ed.), p. 427-450. John Wiley and Sons.

Berner, R.A. (1978) Sulfate reduction and the rate of deposition of marine sediments. Earth Planet. Sci. Lett. v. 37, 492-498.

Berner, R.A. (1980) Early Diagenesis: A Theoretical Approach. Princeton Univ. Press.

Bird, R.B., Stewart, W.E. and Lightfoot, E.D. (1960) Transport Phenomena. John Wiley.

Blackburn, T.H., Blackburn, N.D., Jensen, K. and Risgaard-Petersen, N. (1994) Simulation model of the coupling between nitrification and denitrification in a freshwater sediment. Appl. Environ. Microbiol. v. 60, 3089-3095.

Blackwelder, R.F. and Haritonidis, J.H. (1983) Scaling of the bursting frequency in turbulent boundary layers. Jour. Fluid Mech. v. 132, 87-103.

Blanch, H.W. (1981) Invited review: Microbial growth kinetics. Chem. Eng. Commun. v. 8, 181-221.

Blom, J.G. and Zegeling, P.A. (1994) Algorithm 731: A moving-grid interface for systems of one-dimensional time-dependent partial differential equations. ACM Trans. Math. Software v. 20, 194-214.

Bockris, J.O'M. and Reddy, A.K.N. (1970) Modern Electrochemistry. Plenum (Rosetta).

Bosatta, E. and Ågren, G.I. (1995) The power and reactive continuum models as particular cases of the q-theory of organic matter dynamics. Geochim. Cosmochim. Acta v. 59, 3833-3835.

Boudreau, B.P. (1981) The Influence of a Diffusive Sublayer on Diagenesis at the Sea Floor. M.S. Thesis. Texas A&M University.

Boudreau, B.P. (1984) On the equivalence of nonlocal and radial-diffusion models for porewater irrigation. Jour. Mar. Res. v. 42, 731-735.

Boudreau, B.P. (1986a) Mathematics of tracer mixing in sediments: I. Spatially-dependent, diffusive mixing. Amer. Jour. Sci. v. 286, 161-198.

Boudreau, B.P. (1986b) Mathematics of tracer mixing in sediments: II. Nonlocal mixing and biological conveyor-belt phenomena. Amer. Jour. Sci. v. 286, 199-238.

Boudreau, B.P. (1987) A steady state diagenetic model for dissolved carbonate species and pH in the porewaters of oxic and suboxic sediments. Geochim. Cosmochim. Acta v. 51, 1985-1996.

Boudreau, B.P. (1989) The diffusion and telegraph equations in diagenetic modelling. Geochim. Cosmochim. Acta v. 53, 1857-1866.

Boudreau, B.P. (1990a) Asymptotic forms and solutions of the model for silica-opal diagenesis in bioturbated sediments. Jour. Geophys. Res. v. 95, 7367-7379.

Boudreau, B.P. (1990b) Modelling early diagenesis of silica in non-mixed sediments. Deep-Sea Res. v. 37, 1543-1567.

Boudreau, B.P. (1991) Modelling the sulfide-oxygen reaction and associated pH gradients in porewaters. Geochim. Cosmochim. Acta v. 55, 145-159.

Boudreau, B.P. (1992) A kinetic model for microbic organic-matter decomposition in marine sediments. FEMS Microbiol. Ecol. v. 102, 1-14.

Boudreau, B.P. (1994) Is burial velocity a master parameter for bioturbation? Geochim. Cosmochim. Acta. v. 58, 1243-1249.

Boudreau, B.P. (1996) A method-of-lines code for Carbon and nutrient diagenesis in aquatic sediments. Computers Geosci. (in press)

Boudreau, B.P. and Canfield, D.E. (1988) A provisional diagenetic model for pH in anoxic porewaters: application to the FOAM site. Jour. Mar. Res. v. 46, 429-455.

Boudreau, B.P. and Canfield, D.E. (1993) A comparison of closed- and open-system models for porewater pH and calcite dissolution. Geochim. Cosmochim. Acta v. 57, 317-334.

Boudreau, B.P. and Guinasso, N.L. (1982) The influence of a diffusive sublayer on accretion, dissolution and diagenesis at the sea floor, *in* The Dynamic Environment of the Sea Floor (Fanning, K.A. and Manheim, F.T., eds), p. 115-145. Lexington Books.

Boudreau, B.P. and Imboden, D.M. (1987) Mathematics of tracer mixing in sediments: III. The theory of nonlocal mixing within sediments. Amer. Jour. Sci. v. 286, 693-719.

Boudreau, B.P. and Marinelli, R.L. (1994) A modelling study of discontinuous irrigation. Jour. Mar. Res. v. 52, 947-968.

Boudreau, B.P. and Ruddick, B.R. (1991) On a reactive continuum representation of organic matter diagenesis. Amer. Jour. Sci. v. 291, 507-538.

Boudreau, B.P. and Scott, M.R. (1978) A model for the diffusion controlled growth of deep-sea manganese nodules. Amer. Jour. Sci. v. 278, 903-929.

Boudreau, B.P. and Taylor, R.J. (1989) A theoretical study of diagenetic concentrations fields near manganese nodules at the sediment-water interface. Jour. Geophys. Res. v. 94, 2124-2136.

Boudreau, B.P. and Westrich, J.T. (1984) The dependence of bacterial sulfate reduction on sulfate concentration in marine sediments. Geochim. Cosmochim. Acta v. 48, 2503-2516.

Bouldin, D.R. (1968) Models for describing the diffusion of oxygen and other mobile constituents across the mud-water interface. Jour. Ecol. 56, 77-87.

Boussinesq, M.J. (1877) Essai: Sur la théorie des eaux courantes. Mem. Prés. Divers Savants, Acad. Sci. Paris Ser. 2 v. 23, p.46.

Boyce, W.E. and DiPrima, R.C. (1977) Elementary Differential Equations and Boundary Value Problems. John Wiley and Sons. (Later editions are available.)

Boyle, W.C. and Berthouex, P.M. (1974) Biological wastewater treatment model building fits and misfits. Biotechnol. Bioeng. v. 16, 1139-1159.

Brady, J.B. (1975) Reference frames and diffusion coefficients. Amer. Jour. Sci. v. 275, 954-983.

Brezonik, P.L. (1994) Chemical Kinetics and Process Dynamics in Aquatic Systems. Lewis Publishers.

Brown, P.N., Byrne, G.D. and Hindmarsh, A.C. (1989) VODE, a variable-coefficient ODE solver. SIAM Jour. Sci. Stat. Comput., v. 10, p. 1038-1051.

Bruggemann, D.A.G. (1935) Berechnung verschiedener physicalischer konstaten von heterogenen substanzen. Ann. Physik v. 24, 636-664.

Brusseau, M.L., Jessup, R.E. and Rao, P.S.C. (1989) Modeling the transport of solutes influenced by multiprocess nonequilibrium. Water Resour. Res., v. 25, 1971-1988.

Bryant, W.R., Hottman, W. and Trabant, P. (1975) Permeability of unconsolidated and consolidated marine sediments, Gulf of Mexico. Mar. Geotechnol. v. 1, 1-14.

Bussian, A.E. (1983) Electrical conductance in a porous medium. Geophys. v. 48, 1258-1268.

Byrne, G.D. and Hindmarsh, A.C. (1987) Stiff ODE solvers: A review of current and coming attractions. Jour. Comput. Phys. v. 70, 1-62.

Campbell, J.B. (1979) Bessel functions $J_\nu(x)$ and $Y_\nu(x)$ of real order and real argument. Computer Physics Commun. v. 18, 133-142.

Campbell, J.B. (1981) Bessel functions $I_\nu(x)$ and $J_\nu(x)$ of real order and complex argument. Computer Physics Commun. v. 24, 97-105.

Canfield, D.E. (1989) Sulfate reduction and oxic respiration in marine sediments: implications for organic carbon preservation in euxinic environments. Deep-Sea Res. v. 36, 121-138.

Canfield, D.E. (1991) Sulfate reduction in deep-sea sediments. Amer. Jour. Sci. v. 291, 177-188.

Canfield, D.E. (1993) Organic matter oxidation in marine sediments, *in* Interactions of C, N, P and S Biogeochemical Cycles and Global Change (Wollast, R., Mackenzie, F.T. and Chou, L., eds), NATO ASI Series v. 14, p. 333-363.

Canfield, D.E. (1994) Factors influencing organic carbon preservation in marine sediments. Chem. Geol. v. 114, 315-329.

Carman, P.C. (1937) Fluid flow through a granular bed. Trans. Inst. Chem. Eng. v. 15, 150-156.

Carslaw, H.S. and Jaeger, J.C. (1959) Conduction of Heat in Solids. Oxford Univ. Press.

Cederberg, G.A., Street, R.L. and Leckie, J.O. (1985) A groundwater mass transport and equilibrium chemistry model for multicomponent systems. Water Resour. Res. v. 21, 1095-1104.

Chalabi, A. (1992) Stable upwind schemes for hyperbolic conservation laws with source terms. IMA Jour. Numer. Anal. v. 12, 217-241.

Chang, C.-S. and Rochelle, G.T. (1982) Mass transfer enhanced by equilibrium reactions. Indus. Eng. Chem. Fundam. v. 21, 379-385.

Chellam, S., Wiesner, M.R. and Dawson, C. (1992) Slip at a uniformly porous boundary: effect on fluid flow and mass transfer. Jour. Eng. Math. v. 26, 481-492.

Chiu, S.Y., Erickson, L.E., Fan, L.T. and Kao, I.C. (1972) Kinetic model identification in mixed populations using continuous culture data. Biotechnol. Bioeng. v. 14, 207-231.

Christensen, E.R. (1982) A model for radionuclides in sediments influenced by mixing and compaction. Jour. Geophys. Res. v. 87, 566-572.

Christensen, E.R. and Goetz, R.H. (1987) Historical fluxes of particle-bound pollutants from deconvolved sedimentary records. Env. Sci. Tech. v. 21, 1088-1096.

Christensen, E.R. and Osuna, J. L. (1989) Atmospheric fluxes of Lead, Zinc, and Cadmium from frequency domain deconvolution of sedimentary records. Jour. Geophys. Res. v. 94, 14585-14597.

Christensen, J.P., Smethie, W.M. and Devol, A.H. (1987) Benthic nutrient regeneration and denitrification on the Washington continental shelf. Deep-Sea Res. v. 34, 1027-1047.

Clukey, E.C. and Silva, A.J. (1981) Permeability of deep-sea clays: Northwestern Atlantic. Mar. Geotechnol. v. 5, 1-26.

Cody, W.J. (1983) Algorithm 597: Sequence of modified Bessel functions of the first kind. ACM Trans. Math. Software v. 9, 242-245.

Cole, J.D. (1968) Perturbation Methods in Applied Mathematics. Blaisdell.

Conte, S.D. and de Boor, C. (1972) Elementary Numerical Analysis. McGraw-Hill.

Corino, E.R. and Brodkey, R.S. (1969) A visual investigation of the wall region in turbulent flow. Jour. Fluid Mech. v. 37, 1-30.

Cornel, P.K., Summers, R.S. and Roberts, P.V. (1986) Diffusion of humic acid in dilute aqueous solution. Jour. Colloid Interface Sci. v. 110, 149-164.

Crank, J. (1975) The Mathematics of Diffusion. Oxford Univ. Press.

Crank, J. (1984) Free and Moving Boundary Problems. Clarendon Press.

CRC (1979) Standard Mathematical Tables (Beyer, W.H., ed.) CRC Press.

CRC (1993) Handbook of Chemistry and Physics. CRC Press.

Crittenden, J.C., Hutzler, N.J., Geyer, D.G., Oravitz, J.L. and Friedman, G. (1986) Transport of organic compounds with saturated groundwater flow: model development and parameter sensitivity. Water Resour. Res. v. 22, 271-284.

Cussler, E.L. (1984) Diffusion. Mass Transfer in Fluid Systems. Cambridge Univ. Press.

Dabes, J.N., Finn, R.K. and Wilke, C.R. (1973) Equations of substrate-limited growth: the case for Blackman kinetics. Biotechnol. Bioeng. v. 15, 1159-1177.

Dade, W.B. (1993) Near-bed turbulence and hydrodynamic control of diffusional mass transfer at the sea floor. Limnol. Oceanogr. v. 38, 52-69.

Danckwerts, P.V. (1951) Absorption by simultaneous diffusion and chemical reaction into particles of various shapes and into falling drops. Trans. Faraday Soc. v. 47, 1014-1023.

Davies, J.T. (1977) Diffusion and heat transfer at the boundaries of turbulent liquids, *in* Physico-Chemical Hydrodynamics (Spalding, D.B., ed.), p. 3-22. Advance Publishers.

Davis, P.J. and Rabinowitz, P. (1984) Methods of Numerical Integration. 2nd Edition. Academic Press.

Dawson, D.A. and Trass, O. (1972) Mass transfer at rough surfaces. Internat. Jour. Heat Mass Transfer v. 15, 1317-1336.

DeCoursey, W.J. and Thring, R.W. (1989) Effects of unequal diffusivities on enhancement factors for reversible and irreversible reactions. Chem. Eng. Sci. v. 44, 1715-1721.

Deemer, A.R. and Slattery, J.C. (1978) Balance equations and structural models for phase interfaces. Internat. Jour. Multiphase Flow v. 4, 171-192.

de Groot, S.R. and Mazur, P. (1984) Non-Equilibrium Thermodynamics. Dover.

de Ligny, C.L. (1970) Coupling between diffusion and convection in radial dispersion of matter by flow through packed beds. Chem. Eng. Sci. v. 25, 1177-1181.

Demiray, H. (1981) A continuum theory of chemically reacting mixtures of fluids and solids. Internat. Jour. Eng. Sci. v. 19, 253-268.

Dhakar, S.P. and Burdige, D.J. (1996) A coupled, non-linear, steady state model for early diagenetic processes in pelagic sediments. Amer. Jour. Sci. v. 296, 296-330.

Do, D.D. (1984) Determination of boundary conditions for zero-order chemical reaction inside a single catalyst pellet under transient condition. Chem. Eng. Commun. v. 25, 251-265.

Dobos, D. (1975) Electrochemical Data. Elsevier.

Domenico, P.A. (1977) Transport phenomena in chemical rates processes in sediments. Ann. Rev. Earth Planet. Sci. v. 5, 287-317.

Domenico, P.A. and Palciauskas, V.V. (1979) The volume-averaged mass-transport equation for chemical diagenetic models. Jour. Hydrol. v. 43, 427-438.

Drumheller, D.S. (1978) The theoretical treatment of a porous solid using a mixture theory. Internat. Jour. Solids Structures v. 14, 441-456.

Drumheller, D.S. and Bedford, A. (1980) A thermomechanical theory for reacting immiscible mixtures. Arch. Rat. Mech. v. 73, 257-284.

Dullien, F.A.L. (1992) Porous Media: Fluid Transport and Pore Structure. 2nd Edition. Academic Press.

Duursma, E.K. and Smies, M. (1982) Sediments and transfer at and in the bottom interfacial layer, *in* Pollutant Transfer and Transport in the Sea v. II (Kullenberg, G., ed.), p. 101-139. CRC Press.

Dykhuizen, R.C. and Casey, W.H. (1989) An analysis of solute diffusion in rocks. Geochim. Cosmochim. Acta v. 53, 2797-2805.

Easteal, A.J., Price, W.E. and Woolf, L.A. (1989) Diaphragm cell for high-temperature diffusion measurements. Jour. Chem. Soc., Faraday Trans. 1 v. 85, 1091-1097.

Einstein, A. (1906, reprinted 1956) On the theory of Brownian movement, *in* Investigations on the Theory of Brownian Movement (Furth, R., ed., Cowper, A.D., transl.), p. 19-35. Dover.

Emerson, S. (1985) Organic carbon preservation in marine sediments, *in* The Carbon Cycle and Atmospheric CO_2: Natural Variations Archean to Present (Sundquist, E., and Broecker, W.S., eds), Amer. Geophys. Union Monogr. no. 32, p. 78-87.

Emerson, S., Fischer, K., Reimers, C. and Heggie, D. (1985) Organic carbon dynamics and preservation in deep-sea sediments. Deep-Sea Res. v. 32, 1-21.

Emerson, S., Jahnke, R. and Heggie, D. (1984) Sediment-water exchange in shallow water estuarine sediments. Jour. Mar. Res. v. 42, 709-730.

Engesgaard, P. and Kipp, K.L. (1992) A geochemical transport model for redox-controlled movement of mineral fronts in groundwater flow systems: a case of nitrate removal by oxidation of pyrite. Water Resour. Res. v. 28, 2829-2843.

Epstein, N. (1989) On tortuosity and the tortuosity factor in flow and diffusion through porous media. Chem. Eng. Sci. v. 44, 777-779.

Erdélyi, A. (1953) Higher Transcendental Functions, v. 1, 2 and 3. McGraw-Hill.

Erdélyi, A. (1954) Tables of Integral Transforms, v. 1 and 2. McGraw-Hill.

Evans, N.T.S. and Gourlay, A.R. (1977) The solution of a two-dimensional time-dependent diffusion problem concerned with oxygen metabolism in tissues. Jour. Inst. Math. Appl. v. 19, 239-251.

Farlow, S.J. (1993) Partial Differential Equations for Scientists and Engineers. Dover.

Felmy, A.R. and Weare, J.H. (1991a) Calculation of multicomponent ionic diffusion from zero to high concentration: I. The system Na-K-Ca-Mg-Cl-SO_4-H_2O at 25°C. Geochim. Cosmochim. Acta v. 55, 113-131.

Felmy, A.R. and Weare, J.H. (1991b) Calculation of multicomponent ionic diffusion from zero to high concentration: II. Inclusion of associated ion species. Geochim. Cosmochim. Acta v. 55, 133-144.

Fiadeiro, M.E. and Veronis, G. (1977) On weighted-mean schemes for the finite-difference approximation to the advection-diffusion equation. Tellus v. 29, 512-522.

Fick, A. (1855) Uber diffusion. Poggendorff's Annalen der Physik v. 94, 58-86.

Fisher, J.B., Lick, W.J., McCall, P.L. and Robbins, J.A. (1980) Vertical mixing of lake sediments by tubificid oligochaetes. Jour. Geophys. Res. v. 85, 3997-4006.

Fornberg, B. (1988) Generation of finite difference formulas on arbitrarily spaced grids. Math. Comput. v. 51, 699-706.

Frank-Kamenetskii, D.A. (1969) Diffusion and Heat Transfer in Chemical Kinetics. Plenum Press.

Frey, R.W., Howard, J.D. and Pryor, W.A. (1978) Ophiomorpha: Its morphologic, taximetric, and environmental significance. Paleogeogr. Paleoclimat. Paleoecol. v. 23, 199-229.

Fukumori, E., Christensen, E.R. and Klein, R. (1992) A model for ^{137}Cs and other tracers in lake sediments considering particle size and the inverse solution. Earth Planet. Sci. Lett. v. 114, 85-99.

Gallagher, P.M., Athayde, A.L. and Ivory, C.F. (1986) The combined flux technique for diffusion-reaction problems in partial equilibrium: application to the facilitated transport of carbon dioxide in aqueous bicarbonate solutions. Chem. Eng. Sci. v. 41, 567-578.

Gardner, L.R. and Lerche, I. (1987) Simulation of sulfate-dependent sulfate reduction using Monod kinetics. Jour. Math. Geol. v. 19, 219-239.

Gardner, L.R. and Lerche, I. (1990) Simulation of sulfur diagenesis in anoxic marine sediments using Rickard kinetics for FeS and FeS_2 formation. Computers Geosci. v. 16, 441-460.

Gardner, L.R., Sharma, P. and Moore, W.S. (1987) A regeneration model for the effect of bioturbation by fiddler crabs on ^{210}Pb profiles in salt marsh sediments. Jour. Env. Radioact. v. 5, 25-36.

Glaister, P. (1993) Second order accurate upwind difference schemes for scalar conservation laws with source terms. Computers Math. Appl. v. 25, 65-73.

Goddard, J.D. (1990) Consequences of the partial equilibrium approximation for chemical reaction and transport. Proc. Royal Soc. London v. A432, 271-284.

Goldhaber, M.B., Aller, R. Cochran, J.K., Rosenfeld, J., Martens, C. and Berner, R.A. (1977) Sulfate reduction, diffusion, and bioturbation in Long Island Sound sediments: report of the FOAM Group. Amer. Jour. Sci. v. 277, 193-237.

Gradshteyn, I.S. and Ryzhik, I.M. (1980) Table of Integrals, Series, and Products (Jeffrey, A., ed.). Academic Press.

Gray, W.G. (1975) A derivation of the equations for multi-phase transport. Chem. Eng. Sci. v. 30, 229-233.

Gray, W.G. and Hassanizadeh, S.M. (1989) Averaging theorems and averaged equa-tions for transport of interface properties in multiphase systems. Internat. Jour. Multiphase Flow v. 15, 81-95.

Gray, W.G., Leijnse, A., Kolar, R.L. and Blain, C.A. (1993) Mathematical Tools for Changing Spatial Scales in the Analysis of Physical Systems. CRC Press.

Grimanis, M. and Abedian, B. (1985) Turbulent mass transfer in rough tubes at high Schmidt numbers. Physico-Chem. Hydrodyn. v. 6, 75-78.

Grundmanis, V. and Murray, J.W. (1982) Aerobic respiration in pelagic sediments. Geochim. Cosmochim. Acta v. 46, 1101-1120.

Guinasso, N.L., Jr. and Schink, D.R. (1975) Quantitative estimates of biological mixing rates in abyssal sediments. Jour. Geophys. Res. v. 80, 3032-3043.

Gundersen, J.K. and Jørgensen, B.B. (1990) Microstructure of diffusive boundary layers and the oxygen uptake of the sea floor. Nature v. 345, 604-607.

Haber, S. and Mauri, R. (1983) Boundary conditions for the Darcy's flow through porous media. Internat. Jour. Multiphase Flow v. 9, 561-574.

Hammond, D.E. and Fuller, C. (1979) The use of radon-222 to estimate benthic exchange and atmospheric exchange rates in San Francisco Bay, *in* San Francisco Bay: The Urbanized Estuary, p. 213-230. Pacific Division, Amer. Assoc. Advance. Sci.

Harrison, W.D., Musgrave, D. and Reeburgh, W.S. (1983) A wave-induced transport process in marine sediments. Jour. Geophys. Res. v. 88, 7617-7622.

Hassanizadeh, S.M. (1986) Derivation of basic equations of mass transport in porous media, Part 1. Macroscopic balance laws. Adv. Water Resour. v. 9, 196-206.

Hassanizadeh, S.M. and Gray, W.G. (1989a) Derivation of conditions describing transport across zones of reduced dynamics within multiphase systems. Water Resour. Res. v. 25, 529-539.

Hassanizadeh, S.M. and Gray, W.G. (1989b) Boundary and interface conditions in porous media. Water Resour. Res. v. 25, 1705-1715.

Hayduk, W. and Laudie, H. (1974) Prediction of diffusion coefficients for nonelectrolytes in dilute aqueous solutions. Amer. Inst. Chem. Eng. Jour. v. 20, 611-615.

Heggie, D., Maris, C., Hudson, A., Dymond, J., Beach, R. and Cullen, J. (1987) Organic carbon oxidation and preservation in NW Atlantic continental margin sediments, *in* Geology and Geochemistry of Abyssal Plains (Weaver, P.P.E. and Thomson, J., eds.), Geol. Soc. Spec. Publ. No. 31, p. 215-236.

Hemker, P.W. (1977) A Numerical Study of Stiff Two-Point Boundary Problems. Mathematical Centre Tracts 80. Mathematisch Centrum, Amsterdam.

Henrichs, S.M. and Reeburgh, W.S. (1987) Anaerobic mineralization of marine organic matter: rates and role of anaerobic processes in the oceanic carbon economy. Geomicrobiol. Jour. v. 5, 191-237.

Hershey, D. (1973) Transport Analysis. Plenum (Rosetta).

Hildebrand, F.B. (1974) Introduction to Numerical Analysis. 2nd Edition. Dover.

Hildebrand, F.B. (1976) Advanced Calculus for Applications. 2nd Edition. Prentice-Hall, Inc.

Himmelblau, D.M. (1964) Diffusion of dissolved gases in liquids. Chem. Rev. v. 64, 527-550.

Ho, C.S., Ju, L.-K., Baddour, R.F. and Wang, D.I.C. (1988) Simultaneous measurement of oxygen diffusion coefficients and solubilities in electrolyte solutions with a polarographic oxygen electrode. Chem. Eng. Sci. v. 43, 3093-3107.

Ho, F.-G. and Strieder, W. (1981) A variational calculation of the effective surface diffusion coefficient and tortuosity. Chem. Eng. Sci. v. 36, 253-258.

Ho, T.C. and Aris, R. (1987) On apparent second-order kinetics. Amer. Inst. Chem. Eng. Jour. v. 33, 1050-1051.

Hsu, C.T. and Cheng, P. (1991) A singular perturbation solution for Couette flow over a semi-infinite porous bed. Jour. Fluid Eng. v. 113, 137-142.

Humphrey, A.E. (1972) The kinetics of biosystems: a review, *in* Chemical Reaction Engineering (Gould, R.F., ed.), ACS Adv. Chemistry Ser. no. 109, p. 630-650.

Hunt, B. (1978) Finite difference approximation of boundary conditions along irregular boundaries. Internat. Jour. Numer. Meth. Eng. v. 12, 229-235.

Huntley, E. (1986) Accurate finite difference approximations for the solution of parabolic partial differential equations by semi-discretization. Internat. Jour. Numer. Methods Eng. v. 23, 2325-2346.

Hutchinson, P. and Luss, D. (1970) Lumping of mixtures with many parallel first order reactions. Chem. Eng. Jour. v. 1, 129-135.

Imboden, D.M. (1975) Interstitial transport of solutes in non-steady state accumulating and compacting sediments. Earth Planet. Sci. Lett. v. 27, 221-228.

Ince, E.L. (1956) Ordinary Differential Equations. Dover.

Isenberg, J. and de Vahl Davis, G. (1975) Finite difference methods in heat and mass transfer, *in* Topics in Transport Phenomena (Gutfinger, C., ed.), p. 457-553. John Wiley and Sons.

Iversen, N. and Jørgensen, B.B. (1993) Diffusion coefficients of sulfate and methane in marine sediments: influence of porosity. Geochim. Cosmochim. Acta v. 57, 571-578.

Jähne, B., Heinz, G. and Dietrich, W. (1987) Measurement of the diffusion coefficients of sparingly soluble gases in water. Jour. Geophys. Res. v. 92, 10767-10776.

Jahnke, R.A. (1985) A model of microenvironments in deep-sea sediments: formation and effects on porewater profiles. Limnol. Oceanogr. v. 30, 956-965.

Jahnke, R.A., Heggie, D., Emerson, S. and Grundmanis (1982) Pore water of the central Pacific Ocean: nutrient results. Earth Planet. Sci. Lett. v. 61, 233-256.

Jahnke, R.A. and Jackson, G.A. (1992) The spatial distribution of sea floor oxygen consumption in the Atlantic and Pacific Oceans, *in* Deep-Sea Food Chains and the Global Carbon Cycle (Rowe, G.T. and Pariente, V., eds), p. 295-307. Kluwer.

Jennings, A.A. (1987) Critical chemical reaction rates for multicomponent groundwater contamination models. Water Resour. Res. v. 23, 1775-1784.

Johnson, K.S. (1981) The calculation of ion pair diffusion coefficients: a comment. Mar. Chem. v. 10, 195-208.

Jørgensen, B.B. (1977) Bacterial sulfate reduction within reduced microniches of oxidized marine sediments. Mar. Biol. v. 41, 7-17.

Jørgensen, B.B. (1978) A comparison of methods for the quantification of bacterial sulfate reduction in coastal marine sediments. II. Calculations from mathematical models. Geomicrobiol. Jour. v. 1, 29-47.

Jørgensen, B.B. and Des Marais, D.J. (1990) The diffusive boundary layer of sediments: oxygen microgradients over a microbial mat. Limnol. Oceanogr. v. 35, 1343-1355.

Jørgensen, B.B. and Revsbech, N.P. (1985) Diffusive boundary layers and the oxygen uptake of sediments. Limnol. Oceanogr. v. 30, 111-122.

Jost, W. (1964) Fundamental aspects of diffusion processes. Angew. Chem. Internat. Edit. v. 3, 713-722.

Jumars, P.A. and Wheatcroft, R.A. (1989) Responses of benthos to changing food quality and quantity, with a focus on deposit feeding and bioturbation, *in* Productivity of the Oceans: Present and Past (W.H. Berger et al., eds.), p. 235-253. John Wiley and Sons.

Kahaner, D., Moler, C. and Nash, S. (1989) Numerical Methods and Software. Prentice-Hall.

Kaluarachchi, J. and Morshed, J. (1995) Critical assessment of the operator-splitting technique in solving the advection-dispersion-reaction equation: 1. First-order reaction. Adv. Water Resour. v. 18, 89-100.

Kamke, E. (1948) Differentialgleichugen, Losungunsmethoden und Losungen, vol. 1, Gewohnliche Differentialgleichugen. Chelsea Publ. Co.

Kaplan, W. (1973) Advanced Calculus. Addison-Wesley.

Katz, A. and Ben-Yaakov, S. (1980) Diffusion of seawater ions. Part II. The role of activity coefficients and ion paring. Mar. Chem. v. 8, 263-280.

Keast, P. and Muir, P.H. (1991) EPDCOL: A more efficient PDECOL code. ACM Trans. Math. Software v. 17, 153-166.

Keir, R. (1982) Dissolution of calcite in the deep-sea: theoretical prediction for the case of uniform size particles settling into a well-mixed sediment. Amer. Jour. Sci. v. 282, 193-236.

Kennett, J.P. (1982) Marine Geology. Prentice-Hall.

Khaliq, A.Q.M. and Twizell, E.H. (1984) Backward difference replacements of the space derivative in first-order hyperbolic equations. Computer Methods Appl. Mech. Eng. v. 43, 45-56.

Kirkwood, J.G., Baldwin, R.L., Dunlop, P.J., Gosting, L.J. and Kegeles, G. (1960) Flow equations and frames of reference for isothermal diffusion in liquids. Jour. Chem. Phys. v. 33, 1505-1513.

Kirwan, A.D. and Kump, L.R. (1987) Models of geochemical systems from mixture theory: diffusion. Geochim. Cosmochim. Acta v. 51, 1219-1226.

Klump, J.V. and Martens, C.S. (1987) Biogeochemical cycling in an organic-rich coastal marine basin. 5. Sedimentary nitrogen and phosphorus budgets based upon kinetic models, mass balances, and the stoichiometry of nutrient regeneration. Geochim. Cosmochim. Acta v. 51, 1161-1173.

Knapp, R.B. (1989) Spatial and temporal scales of local equilibrium in dynamic fluid-rock systems. Geochim. Cosmochim. Acta v. 53, 1955-1964.

Krom, M.D. and Berner, R.A. (1980) The diffusion coefficients of sulfate, ammonia, and phosphate ions in anoxic marine sediments. Limnol. Oceangr. v. 25, 327-337.

Kubicek, M. and Hlavacek, V. (1983) Numerical Solution of Nonlinear Boundary Value Problems with Applications. Prentice-Hall.

Kukulka, D.J., Gebhart, B. and Mollendorf, J.C. (1987) Thermodynamic and transport properties of pure and saline water. Adv. Heat Transfer v. 18, 325-363.

Lambert, J.D. (1973) Computational Methods in Ordinary Differential Equations. John Wiley and Sons.

Landahl, M.T. and Mollo-Christensen, E. (1986) Turbulence and Random Processes in Fluid Mechanics. Cambridge University Press.

Lankin, W.D. and Sanchez, D.A. (1970) Topics in Ordinary Differential Equations. Dover.

Lapidus, L. and Pinder, G.F. (1982) Numerical Solution of Partial Differential Equations in Science and Engineering. Wiley.

Lasaga, A.C. (1979) The treatment of multi-component diffusion and ion pairs in diagenetic fluxes. Amer. Jour. Sci. v. 279, 324-346.

Lasaga, A.C. (1981) Rate laws of chemical reactions, *in* Kinetics of Geochemical Processes (Lasaga, A.C. and Kirkpatrick, R.J., eds), Reviews in Mineralogy v. 8, p. 1-86. Mineralogical Society of America.

Lasaga, A.C. and Holland, H.D. (1976) Mathematical aspects of non-steady-state diagenesis. Geochim. Cosmochim. Acta v. 40, 257-266.

Lau, P.C.M. (1981) Finite difference approximation for ordinary derivatives. Internat. Jour. Numer. Methods Eng. v. 17, 663-678.

Leaist, D.G. (1987) Diffusion of aqueous carbon dioxide, sulfur dioxide, and ammonia at very low concentrations. Jour. Phys. Chem. v. 81, 4635-4638.

Lebedev, N.N. (1972) Special Functions and their Applications. (Silverman, R.A., transl.) Dover.

Leij, F.J. and Dane, J.H. (1990) Analytical solutions of the one-dimensional advection equation and two- or three-dimensional dispersion equation. Water Resour. Res. v. 26, 1475-1482.

Leij, F.J., Skaggs, T.H. and van Genuchten, M.Th. (1991) Analytical solution for solute transport in three-dimensional semi-infinite porous media. Water Resour. Res. v. 27, 2719-2733.

Lerman, A. (1975) Maintenance of steady state in oceanic sediments. Amer. Jour. Sci. v. 275, 609-635.

Lerman, A. (1977) Migrational processes and chemical reactions in interstitial waters, *in* The Sea: Ideas and Observations on Progress in the Study of the Sea

(Goldberg, E.D., McCave, I.N., O'Brien and Steele, J.H., eds), v. 6, p. 695-738. Macmillan.

Lerman, A. (1978) Chemical exchange across sediment-water interface. Ann. Rev. Earth Planet. Sci. v. 6, 281-303.

Lerman, A. (1979) Geochemical Processes: Water and Sediment Environments. Wiley Interscience.

Lerman, A. and Jones, B.F. (1973) Transient and steady-state salt transport between sediments and brine in saline lakes. Limnol. Oceanogr. v. 18, 75-85.

Lerman, A. and Weiler, R.R. (1970) Diffusion and accumulation of chloride and sodium in Lake Ontario sediments. Earth Planet. Sci. Lett. v. 10, 150-156.

LeVeque, R.J. and Yee, H.C. (1990) A study of numerical methods for hyperbolic conservation laws with stiff source terms. Jour. Comput. Physics v. 86, 187-210.

Li, Y.-H., and Gregory, S. (1974) Diffusion of ions in seawater and in deep-sea sediments. Geochim. Cosmochim. Acta v. 38, 703-714.

Lichtner, P.C. (1993) Scaling properties of time-space kinetic mass transport equations and the local equilibrium limit. Amer. Jour. Sci. v. 293, 257-296.

Lick, W.J. (1989) Difference Equations from Differential Equations. Lecture Notes in Engineering v. 41. Springer-Verlag.

Lin, C.C. and Segel, L.A. (1974) Mathematics Applied to Deterministic Problems in the Natural Sciences. Macmillan Publishing Co. (Now reprinted by SIAM.)

Lin, C.S., Moulton, R.W. and Putnam, G.L. (1953) Mass transfer between solid wall and fluid streams. Industr. Eng. Chem. v. 45, 636-646.

Liu, C.W. and Narasimhan, T.N. (1989a) Redox-controlled multiple species reactive chemical transport. 1. Model development. Water Resour. Res. v. 25, 869-882.

Liu, C.W. and Narasimhan, T.N. (1989b) Redox-controlled multiple species reactive chemical transport. 2. Verification and application. Water Resour. Res. v. 25, 883-910.

Liu, P.L.-F. (1979) Permeable wall effects on Poiseuille flow. ASCE Jour. Eng. Mech. v. 105, 470-476.

Lobo, V.M.M. (1993) Mutual diffusion coefficients in aqueous electrolyte solutions. Pure Appl. Chem. v. 65, 2613-2640.

Loureiro, J.M. and Rodrigues, A.E. (1991) Two solution methods for hyperbolic systems of partial differential equations in chemical engineering. Chem. Eng. Sci. v. 46, 3259-3267.

Low, P.F. (1981) Principles of ion diffusion in clays, *in* Chemistry in the Soil Environment, Amer. Soc. Agronomy Spec. Publ. no. 40, p. 31-45.

Luke, Y.L. (1977) Algorithms for the Computation of Mathematical Functions. Academic Press.

Lynch, D.R. and Officer, C.B. (1984) Nonlinear parameter estimation for sediment cores. Chem. Geol. v. 44, 203-225.

Mackie, J.S. and Meares, P. (1955) The diffusion of electrolytes in a cation-exchange resin membrane. Proc. Royal Soc. v. A232, 498-509.

Mackin, J.E. (1986) The free-solution diffusion coefficient of Boron: influence of dissolved organic matter. Mar. Chem. v. 20, 131-140.

Mackin, J.E. and Aller, R.C. (1983) The infinite dilution diffusion coefficient for $Al(OH)_4^-$ at 25°C. Geochim. Cosmochim. Acta v. 47, 959-961.

Madsen, N.K. and Sincovec, R.F. (1979) Algorithm 540: PDECOL, general collocation software for partial differential equations. ACM Trans. Math. Software v. 5, 326-351.

Manheim, F.T. and Waterman, L.S. (1974) Diffusimetry (diffusion constant estimation) on sediment cores by resistivity probe, *in* Initial Reports of the Deep Sea Drilling Project (von der Borch, C.C. and Sclater, G.C., eds), v. 22, p. 663-670. U.S. Printing Office.

Manteuffel, T.A. and White, A.B. (1986) The numerical solution of second-order boundary value problems on nonuniform meshes. Math. Comput. v. 47, 511-535.

Marchuk, G.I. (1990) Splitting and alternating direction methods, *in* Handbook of Numerical Analysis (Ciarlet, P.G. and Lions, J.L., eds.), v. 1, Finite Difference Methods, p. 198-462. North Holland.

Marle, C.M. (1982) On macroscopic equations governing multiphase flow with diffusion and chemical reaction in porous media. Intern. Jour. Eng. Sci. v. 20, 643-662.

Martens, C.S. and Klump, J.V. (1984) Biogeochemical cycling in an organic-rich coastal marine basin. 4. An organic carbon budget for sediments dominated by sulfate reduction and methanogenesis. Geochim. Cosmochim. Acta v. 48, 1987-2004.

Martin, M. (1989) The source solution for diffusion with a linearly position dependent diffusion coefficient. Zeitschr. Physik. Chemie Neue Folge v. 162, 245-253.

Martin, W.R. and Banta, G.T. (1992) The measurement of sediment irrigation rates: a comparison of the Br^- tracer and $^{222}Rn/^{226}Ra$ disequilibrium techniques. Jour. Mar. Res. v. 50, 125-154.

Martin, W.R. and Bender, M.L. (1988) The variability of benthic fluxes and sedimentary remineralization rates in response to seasonally variable organic carbon rain rates in the deep sea: a modeling study. Amer. Jour. Sci. v. 288, 561-574.

Martin, W.R. and Sayles, F.L. (1987) Seasonal cycles of particulate and solute trans-port processes in nearshore sediments: $^{222}Rn/^{226}Ra$ and $^{234}Th/^{238}U$ disequili-brium at a site in Buzzards Bay, MA. Geochim. Cosmochim. Acta v. 51, 927-943.

Matisoff, G. (1980) Time dependent transport in Chesapeake Bay sediments: Part 1. Temperature and chloride. Amer. Jour. Sci. v. 280, 1-25.

Maxwell, J.C. (1881) Treatise on Electricity and Magnetism. 2nd Edition. Clarendon Press.

McClain, C.R., Huang, N.E. and Pietrafesa, L.J. (1977) Application of a "radiation-type" boundary condition to the wave, porous bed problem. Jour. Phys. Oceanogr. v. 7, 823-835.

McDuff, R.E. and Ellis, R.A. (1979) Determining diffusion coefficients in marine sediments: a laboratory study of the validity of resistivity techniques. Amer. Jour. Sci. v. 279, 666-675.

Meldon, J.H., Stroeve, P. and Gregoire, C.E. (1982) Facilitated transport of Carbon Dioxide. A Review. Chem. Eng. Commun. v. 16, 263-300.

Melgaard, D.K. and Sincovec, R.F. (1981) Algorithm 565: PDETWO/PSETM/ GEARB, Solution of systems of two-dimensional nonlinear partial differential equations. ACM Trans. Math. Software v. 7, 126-135.

Middelburg, J.J. (1989) A simple rate model for organic matter decomposition in marine sediments. Geochim. Cosmochim. Acta v. 53, 1577-1581.

Middelburg, J.J. and De Lange, G.J. (1989) The isolation of Kau Bay during the last glaciation: direct evidence from interstitial water chlorinity. Neth. Jour. Sea Res. v. 24, 615-622.

Middelburg, J.J., Soetaert, K. and Herman, P.M.J. (in press) Empirical relationships for use in global diagenetic models. Deep-Sea Res.

Mikhailov, M.D. and Özisik, M.N. (1984) Unified Analysis and Solutions of Heat and Mass Diffusion. Dover.

Miller, C.W., and Benson, L.V. (1983) Simulation of solute transport in a chemically reactive heterogeneous system: model development and application. Water Resour. Res. v. 19, 381-391.

Miller, D.G. (1982) Estimation of tracer diffusion coefficients of ions in aqueous solution. Lawrence Livermore Laboratory Report UCRL-53319.

Millero, F.J. (1995) Thermodynamics of the carbon dioxide system in the oceans. Geochim. Cosmochim. Acta v. 59, 661-677.

Millington, R.J. (1959) Gas diffusion in porous media. Science v. 130, 100-102.

Mitchell, A.R. and Griffiths, D.F. (1980) The Finite Difference Method in Partial Differential Equations. John Wiley and Sons.

Moré, J.J. and Cosnard, M.Y. (1980) Algorithm 554: BRENTM, A fortran subroutine for the numerical solution of systems of nonlinear equations. ACM Trans. Math. Software v. 6, 240-251.

Morshed, J. and Kaluarachchi, J.J. (1995) Critical assessment of the operator-splitting technique in solving the advection-dispersion-reaction equation: 2. Monod kinetics and coupled transport. Adv. Water Resour. v. 18, 101-110.

Mucci, A. (1983) The solubility of calcite and aragonite in seawater at various salinities, temperatures and one atmosphere total pressure. Amer. Jour. Sci. v. 283, 780-799.

Murphy, G.M. (1960) Ordinary differential equations and their solutions. D. van Nostran Co.

Murray, J.W., Grundmanis, V. and Smethie, W.J. (1978) Interstitial water chemistry in sediments of Saanich Inlet. Geochim. Cosmochim. Acta v. 42, 1011-1026.

Murray, J.W. and Kuivila, K.M. (1990) Organic matter diagenesis in the northeast Pacific: transition from aerobic red clay to suboxic hemipelagic sediments. Deep-Sea Res. v. 37, 59-80.

Nagaoka, H. and Ohgaki, S. (1990) Mass transfer mechanism in a porous riverbed. Water Res. v. 24, 417-425.

Neale, G.H. and Nader, W.K. (1973) Prediction of transport processes in porous media. Amer. Inst. Chem. Eng. Jour. v. 19, 112-119.

Nguyen, V.V., Gray, W.G., Pinder, G.F., Botha, J.F. and Crerar, D.A. (1982) A theoretical investigation on the transport of chemicals in reactive porous media. Water Resour. Res. v. 18, 1149-1156.

Nicolis, C. (1995) Tracer dynamics in ocean sediments and deciphering of past climates. Math. Comput. Modelling v. 21, 27-38.

Nield, D.A. (1983) The boundary correction for the Rayleigh-Darcy problem: limitations of the Brinkman equation. Jour. Fluid Mech. v. 128, 37-46.

Nigrini, A. (1970) Diffusion in rock alteration systems: I. Prediction of limiting equivalent ionic conductances at elevated temperatures. Amer. Jour. Sci. v. 269, 65-91.

Nihtilä, M. and Virkkunen, J. (1977) Practical identifiability of growth and substrate consumption models. Biotechnol. Bioeng. v. 19, 1831-1850.

Noble, R.D. (1982) Mathematical modelling in the context of problem solving. Math. Model. v. 3, 215-219.

Nogotov, E.F. (1978) Applications of Numerical Heat Transfer. UNESCO, Hemisphere Publ. Corp.

Nowak, A., Bialecki, R. and Kurpisz K. (1987) Evaluating eigenvalues for boundary value problems of heat conduction in rectangular and cylindrical coordinate systems. Internat. Jour. Numer. Meth. Eng. v. 24, 419-445.

Noye, B.J. and Arnold, R.J. (1990) Accurate finite difference approximations for the Neumann condition on a curved boundary. Appl. Math. Model. v. 14, 2-13.

Nozaki, Y. (1977) Distribution of natural radionuclides in sediments influenced by bioturbation (in Japanese). Geol. Soc. Japan Jour. v. 83, 699-706.

Nunziato, J.W. and Walsh, E.K. (1980) On ideal multiphase mixtures with chemical reaction and diffusion. Arch. Rat. Mech. v. 73, 285-311.

Oelkers, E.H. (1991) Calculation of diffusion coefficients for aqueous organic species at temperatures from 0 to 350°C. Geochim. Cosmochim. Acta v. 55, 3515-3529.

Officer, C.B. and Lynch, D.R. (1983) Determination of mixing parameters from tracer distributions in deep-sea sediment cores. Mar. Geol. v. 52, 59-74.

Ohsumi, T. and Horibe, Y. (1984) Diffusivity of he and Ar in deep-sea sediments. Earth Planet. Sci. Lett. v. 70, 61-68.

Okubo, A. (1980) Diffusion and Ecological Problems: Mathematical Models. Biomathematics v. 10. Springer-Verlag.

Olander, D.R. (1960) Simultaneous mass transfer and equilibrium chemical reaction. Amer. Inst. Chem. Eng. Jour. v. 6, 233-239.

Opdyke, B.N., Gust, G. and Ledwell, J.R. (1987) Mass transfer from a smooth alabaster surface in turbulent flow. Geophys. Res. Lett. v. 14, 1131-1134.

Osher, S. and Chakravarthy, S. (1984) High resolution schemes and the entropy condition. SIAM Jour. Numer. Anal. v. 21, 955-984.

Otto, N.C. and Quinn, J.A. (1971) The facilitated transport of carbon dioxide through bicarbonate solutions. Chem. Eng. Sci. v. 26, 949-961.

Ozisik, N.M. (1980) Heat Conduction. John Wiley and Sons.

Patel, M.K., Markatos, N.C. and Cross, M. (1985) A critical evaluation of seven discretization schemes for convection-diffusion equations. Internat. Jour. Numer. Methods Fluids v. 5, 225-244.

Pearson, C.E. (1968) On a differential equation of boundary layer type. Jour. Math. Phys. v. 47, 134-154.

Pemberton, G.S., Risk, M.J. and Bucklet, D.E. (1976) Supershrimp: Deep bioturbation in the Strait of Canso, Nova Scotia. Science v. 192, 790-791.

Petersen, E.E. (1958) Diffusion in a pore of varying cross section. Amer. Inst. Chem. Eng. Jour. v. 4, 343-345.

Petersen, E.E. (1965) Chemical Reaction Analysis. Prentice-Hall.

Petzold, L.R. (1982a) Differential-algebraic equations are not ODEs. SIAM Jour. Sci. Math. Comput. v. 3, 367-384.

Petzold, L.R. (1982b) A description of DASSL: a differential/algebraic system solver. Report SAND82-8637, Sandia National Laboratory, USA.

Phillips, O.M. (1991) Flow and Reaction in Permeable Rocks. Cambridge University Press.

Pikal, M.J. (1971) Ion-pair formation and the theory of mutual diffusion coefficients in a binary electrolyte. Jour. Phys. Chem. v. 75, 663-675.

Pinczewski, W.V. and Sideman, S. (1974) A model for mass (heat) transfer in turbulent tube flow. Moderate and high Schmidt (Prandtl) numbers. Chem. Eng. Sci. v. 29, 1969-1976.

Plumb, O.A. and Whitaker, S. (1990) Diffusion, adsorption and dispersion in porous media: small-scale averaging and local volume averaging, *in* Dynamics of Fluids in Hierarchical Porous Media, p. 97-148. Academic Press.

Postlethwaite, J. and Lotz, U. (1988) Mass transfer at erosion-corrosion roughened surfaces. Can. Jour. Chem. Eng. v. 66, 75-78.

Postma, D. (1993) The reactivity of iron oxides in sediments: a kinetic approach. Geochim. Cosmochim. Acta v. 57, 5027-5034.

Press, W.H., Teukolsky, S.A., Vetterling, W.T., and Flannery, B.P. (1992) Numerical Recipes (in FORTRAN). 2nd Edition. Cambridge University Press.

Putnam, J.A. (1949) Loss of wave energy due to percolation in a permeable sea bottom. Trans. Amer. Geophys. Union v. 30, 349-356.

Rabouille, C. and Gaillard J.-F. (1990) The validity of steady-state flux calculations in early diagenesis: a computer simulation of deep-sea silica diagenesis. Deep-Sea Res. v. 37, 625-646.

Rabouille, C. and Gaillard, J.-F. (1991a) A coupled model representing the deep-sea organic carbon and oxygen consumption in surficial sediments. Jour. Geophys. Res. v. 96, 2761-2776.

Rabouille, C. and Gaillard, J.-F. (1991b) Towards the EDGE: early diagenetic global explanation. A model depicting the early diagenesis of organic matter, O_2, NO_3, Mn, and PO_4. Geochim. Cosmochim. Acta v. 55, 2511-2525.

Rahm, L. and Svensson, U. (1989) On mass transfer properties of the benthic boundary layer with an application to oxygen fluxes. Neth. Jour. Sea Res. v. 24, 27-35.

Raiswell, R., Whaler, K., Dean, S., Coleman, M.L. and Briggs, D.E.G. (1993) A simple three-dimensional model of diffusion-with-precipitation applied to localized pyrite formation in framboids, fossils and detrital iron minerals. Mar. Geol. v. 113, 89-100.

Ramos, J.I. (1986a) Numerical solution of reaction-diffusive systems. Part 2: Method of lines and implicit algorithms. Internat. Jour. Computer Math. v. 18, 141-161.

Ramos, J.I. (1986b) Numerical solution of reaction-diffusive systems. Part 3: Time linearization and operator splitting techniques. Internat. Jour. Computer Math. v. 18, 289-309.

Ramos, J.I. (1987) Modified equation techniques for reaction-diffusion systems. Part 2: Time-linearization and operator splitting methods. Computer Methods Appl. Mech. Eng. v. 64, 221-236.

Rayleigh, L. (1892) On the influence of obstacles arranged in rectangular order upon the properties of a medium. Phil. Mag. v. 34, 481-489.

Reible, D.D., Valsaraj, K.T. and Thibodeaux, L.J. (1991) Chemodynamic models for transport of contaminants from sediment beds, *in* The Handbook of Environmental Chemistry (Hutzinger, O., ed.), v. 2, p. 185-228.

Reimers, C.E. and Suess, E. (1983) The partitioning of organic carbon fluxes and sedimentary organic matter decomposition rates in the ocean. Mar. Chem. v. 13, 141-168.

Rein, M. (1992) The partial-equilibrium approximation in reacting flows. Phys. Fluids v. A4, 873-886.

Reynolds, O. (1895) On the dynamical theory of incompressible viscous fluids and the determination of the criterion. Royal Soc. London Phil. Trans. v. A186, 123-164.

Rice, D.L. (1986) Early diagenesis in bioadvective sediments: relationships between the diagenesis of Beryllium-7, sediment reworking rates, and the abundance of conveyor-belt deposit-feeders. Jour. Mar. Res. v.44, 149-184.

Richter, R. (1952) Fluidal-texture, *in* Sediment-Geseinen und uber Sedifluktion uberhaupt: Notizbl. Hess. L.-Amt. Bodenforsch., v. 3, p. 67-81.

Robbins, J.A. (1986) A model for particle-selective transport of tracers in sediments with conveyor belt deposit feeders. Jour. Geophys. Res. v. 91, 8542-8558.

Rogers, J.C.W. (1977) A free boundary problems as diffusion with nonlinear adsorption. Jour. Inst. Math. Appl. v. 20, 261-268.

Rogers, J.C.W., Berger, A.E. and Ciment, M. (1979) The alternating phase truncation method for numerical solution of a Stefan problem. SIAM Jour. Numer. Anal. v. 16, 563-587.

Rood, R.B. (1987) Numerical advection algorithms and their role in atmospheric transport and chemistry. Rev. Geophys. v. 25, 71-100.

Rubin, J. (1983) Transport of reacting solutes in porous media: relation between mathematical nature of problem formulation and chemical nature of reactions. Water Resour. Res. v. 19, 1231-1252.

Rutgers van der Loeff, M.M. (1981) Wave effects on sediment water exchange in a submerged sand bed. Neth. Jour. Sea Res. v. 15, 100-112.

Rutherford, J.C., Boyle, J.D., Elliot, A.H., Hatherell, T.V.J. and Chiu, T.W. (1995) Modeling benthic oxygen uptake by pumping. Jour. Envir. Eng. v. 121, 84-95.

Saffman, P.G. (1971) On the boundary condition at the surface of a porous medium. Studies Appl. Math. v. 50, 93-101.

Sahraoui, M. and Kaviany, M. (1992) Slip and no-slip velocity boundary conditions at interface of porous, plain media. Internat. Jour. Heat. Mass Transfer v. 35, 927-943.

Saltzman, E.S., King, D.B., Holmen, K. and Leck, C. (1993) Experimental determination of the diffusion coefficient of dimethylsulfide in water. Jour. Geophys. Res. v. 98, 16481-16486.

Santschi, P.H., Anderson, R.F., Fleisher, M.Q. and Bowles, W. (1991) Measurements of diffusive sublayer thicknesses in the ocean by alabaster dissolution, and their implications for the measurements of benthic fluxes. Jour. Geophys. Res. v. 96, 10641-10657.

Santschi, P.H., Bower, P., Nyffeler, U.P., Azevedo, A. and Broecker, W.S. (1983) Estimates of the resistance to chemical transport posed by the deep-sea boundary layer. Limnol. Oceanogr. v. 28, 899-912.

Savant, S.A., Reible, D.D. and Thibodeaux. L.J. (1987) Convective transport within stable river sediments. Water Resour. Res. v. 23, 1763-1768.

Schiesser, W.E. (1991) The Numerical Method of Lines: Integration of Partial Differential Equations. Academic Press.

Schiesser, W.E. (1994) Computational Mathematics in Engineering and Applied Science: ODEs, DAEs and PDEs. CRC Press.

Schiffelbein, P. (1985) Extracting the benthic mixing impulse response function: a constrained deconvolution technique. Mar. Geol. v. 64, 313-336.

Schink, D.R. and Guinasso, N.L. (1977a) Modelling the influence of bioturbation and other processes on calcium carbonate dissolution at the sea floor, *in* The Fate of Fossil Fuel CO2 in the Oceans (Andersen, N.R. and Malahoff, A., eds), p. 375-399. Plenum.

Schink, D.R. and Guinasso, N.L. (1977b) Effects of bioturbation on sediment-seawater interaction. Mar. Geol. v. 23, 133-154.

Schink, D.R. and Guinasso, N.L. (1978) Redistribution of dissolved and adsorbed materials in abyssal marine sediments undergoing biological stirring. Amer. Jour. Sci. v. 278, 687-702.

Schink, D.R. and Guinasso, N.L. (1980) Processes affecting silica at the abyssal sediment-water interface, *in* Biogéochimie de la Matière Organique à L'Interface Eau-Sédiment Marin, Colloques Internationaux du CNRS, No. 293, p. 81-92.

Schlichting, H. (1968) Boundary Layer Theory. 6th Edition. McGraw-Hill.

Sen, A.K. (1988) Enhancement factors for gas absorption with chemical reaction: approximate analytical solutions. Chem. Eng. Sci. v. 43, 641-648.

Shampine, L.F. and Gear, C.W. (1979) A user's view of solving stiff ordinary differential equations. SIAM Review v. 21, 1-17.

Shaw, D.A. and Hanratty, T.J. (1977) Turbulent mass transfer rates to a wall for large Schmidt numbers. Amer. Inst. Chem. Eng. Jour. v. 23, 28-37.

Sherwood, T.K., Pigford, R.L. and Wilke, C.R. (1975) Mass Transfer. McGraw-Hill.

Shum, K.T. (1992) Wave-induced advective transport below a rippled water-sediment interface. Jour. Geophys. Res. v. 97, 789-808. (see also Correction to Wave-induced advective transport below a rippled water-sediment interface. Jour. Geophys. Res. v. 97, 14475-14477.)

Shum, K.T. (1993) The effects of wave-induced pore water circulation on the transport of reactive solutes below a rippled sediment bed. Jour. Geophys. Res. v. 98, 10289-10301.

Shyy, W. (1985) A study of finite difference approximations to steady-state, convection-dominated flow problems. Jour. Comput. Physics v. 57, 415-438.

Sideman, S. and Pinczewski, W.V. (1975) Turbulent heat and mass transfer at interfaces: transport models and mechanisms, *in* Topics in Transport Phenomena (Gutfinger, C., ed.), p. 47-207. John Wiley and Sons.

Slattery, J.C. (1967) General balance equation for a phase interface. Indus. Eng. Chem. Fund. v. 6, 108-115.

Slattery, J.C. (1980) Interfacial transport phenomena. Chem. Eng. Commun. v. 4, 149-166.

Slattery, J.C. (1981) Momentum, Energy, and Mass Transfer in Continua. 2nd Edition. Robert E. Krieger Publ. Co.

Sleath, J.F.A. (1984) Sea Bed Mechanics. John Wiley and Sons.

Smith, C.R., Pope, R.H., DeMaster, D.J. and Magaard, L. (1993) Age-dependent mixing of deep-sea sediments. Geochim. Cosmochim. Acta v. 57, 1473-1488.

Smith, G.D. (1975) Numerical Solution of Partial Differential Equations. Oxford Univ. Press.

Soetaert, K., Herman, P.M.J. and Middelburg, J.J. (1996) A model of early diagenetic processes from the shelf to abyssal depths. Geochim. Cosmochim. Acta. v. 60, 1019-1040.

Sommeijer, B.P. and van der Houwen, P.J. (1984) Algorithm 621: Software with low storage requirements for two-dimensional, nonlinear, parabolic differential equations. ACM Trans. Math. Software v. 10, 378-396.

Spanier, J. and Oldham, K.B. (1987) An Atlas of Functions. Hemisphere Publishing Corp.

Squire, W. (1976) An efficient iterative method for numerical evaluation of integrals over semi-infinite range. Internat. Jour. Numer. Methods Eng. v. 10, 478-484.

Standart, G. (1964) The mass, momentum and energy equations for heterogeneous flow systems. Chem. Eng. Sci. v. 19, 227-236.

Stumm, W. and Morgan, J.J. (1996) Aquatic Chemistry. 3rd Edition. John Wiley and Sons.

Svensson, U. and Rahm, L. (1991) Toward a mathematical model of oxygen transfer to and within bottom sediments. Jour. Geophys. Res. v. 96, 2777-2783.

Swan, G.W. (1974) Derivation of the equation for concentration profiles in a binary diffusing system, *in* Mathematical Problems in Biology (van den Driessche, P., ed.), Lecture Notes in Biomathematics v. 2, p.226-235. Springer-Verlag.

Swanson, R.C. and Turkel, E. (1992) On central-difference and upwind schemes. Jour. Comput. Physics v. 101, 292-306.

Sweby, P.K. (1984) High resolution schemes using flux limiters for hyperbolic conservation laws. SIAM Jour. Numer. Anal. v. 21, 995-1011.

Sweerts, J.-P. R.A., Kelly, C.A., Rudd, J.W.M., Hesslein, R. and Cappenberg, T.E. (1991) Similarity of whole-sediment molecular diffusion coefficients in freshwater sediments of low and high porosity. Limnol. Oceanogr. v. 36, 335-342.

Tan, K.K. and Thorpe, R.B. (1992) Gas diffusion into viscous and non-Newtonian liquids. Chem. Eng. Sci. v. 47, 3565-3572.

Thibault, J., Bergeron, S. and Bonin, H.W. (1987) On finite-difference solutions of the heat equation in spherical coordinates. Numer. Heat Transfer v. 12, 457-474.

Thibodeaux, L.J. (1979) Chemodynamics. Environmental Movement of Chemicals in Air, Water and Soil. John Wiley and Sons.

Thibodeaux, L.J. and Boyle, J.D. (1987) Bedform-generated convective transport in bottom sediments. Nature v, 325, 341-343.

Tomadakis, M.M. and Sotirchos, S.V. (1993) Transport properties of random arrays of freely overlapping cylinders with various orientation distributions. Jour. Chem. Phys. v. 98, 616-626.

Toor, H.L. and Arnold, K.R. (1965) Nature of the uncoupled multicomponent diffusion equations. Industr. Eng. Chem. Fundamentals v. 4, 363-364.

Toride, N. Leij, F.J. and van Genuchten, M.Th. (1993) A comprehensive set of analytical solutions for nonequilibrium solute transport with first-order decay and zero-order production. Water Resour. Res. v. 29, 2167-2182.

Toth, D.J. and Lerman, A. (1977) Organic matter reactivity and sedimentation rates in the ocean. Amer. Jour. Sci. v. 277, 465-485.

Townsend, A.A. (1956) The Structure of Turbulent Shear Flow. Cambridge Univ. Press.

Tromp, T.K., Van Cappellen, P. and Key, R.M. (1995) A global model for early dia-genesis of organic carbon and organic phosphorus in marine sediments. Geochim. Cosmochim. Acta v. 59, 1259-1284.

Truesdell, C. and Toupin, R. (1960) The classical field theories, *in* Encyclopedia of Physics v. 1 (Flugge, S., ed.), p. 226-793. Springer-Verlag.

Tyrrell, H.J.V. (1964) The origin and present status of Fick's diffusion law. Jour. Chem. Edu. v. 41, 397-400.

Tyrrell, H.J.V. and Harris, K.R. (1984) Diffusion in Liquids. Butterworths.

Ullman, W.J. and Aller, R.C. (1982) Diffusion coefficients in nearshore marine sediments. Limnol. Oceanogr. v. 27, 552-556.

Valocchi, A.J. and Malmstead, M. (1992) Accuracy of operator splitting for advection-dispersion-reaction problems. Water Resour. Res. v. 28, 1471-1476.

van Brakel, J. and Heertjes, P.M. (1974) Analysis of diffusion in macroporous media in terms of a porosity, a tortuosity and a constrictivity factor. Internat. Jour. Heat Mass Transfer v. 17, 1093-1103.

Van Cappellen, P., Gaillard, J.-F. and Rabouille, C. (1993) Biogeochemical transformation in sediments: kinetic models of early diagenesis, *in* Interactions of C, N, P and S Biogeochemical Cycles and Global Change (Wollast, R., Mackenzie, F.T. and Chou, L., eds), NATO ASI Series v. 14, p. 401-445.

Van Cappellen, P. and Wang, Y. (1995) Metal cycling in surface sediments: modelling the interplay between transport and reaction, *in* Metal Speciation and Contamination of Aquatic Sediments (Allen, H.E., ed.), p. 21-64. Ann Arbor Press.

Van Cappellen, P. and Wang, Y. (1996) Cycling of iron and manganese in surface sediments: a general theory for the coupled transport and reaction of Carbon, Oxygen, Nitrogen, Sulfur, Iron and Manganese. Amer. Jour. Sci. v. 296, 197-243.

Vanderborght, J.P., Wollast, R. and Billen, G. (1977) Kinetic models of diagenesis in disturbed sediments. Part 1. Mass transfer properties and silica diagenesis. Limnol. Oceanogr. v. 22, 787-793.

van der Weijden, C.H. (1992) Early diagenesis and marine porewater, *in* Diagenesis III (Wolf, K.H. and Chilingarian, G.V., eds), Developments in Sedimentology 47, p. 13-134.

van Genuchten, M.Th. (1981) Analytical solutions for chemical transport with simultaneous adsorption, zero-order production and first-order decay. Jour. Hydrol. v. 49, 213-233.

van Genuchten, M.Th. (1985) Convective-dispersive transport of solutes involved in sequential first-order decay reactions. Computers Geosci. v. 11, 129-147.

van Genuchten, M. Th. and Alves, W.J. (1982) Analytical Solutions of the One-Dimensional Convective-Dispersive Solute Transport Equations. U.S. Dept. Agriculture, Technical Bulletin 1661.

van Leer, B. (1974) Towards the ultimate conservative difference scheme. II. Monoclicity and conservation combined in a second-order scheme. Jour. Comput. Physics v. 14, 361-370.

Verhallen, P.T.H.M., Oomen, L.J.P., Elsen, A.J.J.M.v.d., Kruger, A.J. and Fortuin, J.M.H. (1984) The diffusion coefficients of helium, hydrogen, oxygen and nitrogen in water determined from the permeability of a stagnant liquid layer in the quasi-steady state. Chem. Eng. Sci. v. 39, 1535-1541.

Verwer, J.G. (1994) Gauss-Seidel iteration for stiff ODEs from chemical kinetics. SIAM Jour. Sci. Comput. v. 15, 1243-1250.

Verwer, J.G. and van Loon, M. (1994) An evaluation of explicit pseudo-steady-state approximation schemes for stiff ODE systems from chemical kinetics. Jour. Comput. Physics v. 113, 347-352.

Weber, W.J., McGinley, P.M. and Katz, L.E. (1991) Sorption phenomena in subsurface systems: concepts, models and effects on contaminant fate and transport. Water Res. v. 25, 499-528.

Webster, I.T. (1992) Wave enhancement of solute exchange within empty burrows. Limnol. Oceanogr. v. 37, 630-643.

Webster, I.T. and Taylor, J.H. (1992) Rotational dispersion in porous media due to fluctuating flows. Water Resour. Res. v. 28, 109-119.

Weissberg, H. (1963) Effective diffusion coefficients in porous media. Jour. Appl. Phys. v. 34, 2636-2639.

Westrich, J.T. and Berner, R.A. (1984) The role of sedimentary organic matter in bacterial sulfate reduction: the G model tested. Limnol. Oceanogr. v. 29, 236-249.

Wheatcroft, R.A., Jumars, P.A., Smith, C.R. and Nowell, A.R.M. (1990) A mechanistic view of the particulate biodiffusion coefficient: step lengths, rest periods and transport directions. Jour. Mar. Res. v. 48, 177-207.

Whitaker, S. (1967) Diffusion and dispersion in porous media. Amer. Inst. Chem. Eng. Jour. v. 13, 420-427.

Whitaker, S. (1985) A simple geometric derivation of the spatial averaging theorem. Chem. Eng. Edu. v. 19, 18-21, 50-52.

Wilke, C.R. and Chang, P. (1955) Correlation of diffusion coefficients in dilute solutions. Amer. Inst. Chem. Eng. Jour. v. 1, 264-270.

Wollast, R. and Garrels, R.M. (1971) Diffusion coefficient of silica in seawater. Nature Physical Science v. 229, 94.

Wong, G.T.F. and Grosch, C.E. (1978) A mathematical model for the distribution of dissolved silicon in interstitial waters - an analytical approach. Jour. Mar. Res. v. 36, 735-750.

Wood, P.E. and Petty, C.A. (1983) New model for turbulent mass transfer near a rigid interface. Amer. Inst. Chem. Eng. Jour. v. 29, 164-167.

Yates, S.R. (1992) An analytical solution for one-dimensional transport in porous media with an exponential dispersion function. Water Resour. Res. v. 28, 2149-2154.

Yeh, G.T. and Tripathy, V.S. (1989) A critical evaluation of recent developments in hydrogeochemical transport models of reactive multichemical components. Water Resour. Res. v. 25, 93-108.

Yeh, G.T. and Tripathy, V.S. (1991) A model for simulating transport of reactive multispecies components: model development and demonstration. Water Resour. Res. v. 27, 3075-3094.

Zauderer, E. (1983) Partial Differential Equations of Applied Mathematics. John-Wiley and Sons.

Zwillinger, D. (1992) Handbook of Differential Equations. 2nd Edition. Academic Press.

12 COMMON AND IMPORTANT SYMBOLS

Symbol	Description	Chapter(s)
A,B	integration constants	6,7
a	age	3
a,b,c	integration constants	5,6,7
a	thermodynamic activity	3,4
a_o	ion size parameter	4
A	area	3,4
c(x,a,t)	age-dependent concentration	3
C	solute or generic concentration	all
C_i	solute concentration at the i-th grid point	8
C_i^k	solute concentration at the i-th grid point and k-th time step	8
cos(x)	Cosine function	all
cotan(x)	Cotangent function	all
coth(x)	Hyperbolic Cotangent function	all
D	generic diffusion coefficient	all
Da	Damkohler number	all
D_B	biodiffusion or mixing coefficient	all
D´	diffusion coefficient corrected for tortuosity	all
D_o	molecular or ionic diffusion coefficient at infinite dilution	all
D_i°	ionic diffusion coefficient of ion i	4
D^*	dispersion coefficient	all
d_p	diameter of a sediment particle	4
$\mathcal{E}$ and $\mathcal{E}'$	portion $\mathcal{E}'$ of and external interface $\mathcal{E}$ within the balancing volume $\mathcal{V}$	5
E_a	activation energy	4
E_{ij}	(nonlocal) rate constant for exchange from finite layer i to finite layer j (time^{-1})	3
e_o	charge on an electron (emu)	4
erf(x)	Error function	all
erfc(x)	Complementary Error function	all
exp(x)	Exponential function	all
F	generic flux vector	all
F_c	flux of organic matter at the sediment-water interface	4
F	Faraday's constant	4
f	formation factor	4
f	friction coefficient	4
$f_p()$	thermodynamic equilibrium relation	8
g	acceleration due to gravity	4
g(x)	inhomogeneous forcing	6
g(x,t,k)	concentration in a continuum of reactivity k	3,4
h	hydraulic head	4
H(t-a)	Heaviside function	7
I and I'	portion I' of and internal interface I within the	

	balancing volume $\mathcal{V}$	5
$I_\nu(x)$	modified Bessel function of the first kind of order ν	all
$J_\nu(x)$	Bessel function of the first kind of order ν	all
k	rate constant	all
k	Boltzmann's constant	4
k_{het}	heterogeneous rate constant	3,4,5
k_b	backwards rate constant	all
k_f	forward rate constant	all
$\mathbf{k}$	permeability	3,4,5
K_{eq}	thermodynamic equilibrium constant	all
K_{ex}	constant symmetric exchange rate-constant	all
K_I	inhibition constant for organic matter oxidation	4
K_{ij}	nonlocal exchange constant from finite layer i to finite layer j (length^{-1} time^{-1})	3
K_{ox}	saturation concentration for an oxidant	4
K(x´,x;t)	exchange function from depth x´ to depth x	all
$K_R(x)$	nonlocal removal rate constant	all
$K_\nu(x)$	modified Bessel function of the second kind of order ν	all
L	special depth, e.g. base of the mixed layer	all
M	total mass of a species in a depth interval	3,7
M_w	molecular weight	4
N	Avogadro's number	4
$N(\beta_m)$	a norm for Eigenfunctions	7
Ox	concentration of an arbitrary oxidant	4
n	a normal vector (to a surface or line)	2,3,5
Pe	Peclet number	all
$P_n(x)$	Legendre polynomial of order n	7
p´	reduced pressure	3
r_j	position of the j-th grid point in the radial direction	8
r_i	crystal radius of ion i	4
r_o	radius of a diffusing ion or molecule	4
R	generic rate of reaction	all
R	gas constant	4
$\mathfrak{R}$	rate of reaction along the surface $\mathcal{E}$ or I	5
R(r)	a separation function for the radial coordinate	7
S	surface area	3,5
$\mathcal{S}$	surface area of the balancing volume $\mathcal{V}$	5
$\bar{s}$	specific surface area	4
S(s)	separation function for transformed azimuthal angle	7
Sc	Schmidt number	all
$sin(x)$	Sine function	all
t	time	all
T	absolute temperature	4
T_d	de-correlation time	4
T(t)	a separation function for time	7
$tan(x)$	Tangent function	all
u	velocity of the porewater	all
U,V,W	transformed concentration	7
$U(a,b,x)$	Tricomi function	6

V	volume	3,5
$\mathcal{V}$	balancing volume used to calculate boundary conditions	5
V_{avg}	averaging volume	3.5
V_b	molar volume of a non-electrolyte at its normal boiling temperature	4
V_m	molar volume	4
$\underline{V}_i$	partial molal volume of species i	4
v	generic velocity	all
v_I	irrigation velocity	4
w	burial velocity of the solids	all
X	electrical field	4
$X(x)$	a separation function for the depth	7
x	depth (Cartesian and cylindrical systems)	all
x_i	depth of the i-th grid point	8
y	a lateral dimension in the Cartesian system	all
Z	generic variable	8
z	a lateral dimension in the Cartesian system	all
z_i	charge on an ion	4
$\alpha(x)$	(nonlocal) porewater irrigation constant	3,4
α	an attenuation constant	2,6,7,8
β	mass-transfer coefficient	5,6
β,μ,λ,κ	(various) separation constants	7
ε	weighting parameter for finite differencing	8
ε_m	eddy diffusion coefficient for momentum	5
ε_c	eddy diffusion coefficient for a solute	5
$\Delta x, \Delta t$	small increments in x and t, respectively	all
δ_v	thickness of the viscous sublayer	7
δ_D	thickness of the diffusive sublayer	7
δ_e	thickness of the effective diffusive sublayer	7
$\delta\mathbf{r}$	(local) position vector	3
$\delta(x-a)$	Dirac Delta function at depth x=a	all
$\Gamma(x)$	Gamma function	all
ξ,ζ	symbols (amongst others) for transformed depth	6,7
γ_i	activity coefficient of species i	4
φ	porosity	all
φ_s	solid volume fraction	all
φ_i	volume fraction of phase i	3
ϕ	coordinate of the spherical system (angle in the horizontal plane, Fig. 3.22)	all
λ	equivalent conductance or ionic conductivity of ion i	4
λ_i°	limiting equivalent conductance of ion i	4
μ	dynamic viscosity	3,4
$\mu(x)$	integration factor for first-order ODEs	6
π	pi = 3.14159...	all
ρ	porewater density	all
ρ_i	intrinsic density of solid species i	all

θ	tortuosity	all
τ	time between mixing events or diffusive jumps	3,4
τ'	momentum flux (stress)	5
ϑ	coordinate in the cylindrical system (angle in the plane perpendicular to the vertical x axis; Fig. 3.20); or,	all
	coordinate in the spherical system	all
σ	shear stress	4
σ	an attenuation constant	6
σ	"blending" constant	8
ν	kinematic viscosity	4
Ψ	electrical potential	4
ψ_i	chemical potential of species i	4
ψ_i°	standard chemical potential per mole of species i	4
$\psi(x)$	Psi or Digamma function	6,7
υ_i	mobility of an ion of species i	4
ω_p	weighting function for DAEs	8
$\mathfrak{J}$	arbitrary flux or production at an interface	5
$\mathcal{K}$	hydraulic conductivity	4
$\mathcal{L}[\,]$	Laplace transform	7
$\mathcal{L}^{-1}[\,]$	Inverse Laplace transform	7
ℓ	mixing or displacement distance	3

Superscripts

~	time-fluctuating quantity	all
′	variation about a mean value	3
•	actual value at a chosen point in the porewaters	3
—	time or spatial average	all
^	value per unit bulk volume (solids+porewater)	3
k	related to the k-th time step	8

Subscripts

B	related to total or dissolvable solids	all
e	related to ions	4
o	related to the sediment-water interface	(selected)
i	related to the i-th grid point in the x-direction	8
I	related to irrigation	4
L	related to the depth x = L	all
M	related to inert solids	all
x,y,z	related ro the Cartesian directions x, y, and z, respectively	all
x,r,ϑ	related ro the Cylindrical directions x, r and ϑ, respectively	all
r,ϑ,ϕ	related ro the Spherical directions r, ϑ, and ϕ, respectively	all

13 INDEX

Springer-Verlag and the Environment

We at Springer-Verlag firmly believe that an international science publisher has a special obligation to the environment, and our corporate policies consistently reflect this conviction.

We also expect our business partners – paper mills, printers, packaging manufacturers, etc. – to commit themselves to using environmentally friendly materials and production processes.

The paper in this book is made from low- or no-chlorine pulp and is acid free, in conformance with international standards for paper permanency.